MATLAB and Simulink Crash Course for Engineers

Eklas Hossain

MATLAB and Simulink Crash Course for Engineers

 Springer

Eklas Hossain
Oregon Institute of Technology
Klamath Falls, OR, USA

ISBN 978-3-030-89764-2   ISBN 978-3-030-89762-8   (eBook)
https://doi.org/10.1007/978-3-030-89762-8

This Springer imprint is published by the registered company Springer Nature Switzerland AG
The registered company address is: Gewerbestrasse 11, 6330 Cham, Switzerland

Preface

MATLAB and Simulink are two programming and simulating tools developed by MathWorks®. They are the go-to tools for solving engineering problems and designing, modelling, and simulating new inventions. They are the present-day engineer's favorite tools and are recognized as the most widely used and standard software used all across the globe in all fields of science and engineering. Numerous researches have been developed based solely on the magic of programming in MATLAB and Simulink. Hence, it is essential for STEM students and academicians, and also professional engineers to be fairly acquainted with the usage of MATLAB and Simulink.

All branches of engineering except computer science have only a handful of programming courses in C/C++ or MATLAB/PSpice at a very few weekly contact hours. However, programming is highly relevant to all branches of engineering study. For undergraduate or graduate level projects, thesis, or assignments, working knowledge on programming is necessary to accomplish tasks, build models, simulate designs, or simply replicate the ideas from others. Programming is so intricately connected in modern life that no engineering student can avoid programming anymore, no matter their major. Therefore, students need to develop adequate programming skills early on in their careers to help them work independently throughout their student and professional life. Although programming is required at all levels of study, it is seldom taught academically, and students are eventually compelled to learn everything on their own or simply surrender their weaknesses.

When I began teaching advanced technologies in electrical engineering at Oregon Tech, I discovered that most students have a sort of phobia in programming. They wander around to find solid material to help them solve their programming requirements. I realized the gravity of the problem even more when in the senior year power system or control system projects, many students are either shy or admit their weakness in programming. In fact, programming is an inherent part of not only academia but also in professional engineering and technical jobs.

Hence, I decided to teach programming to students and approached the head of the department for approval. Getting approval from the department, I worked an entire summer to build the right course material for students to build their programming skills from scratch and develop the course outline based on real-world applications. The students were involved in providing their feedback on the course, and they really appreciated the course curriculum.

After teaching the course on MATLAB and Simulink for three to four consecutive years and having solid proof of its efficacy, the course curriculum is now being presented in the form of a book to aid most engineering students worldwide, if not all. This book will be an independent study guide for engineering students. It is designed from an instructor's perspective to help them teach students, to help students learn on their own, and also for helping out independent researchers and professional engineers. The book is rich in case studies that will be relatable to many fields of work.

As an author, I will consider this book a success only if it is able to help students overcome their fear of programming and use this book as a guide to solving engineering problems using MATLAB and Simulink. Thus, the main aim of publishing this book is to disseminate the knowledge that I have gleaned over the years from experiences in teaching and witnessing the problems that engineering students frequently face.

This book comprising 18 chapters is divided into 2 sections. Chapters 1–11 cover MATLAB programming, while Chaps. 12–18 cover Simulink. Each section begins with chapters encompassing the fundamentals of MATLAB and Simulink, and then proceeds towards the applications of these two tools in various fields, particularly in the electrical engineering arena. Since the main research interest of the author is on energy systems, the book is more biased towards the applications of MATLAB and Simulink in electrical circuits, electronics, power electronics, power systems, control systems, renewable energy, etc. In addition to helping develop programming skills, this book will help engineering students to brush up on concepts related to electrical engineering and build projects based on the case studies included in the chapters.

This book caters to the needs of both students and instructors. Students and professional engineers will be able to use this book as self-study material, and instructors will find this book utterly useful to deliver their lectures based on the outlines of this book and also may use any part of this book directly as course material. By the completion of this book, the reader is expected to develop at least an intermediate level of expertise in programming in MATLAB and Simulink. The advanced level will gradually grow with practice and more exposure to the software.

To all those willing to develop skills on MATLAB and Simulink and set out on the path to discovery and learning, good luck on your journey!

Klamath Falls, OR, USA   Eklas Hossain

Acknowledgments

I am thankful to everyone at Springer Nature who has patiently helped me complete the manuscript of this book. Besides, my colleagues and fellow students at Oregon Tech deserve a heartfelt appreciation for their never-ending faith in me and their amiable cooperation. This book reflects my time as an instructor at Oregon Tech, and my experience and knowledge to develop this book would have been zero without Oregon Tech. I thank each and every member of the Oregon Tech family for helping shape my career as an academician.

This book has been designed and drafted with the help of many people. I am thankful to all those who have supported me throughout the process of writing this book. I am grateful to MathWorks® for developing MATLAB and Simulink, and for providing free guidelines on their website to help people learn the two software. Those resources have profoundly helped to develop this book. In addition, I would like to express my appreciation for all the other books and resources available freely on the Internet that have enriched this book. It would be wrong not to mention the reviewers of the book proposal for helping improve the outline and contents of this book. I am thankful to their suggestions to update the contents of this book.

I am indebted to my family, for they have always been there for me in all ups and downs and showered me with their constant love and support. I thank my friends and collaborators for cheering for me and constantly pushing me to improve myself.

And at last, I am infinitely grateful to my Lord, for making me, this world, and all that is there in the Universe. This book, or even I, would be nothing without the Mercy and Blessings of the Creator.

Contents

About the Author

Eklas Hossain is an associate professor in the Department of Electrical Engineering and Renewable Energy and an Associate Researcher with the Oregon Renewable Energy Center (OREC) at the Oregon Institute of Technology (OIT), which is home to the only ABET-accredited BS and MS programs in renewable energy. He has been working in distributed power systems and renewable energy integration for the last 10 years and has published a number of research papers and posters in this field. He is currently involved with several research projects on renewable energy and grid-tied microgrid systems at OIT. He received his PhD from the College of Engineering and Applied Science at the University of Wisconsin Milwaukee (UWM), his MS in Mechatronics and Robotics Engineering from International Islamic University of Malaysia, and a BS in Electrical & Electronic Engineering from Khulna University of Engineering and Technology, Bangladesh. Dr. Hossain is a registered Professional Engineer (PE) in the state of Oregon and is also a Certified Energy Manager (CEM) and Renewable Energy Professional (REP). He is a senior member of the Association of Energy Engineers (AEE) and an Associate Editor for *IEEE Access, IEEE Systems Journal*, and *IET Renewable Power Generation.* His research interests include modelling, analyzing, designing, and controlling power electronic devices; energy storage systems; renewable energy sources; integration of distributed generation systems; microgrid and smart grid applications; robotics, and advanced control system.

Dr. Hossain has authored the book *Excel Crash Course for Engineers*, coauthored the book *Renewable Energy Crash Course: A Concise Introduction*, and is working on several other book projects. He is the winner of the Rising Faculty Scholar Award in 2019 from Oregon Institute of Technology for his outstanding contribution to teaching. Dr. Hossain, with his dedicated research team, is looking forward to exploring methods to make electric power systems more sustainable, cost-effective, and secure through extensive research and analysis on energy storage, microgrid system, and renewable energy sources.

Chapter 1
Introduction to MATLAB

1.1 Introduction

MATLAB is a highly useful tool in engineering for solving problems, designing systems, and simulating models. The versatile nature of MATLAB makes it suitable for numerous applications catering to the needs of almost all engineering fields. This book is particularly designed for electrical engineering problems. This chapter will provide a basic overview of MATLAB and help the readers get familiarized with the software to be able to gradually build the complete idea given in the next chapters.

1.2 What Is MATLAB?

MATLAB provides an environment for researchers and engineers of all domains to create models and algorithms and compute and analyze numerical data with programming capability. The acronym MATLAB came from Matrix Laboratory, as initially the software was built with a goal to perform numerous operations on matrices and vectors. Over the years, MATLAB has developed several toolboxes that facilitate research on control systems, signal/image processing, deep learning, robotics, and so on. It provides the high graphic capability to visualize data in both 2D and 3D formats. MATLAB is also a high-level programming language, and the platform offers great flexibility to use with other programming languages, e.g., Python, C/C++, Java, Fortran, etc. Another distinguishing feature of MATLAB is that it can run in a public cloud environment outside the MathWorks cloud domain. The features of parallel computing, hardware interfacing, embedded applications, and app building have elevated MATLAB to a much higher level in the field of scientific research and engineering applications.

© The Author(s), under exclusive license to Springer Nature Switzerland AG 2022
E. Hossain, *MATLAB and Simulink Crash Course for Engineers*,
https://doi.org/10.1007/978-3-030-89762-8_1

1.3 History, Purpose, and Importance

The historical development of MATLAB, the purpose of introducing MATLAB, and the importance of MATLAB in the present engineering world are described in the following sections.

1.3.1 History

The earliest origin of MATLAB can be traced back to the invention of EISPAC (Matrix Eigensystem Package) software, which was developed to solve eigenvalue problems. The fundamental of this software was the procedure followed by ALGOL 60 to solve such problems. It was first developed at the Argonne National Laboratory around 1970, and the first version of the software was released in 1971, followed by the release of the second update in 1976. Later, another software for mathematical analysis named LINPACK (Linear Equation Package) was developed as a byproduct in 1975 in the same lab by Cleve Moler, Jack Dongarra, Pete Stewart, and Jim Bunch. EISPACK and LINPACK both were formed in Fortran and can be considered as the very primal stage before the appearance of MATLAB.

Although EISPACK and LINPACK were capable of performing numerical analysis and solving linear algebra problems, Moler wanted to upgrade both of them to reduce the complexity of access for his students. With that aim, Moler first came up with the idea of MATLAB, which was named after the Matrix Laboratory. MATLAB was simply a matrix calculator, where the data type of input was matrix. This version was created by Moler only for the usage of his students, which is later regarded as classical MATLAB.

The idea of the first commercial MATLAB was proposed by a graduate student of Stanford named Jack Little in 1983. Jack Little, Steve Bangert, and Cleve Moler took the initiative to bring IBM PC-based MATLAB, which was translated in C from Fortran. The first PC-based commercial MATLAB made its appearance in December 1984, followed by the first marketization in 1985. This new version of MATLAB was updated and modified significantly by both Jack Little and Steve Bangert, where they added many mathematical functions, graphics, and toolboxes based on various applications. MATLAB has been upgraded multiple times for adjusting itself with the new applications and requirements in the domain of engineering and science. The major changes that have occurred over time in different versions of MATLAB are summarized in Table 1.1.

Table 1.1 Chronological development of MATLAB

MATLAB version	Major features	Year
Classical MATLAB	• Input data type matrix	Around 1981
	• Used as a simple matrix calculator	
	• Written in Fortran	
MATLAB 1.0 (PC-MATLAB)	• Translated in C from Fortran	1984
	• IBM PC-based software	
	• Multiple mathematical functions, graphics, and toolboxes	
MATLAB-3	• Ordinary differential equation toolbox	1987
	• Signal processing toolbox	
MATLAB-4	• Simulink	1992
	• Sparse matrix	
	• 2D and 3D color graphics	
MATLAB-5	• Data types	1996
	• Visualization (advanced)	
	• Cell arrays and structure	
	• Graphical user interface	
MATLAB-6	• Desktop MATLAB	2000
	• Linear Algebra Package (LAPACK)	
MATLAB-7	• Parallel computing toolbox	2004
	• Anonymous function	
	• Nested function	
	• Integer data types	
MATLAB-8	• MATLAB app	2012
	• Toolstrip interface	
MATLAB-9	• Live editor	2016
	• App designer	
MATLAB-9.9	• Simulink online	2020
MATLAB-9.10	• Satellite communication toolbox	2021
	• Radar toolbox	
	• DDS blockset	

1.3.2 Purpose and Importance

The original purpose of MATLAB is to provide a programming platform where mathematical analysis, along with different applications in the engineering and science domain, can be performed in the most optimized and user-friendly way. As mentioned earlier, MATLAB can take input in the form of a matrix; hence, the vector, array, and matrix operations can be performed more easily by writing minimal code compared to other programming languages. In MATLAB, algorithm development and advanced mathematics problems can be solved using numerous built-in functions. One of the most important features of MATLAB is its toolbox, which can be used to enhance the performance of MATLAB in any desired domain.

MATLAB covers control systems, signal processing, image processing, robotics, communications, mechatronics, biology, data analytics, and other numerous fields in different domains. Such widespread versatility and functionality of MATLAB have made it one of the most widely used scientific platforms in the world.

1.4 Installation and Dependencies

Installation of MATLAB software can be conducted in three ways [1], as described hereunder.

Method 1: Installation with Internet Connection

Step 1: Sign in to the MathWorks account and download the installer.
Step 2: Run the installer and accept the license agreements, which will create the appearance of the following window in Fig. 1.1.
Step 3: Use the activation key, or navigate to the relevant license file by selecting the *Advanced Options > I want to install network license manager.*
Step 4: Navigate to the directory, where the MATLAB will be installed.
Step 5: MATLAB provides a variety of products, and it provides the option to the user to only install products that are necessary, or relevant for a particular user. Thus, it facilitates the users to save enough space by avoiding the installation of irrelevant products. Choose the products based on your work or interest.
Step 6: Choose any convenient option and make the confirmation.
Step 7: Click on the "Begin Install" option, and wait for a while.
Step 8: After the installation is completed, a "Finish" option will appear. By clicking on it, finalize the installation of MATLAB.

Method 2: Installation Using File Activation Key

Step 1: Start the installer by clicking on the installer file.
Step 2: Accept all the terms and agreements.

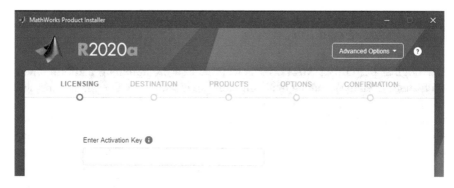

Fig. 1.1 MathWorks Product Installer

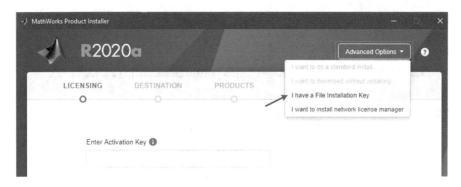

Fig. 1.2 MathWorks Product Installer dropdown menu

Step 3: Find the "Advanced Option" dropdown box, and choose the option "I have a File Installation Key." This is shown in Fig. 1.2.

Step 4: Enter the key.

Step 5: Navigate to the license file location.

Step 6: Navigate to the directory, where MATLAB will be installed.

Step 7: Choose the products for the installation.

Step 8: Choose the options that are convenient.

Step 9: Click on the "Begin Install" option, and wait for a while.

Step 10: After the installation is completed, a "Finish" option will appear. By clicking on it, finalize the installation of MATLAB.

Method 3: Download Installation Package

Step 1: Download the installation file and run the installer.

Step 2: MathWorks sign-in option will appear; hence, sign in to the account. From the dropdown "Advanced Options," select the "I want to download without installing" option.

Step 3: Navigate to the directory where MATLAB will be installed.

Step 4: Choose the platform of installation and also decide the products that are relevant to be installed.

Step 5: After proceeding further, a "Begin Download" option will appear. Click on that option, and wait for a while.

Step 6: When the download will be finished, click on the "Finish" button, which will appear right after the completion of the download.

Step 7: Move the file to the desired destination, and run the installer file. For windows, the installer file will be found as "setup.exe." For Linux and MAC, the name of the installer file will be "install" and "InstallForMacOSX," respectively.

Step 8: After running the installer file, follow Method 1, if the Internet connection is available. Otherwise, follow Method 2, if a file activation key is available for completing the installation.

1.4.1 Dependencies

Before installing MATLAB, the dependencies, or the system requirements, need to be checked. The dependencies for different platforms are listed in Table 1.2.

Table 1.2 Dependencies of MATLAB

	Windows	Mac	Linux
Operating systems	• Windows Server 2016 • Windows Server 2019 • Windows 7 Service Pack 1 • Windows 10 (version 1803 or higher)	• macOS Mojave (10.14) • macOS Catalina (10.15) • macOS Big Sur (11)	• Red Hat Enterprise Linux 7 (minimum 7.6) • Red Hat Enterprise Linux 8 (minimum 8.1) • Ubuntu 16.04 LTS • Ubuntu 18.04 LTS • Ubuntu 20.04 LTS • SUSE Linux Enterprise Server 12 (minimum SP2) • SUSE Linux Enterprise Desktop 12 (minimum SP2) • SUSE Linux Enterprise Server 15 • SUSE Linux Enterprise Desktop 15
RAM	Min. 4 GB Recommended: 8 GB For Polyspace: 4 GB/core (recommended)	Min. 4 GB Recommended: 8 GB For Polyspace: 4 GB/core (recommended)	Min. 4 GB Recommended: 8 GB For Polyspace: 4 GB/core (recommended)
Processors	Min. any Intel or AMD x86-64 processor Recommended: any Intel or AMD x86-64 processor with four logical cores and AVX2 instruction set support	Min. any Intel x86-64 processor Recommended: any Intel x86-64 processor with four logical cores and AVX2 instruction set support	Min. any Intel x86-64 processor Recommended: any Intel x86-64 processor with four logical cores and AVX2 instruction set support
Disk	3.4 GB of HDD space for MATLAB only, 5–8 GB for a typical installation A full installation of all MathWorks products may require 29 GB	3 GB of HDD space for MATLAB only, 5–8 GB for a typical installation A full installation of all MathWorks products may require 22 GB	3.3 GB of HDD space for MATLAB only, 5–8 GB for a typical installation A full installation of all MathWorks products may require 27 GB
Graphics	Graphics card: not limited Recommended: hardware accelerated graphics card supporting OpenGL 3.3 with 1 GB GPU memory	Graphics card: not limited Recommended: hardware accelerated graphics card supporting OpenGL 3.3 with 1 GB GPU memory	Graphics card: not limited Recommended: hardware accelerated graphics card supporting OpenGL 3.3 with 1 GB GPU memory

1.5 Starting MATLAB

Windows: Open the program menu and find the MATLAB application file (matlab. exe) from the directory. Double-click on it to start the MATLAB.

Linux: Open the terminal and write "matlab." After pressing enter, the MATLAB will start.

Mac: On the dock, find the MATLAB icon. By clicking on it, MATLAB can be opened.

1.6 MATLAB Environment

In the MATLAB environment, there are multiple windows that can be set as the starting layout. In a starting layout, the basic windows that are docked in default form are Command Window, Editor, Current Folder, and Workspace. In the header tool strip of MATLAB, the "Layout" option is available, by clicking on which the layout can be changed according to our convenience. In Fig. 1.3, the layout of the MATLAB starter page in default mode is shown. The "Command Window" pro- vides a window where any MATLAB command can be written for execution. The command can be run by pressing the "Enter" button on the keyboard. For writing a long program, the command window probably is not the best place. For that, the "Editor" window is available, where a complete program can be written and run for execution. The output of the program will appear on the command window. In the default mode, there are also two other windows: Current Folder or Directory and

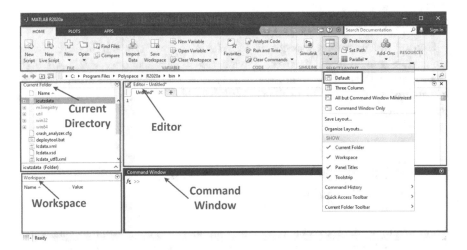

Fig. 1.3 MATLAB environment

Workspace. The current directory indicates the directory from where any MATLAB file can be exported or imported. In the workspace, all the variables defined in the command window, or the editor, will appear along with their values after the execution of the program. From the "Layout" menu, "Command History" window can also be docked in the starter page, which lists all the previous commands in an ordered manner. Therefore, if any previous command needs to be reused, it can be selected from the command history. By clicking the "Up" and "Down" key of the keyboard, the previous commands can also be navigated. In MATLAB, there is a header tool strip, where multiple menus and tools are available, which are also parts of the MATLAB environment.

1.7   Features of MATLAB

The features of MATLAB are crucial in the field of engineering. Some of the most important features of MATLAB can be listed as follows:

1. High-level programming language
2. Built-in graphics
3. Interactive environment
4. High computational capability
5. Numerous mathematical functions
6. Numerous toolboxes dedicated to separate applications
7. Compatibility with other languages
8. Parallel computing
9. App designing
10. Algorithm formation
11. Hardware interfacing
12. Deployment capability, etc.

One of the important aspects of MATLAB is that it is not limited to one application area; instead, it covers a wide area of applications. MATLAB 9.10 version has a total of 82 toolboxes dedicated to numerous application fields in various domains. A list of available toolboxes in different application fields is given in Table 1.3.

1.8   Variables in MATLAB: Categories and Conversion Between Variables

While writing a MATLAB code, variables are used to define different parameters that are needed to be used in the same program multiple times. In variable definition, the main task is to assign data against each variable, and this assigned data can be of

Table 1.3 Number of toolboxes available in MATLAB for various applications

Applications	Number of toolboxes
Aerospace	04
Automotive	10
Code verification	07
Computational biology	02
Computational finance	07
Control systems	10
FPGA, ASIC, and SoC development	08
Image processing and computer vision	03
RF and mixed signal	05
Robotics and autonomous systems	09
Signal processing	06
Test and measurement	06
Wireless communications	05
Total	82

different types, such as integer, float, string, etc. When a variable is defined, it can be observed in the Workspace. The name of the variables can be any letter, or any combination of multiple letters and numbers, such as "var1." However, a variable's name cannot be started with a number; for example, "1var" is not a valid variable name. In addition, special characters such as "@," "#," "$," "&," and "-" cannot be included in a variable's name. Underscore can be allowed to be used in the variable's name except at the starting place. For example, "var_1" is a valid variable name, but "_var" is invalid in MATLAB.

1.8.1   Categories of Data Types

The mostly used categories of data types of variables can be listed as follows:

(a) Numeric data type: integer, float (single and double), and logical
(b) Character and string type: character, string, and cell array
(c) Date and time

An example of variable definition for each of the above-mentioned data types is given below using MATLAB.

1.8.2   MATLAB Example 1.1: Different Data Types

The MATLAB code demonstrating different data types is given in Fig. 1.4, and its output is given in Fig. 1.5.

```
% Variable definiton of different data types
clc;clear;
% Numeric data types: Integer, Float, Logical
var1 = int8(2);                 % Integer of signed 8-bit
var2 = 10.5;                    % Float: Double
var3 = true;                    % Logical
% Character and string type: Character, String, Cell array
var4 = 'MATLAB';                % Character
var5 = ["E","Hossain"];         % String
var6 = {'E','Hossain'};         % Cell array
% Date and time
var7 = datetime('13/05/2021','InputFormat','dd/MM/yyyy');
fprintf('var1 =');  disp(var1)
fprintf('var2 =');  disp(var2)
fprintf('var3 =');  disp(var3)
fprintf('var4 = '); disp(var4)
fprintf('var5 =');  disp(var5)
fprintf('var6 =');  disp(var6)
fprintf('var7 =');  disp(var7)
whos
```

Fig. 1.4 Code—Different data types in MATLAB

```
Command Window                                                          ⊙

  var1 =    2

  var2 =    10.5000

  var3 =    1

  var4 = MATLAB
  var5 =    "E"      "Hossain"

  var6 =    {'E'}     {'Hossain'}

  var7 =    13-May-2021

    Name        Size            Bytes  Class      Attributes

    var1        1x1                 1  int8
    var2        1x1                 8  double
    var3        1x1                 1  logical
    var4        1x6                12  char
    var5        1x2               204  string
    var6        1x2               224  cell
    var7        1x1                 8  datetime
```

Fig. 1.5 Output—Different data types in MATLAB

Output:

In MATLAB *whos* command can be used to print the details of all the defined variables stored in the Workspace. More precisely, it provides the name, size, bytes, class, and attributes of the defined variables. Here, the class indicates the data types of the variable. For example, the class of *var*1 is *int*8, which signifies that *var*1 is a positive signed 8-bit integer. The data types of the rest of the variables can also be found from the above-mentioned output.

After defining a variable, the assigned data of the variable can be reused any time throughout the program, unless the variable is reassigned with new data, or cleared by MATLAB command. To erase any defined variable from MATLAB, the *clear* command can be utilized. Variables can also be cleared by selecting them in the Workspace window and selecting the delete option.

- **If you want to erase all variables from the MATLAB memory, type** *clear* **or** *clear all*
- **To erase a specific variable, say x, type** *clear x*
- **To clear two specific variables, say x and y, type** *clear x y*
- **To clear only the command window, type** *clc*

1.8.3   Conversions

The data types of the defined variables can be converted in MATLAB. The most used conversions that are required are number-to-text conversion and text-to-number conversion. The commands for such conversions are listed in Table 1.4.

A MATLAB example is given below for further illustration.

Table 1.4   MATLAB functions for data type conversion

Number to text	Text to number

• *int2str(number)*
  Convert any integer number to a character. Here, *number* indicates any integer number
• *num2str(number)*
  Convert any number to a character. Here, *number* indicates any number
• *char(number)*
  Convert any number to a character. Here, *number* indicates any number
• *string(number)*
  Convert any number to a string. Here, *number* indicates any number
	• *str2num(text)*
	  Convert any character or string into a double number. Here, *text* indicates any character or string
	• *str2double(text)*.
	  Convert any character or string into a double number. Here, *text* indicates any character or string

1.8.4   MATLAB Example 1.2: Conversion of Data Types

The MATLAB code demonstrating the conversion of data types is given in Fig. 1.6, and its output is given in Fig. 1.7.

Output:

Here, all the conversions can be verified by observing the "Class" of all the variables.

```
% Conversion of data types
clear;clc;
% Number to text
var1 = int8(5);                % Integer
var1_conv = int2str(var1);     % Conversion into character

var2 = 2;                      % Double
var2_conv = num2str(var2);     % Conversion into character

var3 = 2.5;                    % Double
var3_conv = char(var3);        % Conversion into character

var4 = 3;                      % Double
var4_conv = string(var4);      % Conversion into string
fprintf('Number to text conversion:\n');
fprintf('-------------------------------\n');
whos

% Text to number
clear;
var5 = '4';                    % Character
var5_conv = str2num(var5);     % Conversion into double

var6 = "3.1416";               % String
var6_conv = str2double(var6); % Conversion into double
fprintf('\nText to number conversion:\n');
fprintf('-------------------------------\n');
whos
```

Fig. 1.6 Code—Conversion of data types in MATLAB

```
Command Window                                                         ⦿

   Number to text conversion:
   --------------------------------

   Name              Size           Bytes  Class      Attributes

   var1              1x1                1   int8
   var1_conv         1x1                2   char
   var2              1x1                8   double
   var2_conv         1x1                2   char
   var3              1x1                8   double
   var3_conv         1x1                2   char
   var4              1x1                8   double
   var4_conv         1x1              150   string


   Text to number conversion:
   --------------------------------

   Name              Size           Bytes  Class      Attributes

   var5              1x6               12   char
   var5_conv         1x1                8   double
   var6              1x1              150   string
   var6_conv         1x1                8   double
```

Fig. 1.7 Output—Conversion of data types in MATLAB

```
% Without semicolon
a=[1 2 3 4]
```

```
% With semicolon
a=[1 2 3 4];
```

```
Command Window                            ⦿

   a =

       1    2    3    4
```

Fig. 1.8 Suppressing output in MATLAB

1.9   Suppressing Output

In MATLAB, after each command, a semicolon is used to suppress the output. Without a semicolon, the output of the command will be printed on the command window. In general, we normally use a semicolon at the end of every line of code, except the final output, or the results we want to see in the command window.

An example of MATLAB code with and without the usage of the semicolon is shown in Fig. 1.8 to understand the distinction between them:

Here, in the first example, a vector *a* has been defined without using a semicolon. As an outcome, the value of *a* is printed in the command window. In the second example, with the usage of the semicolon, the output is suppressed.

1.10   Recording a MATLAB Session

In MATLAB, *diary* function can be useful while recording a MATLAB session that creates a file containing the keyboard inputs and the outputs. The MATLAB command for the usage of the *diary* function is as follows:

MATLAB command for recording a session:

$$diary('Name')$$

Here, *'Name'* indicates the name of the file in which the session will be recorded.

A MATLAB example is provided in Fig. 1.9 with its output in Fig. 1.10 for further illustration.

Output:

Here the output including the keyboard inputs is recorded in a text file named "DiaryFile.txt," as shown above in the output. To record a particular part of the MATLAB session, *diary on* and *diary off* command placed at the start and the end of a session, respectively, can be very useful. Later, the *diary('Name')* command can be used to define the file name on which the recorded version will be saved.

```
% diary function
clc;clear;
a = input('Enter a:');
b = input('Enter b:');
sum = a + b;
fprintf('Summation: %d\n',sum);
diary('diaryFile.txt');
```

Fig. 1.9 Code—diary function in MATLAB

Fig. 1.10 Output—diary function in MATLAB

1.11   Printing Output

The output of the MATLAB program can be printed in the command window by using two MATLAB built-in functions *fprintf*() and *disp*(). It provides the users great flexibility to choose which output they want to print. Both *fprintf*() and *disp*() commands are explained below:

MATLAB *fprintf* command printing:

$$fprintf('text')$$

$$fprintf('text\%d', var)$$

$$fprintf('text\%f', var)$$

$$fprintf('\backslash n')$$

Here, *'text'* can be any string or character, and *var* indicates the value of a variable that is required to be printed. Finally, *'\n'* is used to shift to the next line.

fprintf() can be used to print both text and numbers in the command window. The input provided within the bracket will be printed in the command window as a string, or characters. If the defined value of any variable is required to be printed with text, the *fprintf*('*text % d*', *var*) command can be utilized where %d will be replaced by the value of the defined variable *var*. Here *var* indicates the name of the variable. However, it is only true if the variable is an integer type. For printing the values of float type variables, %f is used in place of %d. To shift to a new line, \n is used.

```
% fprintf function
clear;clc;
var = int8(5);
var1 = 5.25;
fprintf('Usage of fprintf() function:');
fprintf('\n');
fprintf('The value of the variable is: %d\n',var);
fprintf('The value of the variable is: %f\n',var1);
```

Fig. 1.11   Code—fprintf function in MATLAB

```
Command Window                                                        ⊙

    Usage of fprintf() function:
    The value of the variable is: 5
    The value of the variable is: 5.250000
```

Fig. 1.12   Output—fprintf function in MATLAB

A MATLAB example regarding the usage of the *fprintf()* is provided in Fig. 1.11 with its output in Fig. 1.12.

Output:

Here, in the first *fprintf()* command, a text is printed. In the second command, \n is used to print the next output in a separate line. In the third command, an integer variable named *var* is printed with text. Finally, in the fourth command, a float type variable printing with text is shown.

disp() is another MATLAB command for printing output. The usage of *disp()* command is shown below:

MATLAB *'disp'* **command printing:**

$$disp('text')$$

$$disp(var)$$

$$disp(['text', num2str(var)])$$

Here, *'text'* **can be any string or character, and** *var* **indicates the value of a variable that is required to be printed.**

In *disp()*, a new line shift occurs by default; hence, \n notation, as used in *fprintf()*, is not required to use here. To print both strings and variables, it is necessary to convert the data types of the variables in a string format while using *disp()*. In the

```
% disp function
clear;clc;
var = 5.25;
disp('Usage of disp() function:');
disp(var);
disp(['The value of the variable is: ',num2str(var)]);
```

Fig. 1.13 Code—disp function in MATLAB

```
Command Window                                            ▼

   Usage of disp() function:
        5.2500

   The value of the variable is: 5.25
```

Fig. 1.14 Output—disp function in MATLAB

above example, *num2str*() function is used to convert the data types of the variable in string format. An example of the usage of *disp*() is shown in Fig. 1.13 with its output in Fig. 1.14 for better understanding.

Output:

1.12   Conclusion

This chapter provides a brief introduction to MATLAB and presents the history, purpose, and importance of MATLAB. The concept of MATLAB is put forward such that the chapter can be interesting for a new audience. To help the readers getting started with MATLAB with a hand-in experience, the chapter provides step-by-step methods of its installation procedures. The readers are advised to implement all the examples and coding in MATLAB simultaneously to understand the contents more rigidly. To accomplish this purpose, this chapter introduces the overall MATLAB environment along with some fundamental features, so that the readers can feel comfortable writing MATLAB code while going through the rest of the book.

Exercise 1

1. Write down some of the notable applications of MATLAB in engineering.
2. What are the major data types in MATLAB? How are they represented in MATLAB programming?
3. Mention the usage of the following commands/functions with examples where applicable:

   (a) clc
   (b) num2str()
   (c) str2double()
   (d) int8()
   (e) disp()

4. Perform the following operations in MATLAB and save the results in a variable. Demonstrate the variables using "whos" command:

   (a) 2*4^2
   (b) (2*4)^2
   (c) 503+224−604
   (d) (10^3)/(9*2)
   (e) 6.25*0.42^3.56
   (f) Save "MATLAB is fun!" in a variable

5. Take two numerical inputs from the user and save them in variables *num1* and *num2*. Perform the following operations and record the session using the *diary* function:

   (i) *num1/num2*
   (ii) *num1\num2*

   Do they produce the same result? If not, why?

Reference

1. https://www.mathworks.com/help/install/

Chapter 2
Vectors and Matrices

2.1   Introduction

Vectors and matrices are two of the data types that can be taken as input directly by
MATLAB. Initially, MATLAB was regarded as a matrix calculator, which has been
extended and modified significantly over the years. A vector is a one-dimensional
matrix that contains multiple values ordered either in a row or a column. If the values
are ordered in a row, the vector is regarded as a row vector. Conversely, in a column
vector, all the values are incorporated in a single column. A matrix has multiple
values placed in both rows and columns. In another sense, a matrix can be formed by
aggregating multiple vectors. The number of rows and columns of a matrix determines its size. As in MATLAB, the input data types can be of both matrix and
vector; it is imperative to learn about different vector and matrix operations that can
be performed directly in MATLAB. Therefore, this chapter aims to demonstrate the
manipulations and operations of vectors and matrices in light of MATLAB
implementations.

2.2   Creating Vectors

A vector can be created in MATLAB by using the values enclosed within square
brackets, [ ]. To differentiate among different values, a space is used. A vector can be
a row vector or column vector. In a row vector, the values are separated using space,
and in the case of a column vector, semicolons are used among the values. For
example, A is a row vector, which contains four values that are separated using space
between two neighboring values (Fig. 2.1). Here, the size of vector A is 1×4, which
represents the vector containing four values in a single row.

For a column vector, the values are arranged over a single column. An example of
a column vector having a size of 4×1 is shown in Fig. 2.2.

© The Author(s), under exclusive license to Springer Nature Switzerland AG 2022   19
E. Hossain, *MATLAB and Simulink Crash Course for Engineers*,
https://doi.org/10.1007/978-3-030-89762-8_2

```
A=[1 5 7 9]

 A = 1×4
        1     5     7     9
```

Fig. 2.1 A row vector

```
B=[1;5;7;9]

 B = 4×1
        1
        5
        7
        9
```

Fig. 2.2 A column vector

```
B=1:2:12

 B = 1×6
        1     3     5     7     9    11
```

Fig. 2.3 A row vector with increments

```
C=12:-2:1

 C = 1×6
       12    10     8     6     4     2
```

Fig. 2.4 A row vector with decrements

A vector having values of equal increment or decrement can be defined in an easier way for avoiding enlisting a large number of values manually. For example, if a vector B contains values starting from 1 to 12 with +2 increment, it can be defined as shown in Fig. 2.3.

The vector B has three entities—the first one is the starting value and the last one is the ending value of the vector. The values are equally spaced and have a +2 increment. The increment value represents the middle entity of the B vector. In case, the increment is +1, the middle entity can be skipped as MATLAB considers +1 as the default increment. For decrement or negative increment, the middle entity will become a negative value. An example of a vector having descending values with negative increment is given in Fig. 2.4.

```
A=[1 3 5;2 4 3;2 8 4;1 6 9]
```

```
A = 4×3
          1      3      5
          2      4      3
          2      8      4
          1      6      9
```

Fig. 2.5 A 4 × 3 matrix

2.3 Creating Matrices

A matrix can be defined in the same way as a vector. A matrix can have multiple rows and columns, based on which its size is determined. The size can be controlled by using space and semicolons. Space indicates a shift from one column to another within a single row, whereas a semicolon represents the end of one row or the start of a new row. The size of any vector is defined by row × column.

An example is shown in Fig. 2.5 to realize the row-column definition of a matrix in MATLAB.

Here, the size of matrix A is 4 × 3, which signifies A has 4 rows and 3 columns. The semicolon is used to shift to a new row, while space corresponds to separate column values.

2.4 Manipulation of Vectors and Matrices

A matrix can be formed by combining multiple vectors. A simple example is given in Fig. 2.6, where three row vectors $V1$, $V2$, and $V3$ formed a new matrix M.

While combining vectors to make a matrix, it is noted that the sizes of all the vectors need to be identical.

After defining a vector or matrix, it is essential to access each value. In a matrix, row and column numbers are used to specify the position of a certain value, which can be used to access a specific value in a matrix. For example, in matrix M, the value "8" situated in the third column of the second row. This value can be accessed by using the following command in Fig. 2.7.

Here, 2 signifies the row number and 3 represents the column number. Using this concept, a matrix can be broken down into separate vectors. For example, the matrix M can be interpreted as three separate row vectors as presented in Fig. 2.8.

Here, $V1$, $V2$, and $V3$ represent the values of the first, second, and third rows of M, respectively. It is noted that a colon (:) is used while accessing multiple values. $V1 = M(1, :)$ signifies all the values that are positioned in the first row of every column of matrix M. If the entities are interchanged as $V1 = M(:, 1)$, it will depict all the values of M residing on the first column and every row. In that case, $V1$ will become a column vector instead of a row vector.

```
V1=[1 2 5]
```

```
V1 = 1×3
        1     2     5
```

```
V2=[4 6 8]
```

```
V2 = 1×3
        4     6     8
```

```
V3=[5 7 9]
```

```
V3 = 1×3
        5     7     9
```

```
M=[V1;V2;V3]
```

```
M = 3×3
        1     2     5
        4     6     8
        5     7     9
```

Fig. 2.6 Manipulation of vectors and matrices

```
M(2,3)
```

```
ans = 8
```

Fig. 2.7 Accessing value from a matrix

2.5   Dimensions of Matrices

The dimension of a matrix can be represented by row $\times$ column, where row indicates the total number of rows and column refers to the number of columns of that matrix. In MATLAB, the dimension of a matrix can be determined by using $size()$.

The MATLAB command for determining the dimension of a matrix, A:

$$size(A)$$

```
M=[1 2 5;4 6 8;5 7 9]
```

```
M = 3×3
           1     2     5
           4     6     8
           5     7     9
```

```
V1=M(1,:)
```

```
V1 = 1×3
           1     2     5
```

```
V2=M(2,:)
```

```
V2 = 1×3
           4     6     8
```

```
V3=M(3,:)
```

```
V3 = 1×3
           5     7     9
```

Fig. 2.8 Accessing vectors from a matrix

```
% Dimension of a matrix
clear;clc;
A = [2 1;4 3;2 1]
dim = size(A);
fprintf('Dimension of matrix A:\n');
disp(dim)% Dimension of a matrix
```

Fig. 2.9 Code—Dimension of a matrix

2.5.1 MATLAB Example 2.10: Dimension of a Matrix

The MATLAB code for determining the dimension of a matrix is given in Fig. 2.9, and its output is given in Fig. 2.10.

Output:

Fig. 2.10 Output—
Dimension of a matrix

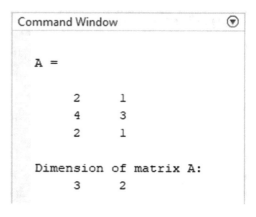

2.6   Operations on Matrices

Several operations may be performed upon a matrix, such as addition, subtraction, multiplication, making transpose, making inverse, etc. These operations are described in the following sections.

2.6.1   Addition and Subtraction

The addition and subtraction of two matrices can be performed by using the "+" and "−" signs, respectively. However, it is to be noted that the sizes of two matrices need to be identical while performing either addition or subtraction.

> **The MATLAB command for addition is "+" sign.**
> **The MATLAB command for subtraction is "−" sign.**

2.6.2   MATLAB Example 2.1: Addition and Subtraction

In Fig. 2.11, both A and B are 3×3 matrices. The addition of these two matrices is the element-wise addition, and the size of the output is the same as matrices A and B. Similarly, the subtraction (sub) is the element-wise subtraction of two matrices A and B.

```
A=[1 4 6;2 5 7;3 6 8]

A = 3×3
        1       4       6
        2       5       7
        3       6       8


B=[2 4 6;3 5 7;4 6 8]

B = 3×3
        2       4       6
        3       5       7
        4       6       8


Add=A+B

Add = 3×3
        3       8      12
        5      10      14
        7      12      16


Sub=A-B

Sub = 3×3
       -1       0       0
       -1       0       0
       -1       0       0
```

Fig. 2.11 Addition and subtraction of matrices

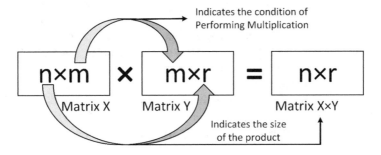

Fig. 2.12 Matrix size in multiplication

2.6.3   *Multiplication*

The multiplication of two matrices is possible only if the number of columns of the first matrix equals the number of rows of the second matrix. Consider two matrices X and Y as shown in Fig. 2.12, whose sizes are $n \times m$ and $m \times r$, respectively. As the number of columns of X and the number of rows of Y are similar, X and Y can be multiplied. The size of the product of these matrices will be $n \times r$.

```
X=[1 2 4;2 5 6]

    X = 2×3
              1       2       4
              2       5       6


Y=[1 2;3 4;5 7]

    Y = 3×2
              1       2
              3       4
              5       7


mul=X*Y

   mul = 2×2
             27      38
             47      66
```

Fig. 2.13 Matrix multiplication

2.6.4   MATLAB Example 2.2: Multiplication

The MATLAB command for multiplication is the "*" symbol.

In Fig. 2.13, the sizes of matrices X and Y are 2×3 and 3×2, respectively. The column number of X and the row number of Y are similar, which is 3. Hence, multiplication can be performed and the size of the output is 2×2 (the row number of $X \times$ the column number of Y).

Note that the multiplication of matrices is noncommutative, i.e., $M \times N$ and $N \times M$ are not the same thing. Therefore, if two matrices fulfill the condition of multiplicativity, it is not necessary that they will do so in the reverse order as well.

2.6.5   Transpose

A transpose matrix is a matrix whose row and column values interchange with each other. For example, if X is a $n \times m$ dimensional matrix, the dimension of the transpose matrix of X will be $m \times n$ by swapping the row and column values with each other.

```
X=[1 2 4;2 5 6]
```

```
X = 2×3
         1     2     4
         2     5     6
```

```
X_T=X'
```

```
X_T = 3×2
         1     2
         2     5
         4     6
```

Fig. 2.14 Transpose of a matrix

2.6.6 *MATLAB Example 2.3: Transpose*

The MATLAB command for transposing a matrix is to use the prime (′) symbol.

In Fig. 2.14, the dimension of the X matrix is 2×3, while the transpose matrix X_T has a dimension of 3×2 due to the swapping of row and column values.

2.6.7 *Determinant*

A determinant of a matrix provides a distinguished scalar value; however, it is only applicable for a square matrix. A square matrix is a matrix whose row and column numbers are identical. Consider a square matrix X, whose dimension is 2×2. The determinant of matrix X can be referred to as $|X|$ or $\det(X)$, which can be determined as follows:

$$X = [a\ b; c\ d], \tag{2.1}$$

$$|X| = ad - bc. \tag{2.2}$$

Consider another square matrix Y with a dimension of 3×3. In this case, the determinant can be derived by utilizing the following method:

$$Y = [a\ b\ c; d\ e\ f; g\ h\ i] \tag{2.3}$$

$$|Y| = a * \det[e\ f\ h\ i] - b * \det[d\ f\ g\ i] + c * \det[d\ e\ g\ h] \tag{2.4}$$

```
A=[1 2 4;3 2 1;2 2 1]

A = 3×3
          1      2      4
          3      2      1
          2      2      1


det(A)

ans = 6
```

Fig. 2.15 Determinant of a matrix

2.6.8   MATLAB Example 2.4: Determinant

Figure 2.15 shows how to create a determinant in MATLAB.

> **The MATLAB command for calculating determinant is to use** *det()*
> **function.**

2.6.9   Identity Matrix

An identity matrix has three distinctive characteristics:

1. It is a square matrix.
2. The diagonal values of an identity matrix are 1. The other values except for the diagonal values are all zero.
3. The determinant of an identity matrix is always 1.

   For example, a 3×3 identity matrix, $I_{3 \times 3}$, or $I_3 = [1\ 0\ 0; 0\ 1\ 0; 0\ 0\ 1\ ]$

2.6.10   MATLAB Example 2.5: Identity Matrix

> **The MATLAB command for identity matrix is to use** *eye(N)* **function,**
> **where N represents the dimension.**

In Fig. 2.16, *I* is a three-dimensional identity matrix.

```
I=eye(3)
```

```
I = 3×3
      1      0      0
      0      1      0
      0      0      1
```

Fig. 2.16 Identity matrix

2.6.11   Inverse Matrix

If the determinant of a certain matrix is zero, it is called a singular matrix. A non-singular matrix always has a nonzero determinant. Consider a non-singular square matrix X having a dimension of $n \times n$. If another matrix Y, having the same dimension as X, can be related as $XY = I$, the second matrix Y can be regarded as the inverse matrix of X. Here, I represents the identity matrix, which has the same dimension as X and Y.

If X is a matrix, the inverse matrix of X can be represented as X^{-1}. The matrix X is invertible only if:

1. X is a square, non-singular matrix
2. $XX^{-1} = I$

The mathematical formula for determining inverse matrix X^{-1} is as follows:

$$X^{-1} = \frac{\mathrm{adj}(X)}{\det(X)} \tag{2.5}$$

Here, $\mathrm{adj}(X)$ is the adjugate of matrix X. The adjugate of a matrix can be determined by transposing the cofactor of that matrix.

2.6.12   MATLAB Example 2.6: Inverse Matrix

The MATLAB command for calculating the inverse of a matrix is to use *inv()* function.

In Fig. 2.17, X is a 3×3 square non-singular matrix. Implementing inv(X), the inverse matrix of X is determined, which has the same dimension as matrix X.

```
X=[1 2 4;3 2 1;2 2 1]

X = 3×3
          1     2     4
          3     2     1
          2     2     1


inv(X)

ans = 3×3
     -0.0000     1.0000    -1.0000
     -0.1667    -1.1667     1.8333
      0.3333     0.3333    -0.6667
```

Fig. 2.17 Inverse matrix

```
% Matrix concatenation
clear;clc;
A = [1 4;2 4;3 2];
B = [2 -4;1 3;7 9];
fprintf('Horizontal concatenation:\n');
C = [A,B]
fprintf('Vertical concatenation:\n');
D = [A;B]
```

Fig. 2.18 Code—Matrix concatenation

2.7 Simple Matrix Concatenation

Multiple matrices can be appended together to form a larger combined matrix using matrix concatenation. Matrix concatenation can be of two types—horizontal and vertical concatenation. When two matrices are concatenated horizontally, the process is called horizontal concatenation; conversely, vertical concatenation occurs when one matrix appends with the other vertically.

A MATLAB example showing both horizontal and vertical concatenation is given below.

2.7.1 MATLAB Example 2.9: Matrix Concatenation

The MATLAB code demonstrating matrix concatenation is given in Fig. 2.18, and its output is given in Fig. 2.19.

Fig. 2.19 Output—Matrix
concatenation

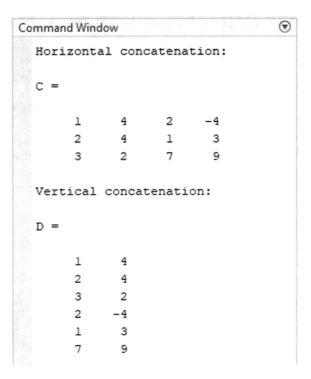

Output:

In horizontal concatenation, two matrices are enclosed within the third bracket by separating them with a comma. For vertical concatenation, the comma is replaced by a semicolon.

2.8   Creating Arrays of Zeros, Ones, and Random Numbers

While writing MATLAB programs, arrays of zeros and ones may become essential in some aspects. In MATLAB the commands for producing each of these arrays are listed below:

> **The MATLAB command for producing arrays of zeros:** *zeros(row, col)*
> **The MATLAB command for producing arrays of ones:** *ones(row, col)*

2.8.1   MATLAB Example 2.7: Arrays of Zeros and Ones

The MATLAB code for demonstrating arrays of zeros and ones is given in Fig. 2.20, and its output is given in Fig. 2.21.

Output:

Here, the sizes of arrays for both zeros and ones are considered as 3×4.

Random number generation is another important task that is usually needed while writing MATLAB code. For verifying any particular program with random input values, random number generation can be very effective. In MATLAB, based on the types of random numbers that are required to be generated, the MATLAB command may vary. Some of the MATLAB commands for generating random number arrays are given below:

```
% Arrays of zeros, ones
clear;clc;
row = 3;
col = 2;
A = zeros(row,col); % array of zeros
B = ones(row,col);  % array of ones
fprintf('Array of zeros:\n');
disp(A);
fprintf('Array of ones:\n');
disp(B);
```

Fig. 2.20 Code—Arrays of zeros and ones

Fig. 2.21 Output—Arrays of zeros and ones

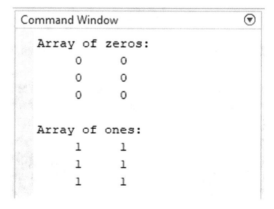

The MATLAB command for producing arrays of random numbers that are uniformly distributed:

$$rand(row, col)$$

The MATLAB command for producing arrays of random numbers that are normally distributed:

$$randn(row, col)$$

The MATLAB command for producing arrays of random pseudo-integer numbers that are uniformly distributed:

$$randi([num_{min}, num_{max}], [row, col])$$

Here, *row* and *col* indicate the array size; num_{min} and num_{max} are the minimum and maximum range of the generated random numbers.

The MATLAB command for producing arrays of random numbers that are uniformly distributed:

$$rand(row, col)$$

The MATLAB command for producing arrays of random numbers that are normally distributed:

$$randn(row, col)$$

The MATLAB command for producing arrays of random pseudo-integer numbers that are uniformly distributed:

$$randi([num_{min}, num_{max}], [row, col])$$

Here, *row* and *col* indicate the array size; num_{min} and num_{max} are the minimum and maximum range of the generated random numbers.

2.8.2   MATLAB Example 2.8: Random Numbers

The MATLAB code to create arrays of random numbers is given in Fig. 2.22, and its output is given in Fig. 2.23.

```
% Arrays of random numbers
clear;clc;
row = 3;
col = 2;
num_min = 2;
num_max = 8;
% uniformly distributed random numbers
A = rand(row,col);
% normally distributed random numbers
B = randn(row,col);
% uniformly distributed random pseudo-integer
C = randi([num_min,num_max],[row,col]);
fprintf('Array of uniformly distributed random numbers:\n');
disp(A);
fprintf('Array normally distributed random numbers:\n');
disp(B);
fprintf('Array uniformly distributed random pseudo-integer:\n');
disp(C);
```

Fig. 2.22   Code—Arrays of random numbers

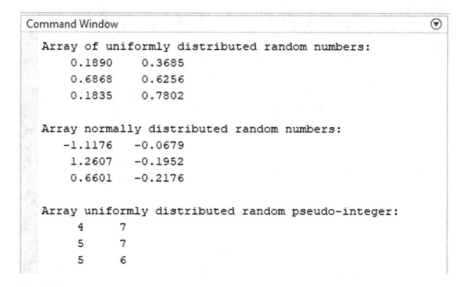

Fig. 2.23   Output—Arrays of random numbers

Output:

From the output, it can be observed that *randi*() can be used to generate an array of pseudo-integers within a predefined range. For the other two functions, *rand*() and *randn*() do not have the features of specifying the range.

2.9 Array Function for One-Dimensional Arrays

Several functions are available in MATLAB to perform operations on a one-dimensional array. Three important and widely used functions will be discussed in this section, namely, *linspace()*, *max ()*, and *min()*.

linspace() is a function that can be used to create a one-dimensional array, which contains equally spaced values within a specific range.

The MATLAB command for *linspace()* function:

$$linspace(Lower_{limit}, Upper_{limit}, point)$$

The above command creates a one-dimensional array that includes values within the range of [*Lower_{limit}, Upper_{limit}*], and *point* indicates the number of evenly spaced values.

In a *linspace()*, the number of values that one requires within a certain range can be given as input. The function automatically creates an array of evenly spaced values maintaining the range and the number of values.

In another version of MATLAB command, the space/range between the upper and lower limits can be specified. This command can create a one-dimensional array maintaining the specified range. The MATLAB command can be written as follows:

Alternative MATLAB command for *linspace()* function:

$$Lower_{limit} : space : Upper_{limit}$$

A MATLAB example is provided below to show a demonstration of both the commands:

2.9.1 MATLAB Example 2.11: Creating Linearly Spaced One-Dimensional Array

The MATLAB code to create a linearly spaced one-dimensional array is given in Fig. 2.24, and its output is given in Fig. 2.25.

Output:

```
% Linearly spaced one-dimensional array
% Using linspace function
clear;clc;
Up_range = 2;
Low_range = 12;
point = 5;
A = linspace(Up_range,Low_range,point);
fprintf('Linearly spaced one-dimensional array:\n');
fprintf('--------------------------------------------\n');
fprintf('Using linspace function:\n');
disp(A)
% Alternative version
space = 2.5;
B = Up_range:space:Low_range;
fprintf('Without using linspace function:\n');
disp(B)
```

Fig. 2.24 Output—Arrays of random numbers

```
Command Window                                                    ⊙

    Linearly spaced one-dimensional array:
    ---------------------------------------
    Using linspace function:
        2.0000    4.5000    7.0000    9.5000    12.0000

    Without using linspace function:
        2.0000    4.5000    7.0000    9.5000    12.0000
```

Fig. 2.25 Output—Linearly spaced one-dimensional array

From the output in Fig. 2.25, it can be observed that the results are the same for both MATLAB commands. However, the difference is the input that the user can provide. For *linspace*(), the number of points is given as an input, which is 5. For the second command, the space or the difference among the values is specified, which is 2.5. As the range is kept the same for both of the instances, the generated arrays are identical in both cases.

To determine the maximum or the minimum values within an array, *max*() and *min*() can be used in MATLAB.

> **The MATLAB command for determining the maximum value of an array, A:** *max(A)*
> **The MATLAB command for determining the minimum value of an array, A:** *min(A)*

```
% Finding Maximum and minimum value from an array
clear;clc;
A = randi([1,30],1,5)
max_A = max(A);
min_A = min(A);
fprintf('Maximum value of the array A:');
disp(max_A);
fprintf('Minimum value of the array A:');
disp(min_A);
```

Fig. 2.26 Code—Finding maximum and minimum value from an array

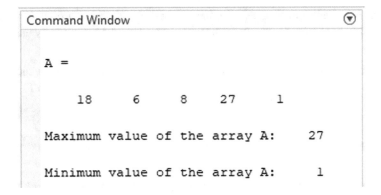

Fig. 2.27 Output—Finding maximum and minimum value from an array

2.9.2   MATLAB Example 2.12: Finding Maximum and Minimum Value from an Array

The MATLAB code to find the minimum and maximum value from an array is given in Fig. 2.26, and its output is given in Fig. 2.27.

Output:

2.10   Mean, Standard Deviation, Variance, and Mode

Mean, standard deviation, and variance are three essential statistical terminologies to understand the distribution of data. In a one-dimensional array, multiple values are available. Determining the mean, standard deviation, and variance of an array can help to understand the characteristics of the data and infer conclusions from the pattern.

Mean: The average of all the values in an array is called the mean or arithmetic mean. In MATLAB, the mean value of an array can be determined using the *mean* () function.

Variance: The variance is the average of the squared difference between each value and the mean of an array, which can be represented as in Eq. (2.6).

$$\text{Variance} = \frac{\sum_{n=1}^{N}(x_n - \underline{x})^2}{N}, \quad \text{where, } n = 1, 2, 3, \ldots, N \qquad (2.6)$$

Here, N is the length of an array or the number of values in an array; $\underline{x}$ is the mean of the array. In MATLAB, the variance of an array can be determined by using *var()* function.

Standard deviation: Standard deviation is the square root of the variance of an array. Therefore, the standard deviation of an array can be written as in Eq. (2.7).

$$\text{Standard deviation} = \sqrt{\frac{\sum_{n=1}^{N}(x_n - \underline{x})^2}{N}}, \quad \text{where, } n = 1, 2, 3, \ldots, N \qquad (2.7)$$

The MATLAB command for determining the standard deviation of an array is *std()*.

Mode: In an array, the value that creates the maximum appearance is regarded as the mode of that array. The MATLAB command for determining the mode of an array is *mode()*.

> **The MATLAB command for determining the mean value of an array, A:**
>   *mean(A)*
> **The MATLAB command for determining the variance of an array, A:**
>   *var(A)*
> **The MATLAB command for determining the standard deviation of an array, A:** *std(A)*
> **The MATLAB command for determining the mode of an array, A:**
>   *mode(A)*
> **It is to be noted that** *mode(A)* **provides the most frequent value of an array. However, if all the values in an array appear only one time,** *mode()* **recognizes the lowest value of that array as its mode.**

```
% Mean, variance, standard deviation, and mode
clear;clc;
A = randi([1,50],1,6);
mean_A = mean(A);
variance_A = var(A);
std_A = std(A);
mode_A = mode(A);
fprintf('One-dimensional array, A:\n');
disp(A);
fprintf('Mean value of A = %.2f\n',mean_A);
fprintf('Variance of A = %.2f\n',variance_A);
fprintf('Standard deviation of A = %.2f\n',std_A);
fprintf('Mode of A = %.2f\n',mode_A);
```

Fig. 2.28 Output—Finding maximum and minimum value from an array

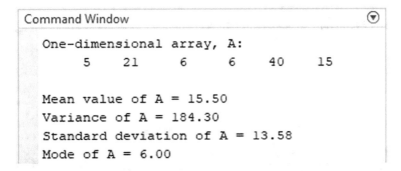

Fig. 2.29 Output—Mean, variance, standard deviation, and mode

2.10.1 *MATLAB Example 2.13: Mean, Variance, Standard Deviation, and Mode*

The MATLAB code demonstrating the use of mean, variance, standard deviation, and mode is given in Fig. 2.28, and its output is given in Fig. 2.29.

Output:

2.11 Dot Operator

The dot operator is utilized to perform element-wise operations. In many cases, the dot operator becomes useful in MATLAB. Some of the useful instances where dot operator can be useful are listed below:

1. Dot multiplication or element-wise multiplication of two arrays having the same size.

    Usage: Dot operator "." is used right before the multiplication sign (*).
2. Element-wise division among two arrays having the same size, or when the numerator is a scalar value and the denominator is an array.

    Usage: Dot operator "." is used right before the division sign (/).
3. Determining power or exponential whenever either the base or the power is an array.

    Usage: Dot operator "." is used right before the power sign (^).
4. For writing complicated and complex equations.

2.11.1   MATLAB Example 2.14: Instances of the Dot Operator

The MATLAB code demonstrating the use of the dot operator is given in Fig. 2.30, and its output is given in Fig. 2.31.

Output:

```
% Some instances of the usage of dot operator
clc;clear;
% A and B are two arrays;
% scalar_val is a scalar value;
A = randi([1,2],2,3);
B = randi([1,2],2,3);
fprintf('Some instances of the usage of dot operator:\n');
fprintf('----------------------------------------------\n');
% Dot multiplication
fprintf('Dot multiplication of two arrays:\n')
disp(A.*B);
% Element-wise division between two arrays
fprintf('Element-wise division of two arrays:\n')
disp(A./B);
% Division: Numerator-scalar and denominator-array
scalar_val = 5;
fprintf('Division when numerator-scalar and denominator-array:\n')
disp(scalar_val./A);
% Power value: Either the base, or power is an array
Base = 10;
fprintf('Power term is an array:\n')
disp(Base.^A)
fprintf('Base term is an array:\n')
disp(A.^scalar_val)
```

Fig. 2.30 Code—Dot operator

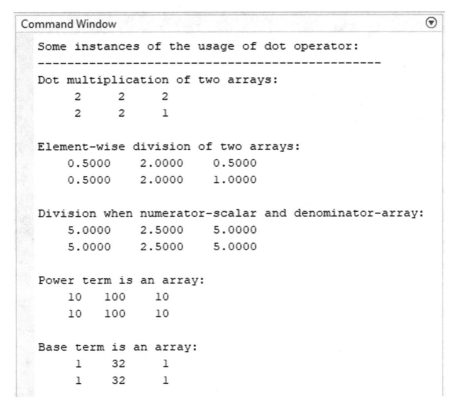

Fig. 2.31 Output—Dot operator

2.12   Table Arrays, Cell Arrays, and Structure Arrays

In MATLAB, a table can be created containing different data types by using the
built-in function *table*(). In a MATLAB-created table, all the variables can be
accessed and operated like separate arrays. The MATLAB command for *table*() is
given below:

> **The MATLAB command for creating a table:** *table*(*variable1*,
>   *variable2*, ...)
> **Here,** *variable1*, *variable2*, . . .**indicate the variables to be incorporated in
>   the table.**

```
% Creating table
% Headers: Battery name, Energy density, Lifecycle, Safety
clear;clc;
Battery_name = {'Li-ion';'Liquid super capacitor';'Lead acid'};
Energy_density = [5;2.5;2];
Life_cycle = [2;5;1.5];
safety = {'High';'Low';'Moderate'};
TABLE = table(Battery_name,Energy_density,Life_cycle,safety);
fprintf('Comparison among different battery types:\n');
fprintf('-----------------------------------------------\n');
disp(TABLE);
% Accessing each column of the table
fprintf('Accessing the data of Battery_name column:\n');
disp(TABLE.Battery_name);
fprintf('Accessing the data of Energy_density column:\n');
disp(TABLE.Energy_density);
fprintf('Accessing the data of Life_cycle column:\n');
disp(TABLE.Life_cycle);
fprintf('Accessing the data of safety column:\n');
disp(TABLE.safety);
```

Fig. 2.32 Code—Creating table in MATLAB

2.12.1   MATLAB Example 2.15: Creating Table

The MATLAB code to create a table is given in Fig. 2.32, and its output is given in Fig. 2.33.

Output:

In the above-mentioned example, a table containing four different variables—Battery_name, Energy_density, Life_cycle, and safety—is generated using the MATLAB command. In the table, "Battery_name" and "safety" are cell arrays, whereas the remaining variables have "double" data type values. Each of these arrays can be accessed separately to perform operations individually, which have been shown in the above examples as well.

2.12.1.1   Cell Array

In a cell array, there are different cells, each of which can contain data of different data types. In MATLAB, the command for creating a cell array is given below:

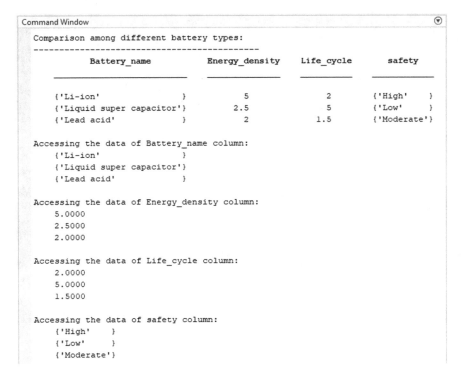

Fig. 2.33 Output—Creating table in MATLAB

The MATLAB command for creating a cell array:

$$cell(row, col)$$

Here, *row*, *col* are used to define the size of the cell array, as well as to access each of the cells.

2.12.2 MATLAB Example 2.16: Cell Array

The MATLAB code demonstrating cell arrays is given in Fig. 2.34, and its output is given in Fig. 2.35.

Output:

```
% Cell array
clear;clc;
A = cell(2,3);
A(1,:)={3,4,4};
A(2,:)={'A','B','C'};
fprintf('Cell array, A:\n')
disp(A)
```

Fig. 2.34 Code—Cell array in MATLAB

Fig. 2.35 Output—Cell
array in MATLAB

```
Command Window                                              ⊙

   Cell array, A:
       {[3]}      {[4]}      {[4]}
       {'A'}      {'B'}      {'C'}
```

In the above cell array, it can be observed that each cell can contain data of different types. In the first row, all the cells contain data of "double" data types, whereas in the second row, all the cells have "character" data types.

2.12.2.1 Structure Array

In a structure array, different data can be grouped in several fields. In each field, there can be data of different types. In MATLAB, the command for creating a structured array is as follows:

The MATLAB command for creating a structured array:

$$struct(Field1, Val1_{Field1}, Field2, Val2_{Field2}, \ldots\ldots)$$

Here, *Field1*, *Field2*, ... **indicates the different fields under which different data will be grouped. The grouped data of each field is represented by** $Val1_{Field1}$, $Val2_{Field2}$, ...

2.12.3 *MATLAB Example 2.17: Structured Array*

The MATLAB code demonstrating structured arrays is given in Fig. 2.36, and its output is given in Fig. 2.37.

```
%% Structure array
clear;clc;
Field1='Battery_Name';
val_Field1 = {'Li-ion','Liquid super capacitor','Lead acid'};
Field2='Energy_Density';
val_Field2 = {5,2.5,2};
Field3='Life_cycle';
val_Field3 = [2,5,1.5];
Field4='safety';
val_Field4 = {'High','Low','Moderate'};
fprintf('Sturcture array of different battery types and properties:\n')
S = struct(Field1,val_Field1,Field2,val_Field2,Field3,...
                  val_Field3,Field4,val_Field4)
fprintf('Accessing first field of the structure:\n\n');
disp(S(1))
fprintf('Accessing second field of the structure:\n\n');
disp(S(2))
fprintf('Accessing third field of the structure:\n\n');
disp(S(3))
```

Fig. 2.36 Code—Structured array in MATLAB

Output:

In this example, a structured array that contains parameters of different battery types is created using the MATLAB command. Here, the fields of the structured array are Battery_name, Energy_density, Life_cycle, and safety. Under each field, there are grouped data of different data types. The above example also shows how to access each field by using the dot operator.

2.13   Conclusion

From this chapter, the readers will be able to learn about different operations of vectors and matrices that can be performed in MATLAB. The manipulations of vectors and matrices along with some special arrays that can be created utilizing MATLAB built-in functions, such as zeros, ones, and random numbers generator, are covered in this chapter. The usage of the dot operator, matrix concatenation, and some widely used array functions applicable for one-dimensional arrays is discussed and demonstrated in MATLAB. At the end of the chapter, some special arrays such as table array, cell array, and structured array are explained due to their importance in the engineering domain. This chapter is prepared in such a way that the readers can be introduced to most of the essential MATLAB commands and functions related to vectors and matrices with relevant MATLAB examples.

```
Command Window                                                           ⊙

   Sturcture array of different battery types and properties:

   S =

     1×3 struct array with fields:

       Battery_Name
       Energy_Density
       Life_cycle
       safety

   Accessing first field of the structure:

         Battery_Name: 'Li-ion'
        Energy_Density: 5
           Life_cycle: [2 5 1.5000]
               safety: 'High'

   Accessing second field of the structure:

         Battery_Name: 'Liquid super capacitor'
        Energy_Density: 2.5000
           Life_cycle: [2 5 1.5000]
               safety: 'Low'

   Accessing third field of the structure:

         Battery_Name: 'Lead acid'
        Energy_Density: 2
           Life_cycle: [2 5 1.5000]
               safety: 'Moderate'
```

Fig. 2.37 Output—Structured array in MATLAB

Exercise 2

1. Define vectors and matrices. What are their applications in engineering?
2. What is the difference between the rand(), randn(), and randi() functions? Explain with examples.
3. What will be the output of the following commands?

   (a) $A = 3:3:15$
   (b) $B = [;]$
   (c) $Z = [143,324,676,432;656,657,987,235;768,876,234,764]; Z(2,:)$

4. Consider three matrices given as follows:

$$MatA = \begin{bmatrix} 4 & 7 & 1 \\ 7 & 2 & 3 \\ 5 & 5 & 9 \end{bmatrix}; MatB = \begin{bmatrix} 6 & 0 & 4 \\ 9 & 8 & 1 \\ 7 & 5 & 2 \end{bmatrix}; MatC = \begin{bmatrix} 2 & 5 & 3 \\ 0 & 17 & 9 \\ 8 & 0 & 1 \end{bmatrix}$$

(i) Calculate the following:

   (a) MatA + MatB
   (b) MatB − MatC
   (c) MatA/MatC
   (d) Transpose of MatB
   (e) Determinant of MatC
   (f) Inverse MatA
   (g) Horizontally concatenate MatB and MatC
   (h) Vertically concatenate MatC and MatA

(ii) Determine MatA ∗ MatB, MatB ∗ MatA, and MatA. ∗ MatB. Do the results vary? If so, why?

5. Given an array $a = linspace(2,20,100)$. What is the mean, variance, standard deviation, and mode of a?

6. Suppose you are working with five semiconductor materials, namely, silicon (Si), germanium (Ge), tin (Sn), carbon (C), and tellurium (Te). Each of them has a bandgap of 1.12, 0.67, 0.08, 5.47, and 0.33 eV, respectively, eV being their unit of measurement.

   (a) Enlist the information in a table with a column for "Serial_Number," "Element_Name," "Element_Symbol," and "Bandgap." Use MATLAB "table" function for the purpose.
   (b) Form and display a structured array from the above information with the same column name as mentioned in 5a. Change the bandgap of tin from 0.08 to 0.07 eV by accessing the specific field and display the array again.

Chapter 3
Programs and Functions

3.1 Introduction

To write a complete program using MATLAB, the creation, saving, running, and publishing of a MATLAB script are the fundamental steps. Similar to other programming languages, MATLAB can also be used to write conditional statements and loops. In this chapter, some examples are shown implementing different conditional statements and loops. In MATLAB, it is possible to create user-defined functions, which becomes important to avoid writing redundant code for performing the same task multiple times within the program. The methods of creating such functions and the implementations are discussed in this chapter with multiple examples.

3.2 Scripts

For compiling a sequence of code, a script is required which allows one to write a complete program and run it sequentially at the end. The script can be saved as a file for a future run by calling the name of the script. In MATLAB, there are multiple ways of creating a new script.

1. A new script can be created by clicking the "New Script" option from the header toolstrip, as shown in Fig. 3.1.
2. Right-click on the "New" button, which will cause an appearance of a scrolling menu. By selecting the "Script" option from that menu bar, a new script can be created (Fig. 3.2).
3. In the command window using the "edit" command, a new script can be created (Fig. 3.3).
4. Using a shortcut—"Ctrl + N"—on the keyboard, a new script can be generated.

© The Author(s), under exclusive license to Springer Nature Switzerland AG 2022
E. Hossain, *MATLAB and Simulink Crash Course for Engineers*,
https://doi.org/10.1007/978-3-030-89762-8_3

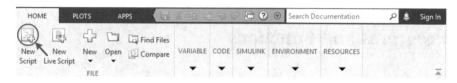

Fig. 3.1 Header Toolstrip

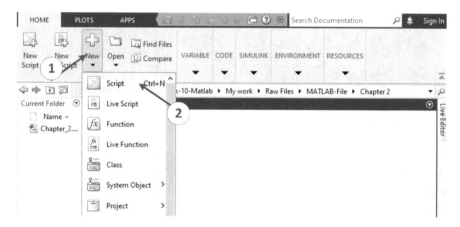

Fig. 3.2 New Script

Fig. 3.3 *edit* command

Fig. 3.4 Live Script

3.2.1 Live Script

A new version of Script called "Live Script" is available from MATLAB 2016 version. It allows accommodating both command and compiled output in the same script. The file can be saved which can have both the programs and the output. The "Live Script" feature facilitates the sharing of the program along with the output. To open a "Live Script," click on the "Live Script" button from the header toolstrip as shown in Fig. 3.4.

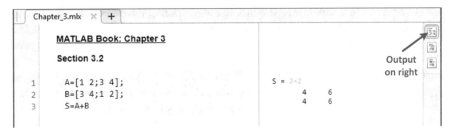

Fig. 3.5 Output on right

Fig. 3.6 Output inline

Fig. 3.7 Hide Code

A Live Script provides three format styles—output on right, output inline, and hide code. A sample version of Live Script along with these three features is shown in Figs. 3.5, 3.6, and 3.7.

3.2.2   Script vs. Live Script

Live script is an interactive platform, where both the code and the output stay in the same environment. The output figures will be generated on the same live script page instead of creating a separate figure file. In the script file, the outputs can be seen in the command window. The figures are produced in a separate figure window. The advantage of a live script is that it can be shared with the produced outputs. However, for a script file, the output cannot be shared. The user needs to run a script file to observe the outputs. Live script is very useful for making lectures or books, as it provides both text and code insert options. A script file is more appropriate and professional to implement in projects or real-time applications.

3.3 Saving, Running, and Publishing a Script

The processes of saving, running, and publishing a script in MATLAB are described in this section.

3.3.1 Saving a Script

After writing a complete program in a script, it is essential to save it for future use. To save a script file, there are multiple ways, such as the following:

1. Click on the "Save" option on the Editor tab from the header toolstrip as shown in Fig. 3.8. Later, navigate to the location and write the appropriate name for the script file to save it. The script file will be saved with an extension of ".m," which can be run in MATLAB in the future. If multiple scripts are open, the "Save" option will save only the active script file. However, MATLAB also provides options to save all the script files opened in the Editor. By opening the scrolling menu bar from the "Save" button, and selecting the "Save All" option will do that trick, as shown in Fig. 3.9.
2. Using the keyboard shortcut "Ctrl + S," a script file can also be saved.

Fig. 3.8 Saving a Script

Fig. 3.9 Saving Multiple Scripts

Fig. 3.10 Running a Script

Fig. 3.11 Running a Script without Saving

3.3.2 Running a Script

1. A script can be run by clicking the "Run" button Editor tab from the header toolstrip as shown in Fig. 3.10. If a script file is not saved before clicking the "Run" button, MATLAB will prompt the save option first by default. Therefore, it is to be noted that the "Run" option works only for the saved script file.
2. Using the keyboard shortcut "F5," a script file can also be run.

MATLAB provides another feature for running a script file without saving upfront. By clicking the "Run and Advance" option (Fig. 3.11), a script file can be compiled without prior saving.

Another useful feature of MATLAB is to create different sections in a script file and run individual sections. For debugging, this option provides great flexibility. To create a new section in a script file, there are two ways:

1. Selecting a place on the script where a new section needs to be created and clicking on the "Insert" button on the Editor tab (Fig. 3.12)
2. By typing "%%"—right from where a new section needs to be created (Fig. 3.13)

Fig. 3.12 Creating Different Sections in a Script File using Insert

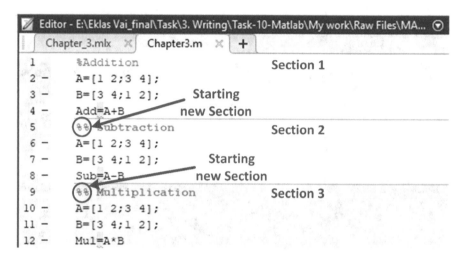

Fig. 3.13 Starting New Section in a Script File using %%

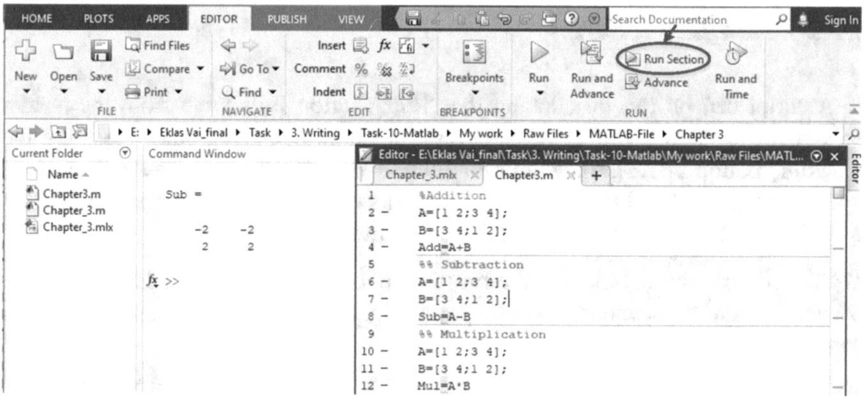

Fig. 3.14 Running a Specific Section from a Script

After creating multiple sections, it may be required to run separate sections. For running individual sections in a script file, "Run Section" on the Editor tab can be utilized after selecting a specific section as shown in Fig. 3.14.

3.3.3 Publishing a Script

After confirming a programming script, it can be published in "html," in "pdf," or in other formats. The purpose of publishing a script is to make the script shareable with others. As the script file can be sharable in "html," "pdf," or other formats, the person with whom the script file will be shared does not require to open the MATLAB application. In the published format, the output will also be included in the document. The steps of publishing a script are listed below and illustrated in Fig. 3.15.

1. After writing a program in a script file, select the "PUBLISH" option from the header menu bar.
2. After that, choose the "Publish" dropdown option from the header toolstrip.
3. Choose the "Edit Publishing Options" from the dropdown box options.
4. Step 3 will create the appearance of the following window in Fig. 3.16, from where you can choose the desired output file format.

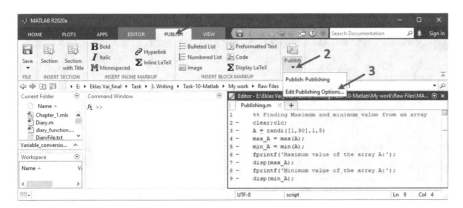

Fig. 3.15 Publishing a Script

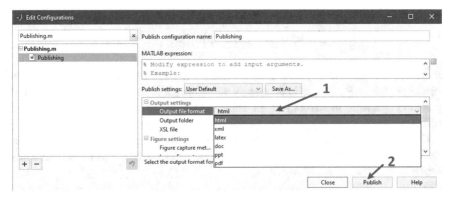

Fig. 3.16 Selecting the Output File Format

Finding Maximum and minimum value from an array

```
clear;clc;
A = randi([1,30],1,5)
max_A = max(A);
min_A = min(A);
fprintf('Maximum value of the array A:');
disp(max_A);
fprintf('Minimum value of the array A:');
disp(min_A);
```

```
A =

     6    10    10     7     8

Maximum value of the array A:    10

Minimum value of the array A:     6
```

Published with MATLAB® R2020a

Fig. 3.17 Maximum and minimum values from an array

5. Click on the "Publish" button. In this example, "html" format is selected for the publishing option, which creates the following publishing page (Fig. 3.17).

3.4 Conditional Statements and Loops

In conditional statements, logical and relational operators play important roles to formulate them. Hence, it is necessary to understand the logical and relational operators available in MATLAB. In Table 3.1, some of the important logical and relational operators along with the meanings and examples are listed.

Conditional statements are essential to execute a particular block of code based on conditions associated with it. The most basic and widely used conditional statements in MATLAB are "if" and "switch."

Table 3.1 Logical and relational operators

Logical and relational operator	Definition	Example
==	Equal $a = = b$ will be true only if both a and b are completely identical	$1 = = 1 : True$ $2 = = 1 : : False$
~=	Not equal $a\sim = b$ will be true only if a and b are unequal	$2\sim = 1 : True$ $2\sim = 2 : False$
&	Logical "and" Condition 1 and Condition 2 will be true only if both Condition 1 and Condition 2 are true	$2 > 1 \ \& \ 5 < 10 : True$ $2 > 1 \ \& \ 5 > 10 : False$
|	Logical "or" Condition 1/Condition 2 will be true if either Condition 1 or Condition 2 is true	$2 > 1 \ \mid 5 < 10 : True$ $2 > 1 \ \mid 5 > 10 : True$ $1 > 2 \ \mid 5 > 10 : False$
>	Greater than $a > b$ will be true, only if a is greater than b	$2 > 1 : True$ $2 > 2 : False$ $1 > 2 : False$
<	Less than $a < b$ will be true, only if a is less than b	$1 < 2 : True$ $1 < 1 : False$ $2 < 1 : False$
>=	Greater than or equal to $a >= b$ will be true, if a is greater than b, or a is equal to b	$5 > = 2 : True$ $2 > = 2 : True$ $5 > = 8 : False$
<=	Less than or equal to $a <= b$ will be true, if a is less than b, or a is equal to b	$2 < = 5 : True$ $2 < = 2 : True$ $8 < = 5 : False$

3.4.1   "If" Statement

The structure of an *if* statement can be represented in three formats as follows:

if **(logical conditions)** **Executable** **command**... **end**	*if* **(logical conditions)** **Executable** **command**... *else* **Executable** **command**... **end**	*if* **(logical conditions)** **Executable** **command**... *elseif* **Executable** **command**... *else* **Executable** **command**... **end**

The first structure is the simplest version of the "if" statement, and the second and the third one provide more extensions. In the above structure, "logical statements" indicate the conditions, which need to be TRUE for the execution of the "Executable command." If the conditions become FALSE, the compiler will skip the "Executable command." An example based on the first structure of the "if" statement is given below:

3.4.2 MATLAB Example 3.1: "If" Statement

The following MATLAB code in Fig. 3.18 shows the use of an *if* statement and Fig. 3.19 shows its output.

Output:

Here, *randi* is a MATLAB function that generates random integer values. The input parameters of the function "20" represent the maximum range of the integer, while 1 signifies the size of the output vector. Therefore, each time, *randi* is run in the above program, a pseudo-integer value of x is randomly generated, within the range of 20.

In the *if* statement within the first bracket, the condition is mentioned as $x > 10$. If this condition is satisfied, the successive command—*disp(x is greater than 10)*—will be executed; otherwise, this command will be skipped. Two outputs generated from two consecutive runs are shown above. In the first output, the initial value of x is 1, which does not satisfy the logical condition. Therefore, the logical condition of the *if* statement in the first run is FALSE, and the next *disp* command is ignored. In the second output, the value of x is 19, which satisfies the logical condition. Hence, the next *disp* command is executed. The *end* command finally truncates the *if* statement.

```
x=randi(20,1);
disp(['The value of x:', num2str(x)])
%% if statement
if (x>10)
    disp('x is greater than 10')
end
```

Fig. 3.18 Code—*if* statement

```
Command Window

   The value of x:1
fx >>
```

```
Command Window

   The value of x:19
   x is greater than 10
fx >>
```

Fig. 3.19 Output—*if* statement

```
x=randi(100,1);
disp(['The value of x:', num2str(x)])
%fprintf('The value of x: %d',x)
if (x>=80)
    disp('Grade: A')
elseif (x>=60 && x<80)
    disp('Grade: B')
elseif (x>=40 && x<60)
    disp('Grade: C')
else
    disp('Grade: F')
end
```

Fig. 3.20 Code—*elseif* and *else* statement

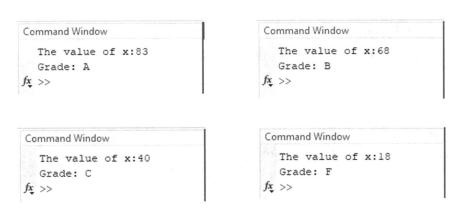

Fig. 3.21 Output—*elseif* and *else* statement

If multiple conditions are available, *else* and *elseif* commands become useful. An example is shown below where both these commands are used. The following MATLAB code in Fig. 3.20 shows the use of an if-else statement and Fig. 3.21 shows its output.

Output for four different runs:

While multiple conditions are needed to be considered for executing different tasks, *elseif* and *else* become more useful. If the first logical condition becomes FALSE, the execution of the first command will be skipped, and sequentially the second logical condition within the *elseif* statement will be checked as before. Thus, using multiple *elseif* statements, multiple conditions can be set forth for different outputs. The *else* statement becomes useful when the program needs to execute a final command in case multiple conditions fail to satisfy. In the above-mentioned code, the outputs have four outcomes—Grade: A, B, C, and F—depending on four conditions, which have been addressed by the one *if* statement, two *elseif* statements, and one *else* statement. The outcomes for four different grades have been shown in the output by executing multiple runs for better comprehensibility of the code.

3.4.3   Switch *Statement*

switch statement is another conditional statement through which multiple cases can be set up for executing different blocks of codes. It acts almost similar to the *if* statement; however, the *switch* statement provides more simplicity in terms of comprehension. However, the *switch* statement cannot be used for inequality; it only works for discrete equality cases. The structure of the *switch* statement is given below for facilitating the understanding:

```
switch Switch Expression
    case Case Expression
            Executable Statement
    case Case Expression
            Executable Statement
    . . . . . . . . . .
    otherwise
            Executable Statement
end
```

To execute a certain command, the Switch Expression needs to be matched with Case Expression. Whenever a specific Case Expression matches with a Switch Expression, the corresponding case statement will be executed, and MATLAB will exit the switch block right away.

3.4.4   *MATLAB Example 3.2:* Switch *Statement*

The following MATLAB code in Fig. 3.22 shows the use of a *switch* statement and Fig. 3.23 shows its output.

Output:

```
x=input('Enter a Month:','s');
switch x
    case 'January'
        disp(['Number of Days in ',x,':31'])
    case 'February'
        disp(['Number of Days in ',x,':28'])
    case 'March'
        disp(['Number of Days in ',x,':31'])
    case 'April'
        disp(['Number of Days in ',x,':30'])
    case 'May'
        disp(['Number of Days in ',x,':31'])
    case 'June'
        disp(['Number of Days in ',x,':30'])
    case 'July'
        disp(['Number of Days in ',x,':31'])
    case 'August'
        disp(['Number of Days in ',x,':31'])
    case 'September'
        disp(['Number of Days in ',x,':30'])
    case 'October'
        disp(['Number of Days in ',x,':31'])
    case 'November'
        disp(['Number of Days in ',x,':30'])
    case 'December'
        disp(['Number of Days in ',x,':31'])
    otherwise
        disp('Enter a Correct Name of Month')
end
```

Fig. 3.22 Code—*switch* statement

Fig. 3.23 Output—*switch* statement

```
Command Window

    Enter a Month:April
    Number of Days in April:30
fx >>
```

3.4.5 For *Loop*

for loop is a repetitive structure, where a certain task will be repeated in a systematic manner. This loop performs a set of tasks defined by the statements for a number of times, which is to be provided in the first line. *for* loop is usually used where there is a range of time for which the operation is needed to be run. Within a *for* loop, other conditional statements, such as *if* statement or nested *if* statement, can be embedded as required. The structure of the *for* loop is shown as follows:

> **for** index = values
>   **statements**
> **end**

An example will help to understand the syntax shown above.

3.4.6   MATLAB Example 3.3: "For" Loop

The following MATLAB code in Fig. 3.24 shows the use of a *for* loop and Fig. 3.25 shows its output.

Output:

In the example, an array *a* is provided, which has to be printed. After the *for* keyword, a range is provided to a variable *i*. The variable is an array from 1 to the length of *a*, which is 5. The statements inside the *for* loop is written after the indentation. The *end* keyword after the indented statements indicates the end of the loop. Upon running the program, the result shows that while the value of *i* is 1, the following two statements are executed, where the number of iteration (the value of *i*) and the value for that iteration (the value in array *a*) are shown. The operation is repeated until the last value of *i* is met.

Multiple *for* loops can also be nested like *if* statements. There are two other statements associated with the loop. A *break* statement within the *for* loop helps to exit the loop without running further. A *continue* statement, on the other hand, helps a program to skip the statements after the *continue* statement, and to start from the next iteration again.

Fig. 3.24 Code—*for* loop

```
a = [2,4,6,8,10];
for i=1:length(a)
    fprintf('Iteration: %d\n', i);
    fprintf('Value: %d\n', a(i))
end
```

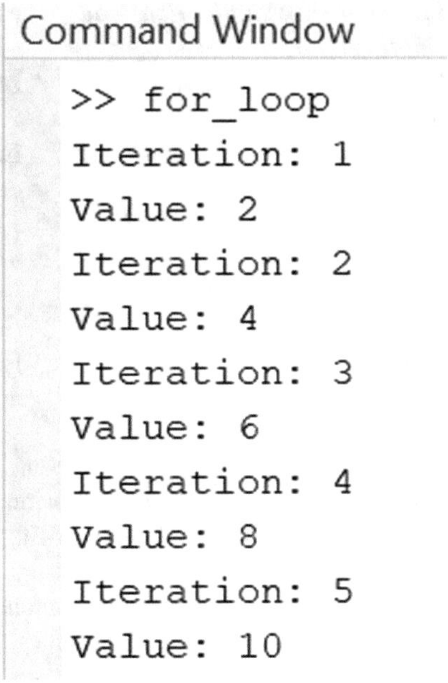

Fig. 3.25 Output—*for* loop

```
Command Window

>> for_loop
Iteration: 1
Value: 2
Iteration: 2
Value: 4
Iteration: 3
Value: 6
Iteration: 4
Value: 8
Iteration: 5
Value: 10
```

3.5 User-Defined Functions

Apart from the built-in functions, MATLAB also provides scope for creating user-defined functions, where the user can create a function to carry out a certain task. The user-defined function consists of three parts—input parameters, output variables, and executable commands—for accomplishing a task. A user-defined function may have multiple inputs and outputs, or even no inputs and outputs at all. This function can be saved as a script file (MATLAB M-file) and can be used in a separate script file by calling out the function's name. It is to be noted that both script files (Function file, Main Script file) need to be in the same path of the directory. MATLAB allows to have another type of user-defined function named Anonymous function, which does not require to be saved as a separate script file. Instead, the function can be created on the same main script file and can be usable in the rest of the code in the same script.

3.6 Creating User-Defined Functions

The structures of creating different user-defined functions based on the number of input and output parameters are given below:

Function with single input and single output:	*Function with multiple inputs and single output:*
function out = functionName (input1)	**function** out = functionName(input1, input2, …)
Executable commands for a task	Executable commands for a task
End	**end**
Function with no input and single output:	*Function with multiple inputs and multiple outputs:*
function out = functionName ()	**function** [out1, out2, …] = functionName (input1, input2, ...)
Executable commands for a task	Executable commands for a task
end	**end**

Here, input1, input2, … represent the input variables of the function, whereas out1, out2, … signify the output parameters. After writing a function in a script file, the file needs to be saved as an M-file in the same path directory of the main script file where the function will be used. By calling the "functionName" along with relevant inputs, the user-defined function will return the outputs. A simple example of such a function is given below:

3.6.1   MATLAB Example 3.4: User-Defined Function

The following MATLAB code in Fig. 3.26 shows how to create a user-defined function and Fig. 3.27 shows its output.

Output:

The structure of the other format of user-defined function—Anonymous function—is given below:

functionName= @ (input1, input2, …..)   Expression

Here, input1, input2, … indicate the input variables as before and the "functionName" represents the name of the function. "Expression" refers to any

```
function [out1, out2]=myfunction(input1, input2)
out1= 2*input1;
out2= 2*input2;
end
```

Fig. 3.26 Code—User defined function

Fig. 3.27 Output—User
defined function

```
Command Window

    >> [out1, out2]=myfunction(10, 20)

    out1 =

          20

    out2 =

          40
```

```
myfunc = @(x) 2*x+3
myfunc(2)

ans = 7
```

Fig. 3.28 User defined function—Anonymous function

mathematical expression with the association of the corresponding input variables.
This type of function can be used inline in the main script file. A simple example of
this type of function is given below:

3.6.2   MATLAB Example 3.5: User-Defined Function–Anonymous Function

The MATLAB code for a user-defined anonymous function is provided in Fig. 3.28.
   Here, $2 * x + 3$ is the expression with a variable x. The name of the function is
"myfunc." By calling "myfunc" with a given value of $x = 2$, the function calculates
the output from the expression, which becomes: 2*2+3=7.

3.6.3   Examples of User-Defined Function

Four examples of user-defined functions for addition, subtraction, multiplication,
and division of two numbers are provided below. Likewise, the functions could be
modified for more than two inputs.

```
function sum=Summation(input1, input2)
sum= input1 + input2;
end
```

Fig. 3.29 Code—Summation function

Fig. 3.30 Output—
Summation function

Command Window

>> Summation(5,10)

ans =

15

```
function sub=Subtraction(input1, input2)
sub= input1 - input2;
end
```

Fig. 3.31 Code—Subtraction function

3.6.3.1   User-Defined Function for Summation

A user-defined function that can perform the summation of two numbers is created in Fig. 3.29.

The above-mentioned function needs to be saved as an M-file named "Summation.m." In the command window, by recalling the function name with two given input numbers, the function will automatically return the summation as an output. The usage of the "Summation" function in the command window is shown in Fig. 3.30.

It is to be noted that the current directory and the directory of the saved file of the function need to be the same to work correctly.

3.6.3.2   User-Defined Function for Subtraction

A user-defined function that can perform a subtraction of any two numbers is formed below with the implementation in the command window as well. The MATLAB code is provided in Fig. 3.31 with its output in Fig. 3.32.

```
Command Window
>> Subtraction(65,37)

ans =

    28
```

Fig. 3.32 Output—Subtraction function

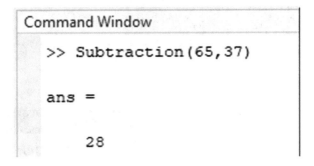

```
function mul=Multiplication(input1, input2)
mul= input1 * input2;
end
```

Fig. 3.33 Code—Multiplication function

```
Command Window
>> Multiplication(12,10)

ans =

   120
```

Fig. 3.34 Output—Multiplication function

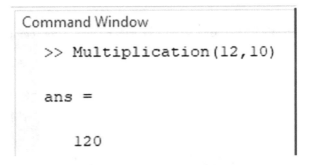

Output:

3.6.3.3 User-Defined Function for Multiplication

The multiplication task performed between any two numbers is written in the forms of a user-defined function in the following example code in Fig. 3.33 with its output in Fig. 3.34.

Output:

```
function div=Division(input1, input2)
div= input1 / input2;
end
```

Fig. 3.35 Code—Division function

Fig. 3.36 Output—
Division function

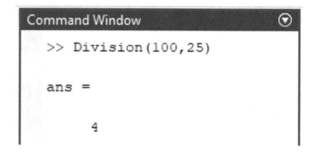

3.6.3.4 User-Defined Function for Division

Finally, a division example is shown using a user-defined function in the following example. The MATLAB code is provided in Fig. 3.35 with its output in Fig. 3.36.

Output:

3.7 Solve Quadratic Equations Using Functions

The general format of a quadratic equation can be represented as in Eq. (3.1).

$$ax^2 + bx + c = 0 \tag{3.1}$$

Here, x is the variable; and a, b, c are real coefficients. In this equation, the highest degree is 2. Hence, the solution to such an equation will have two roots.

The formula used to calculate the roots is shown in Eq. (3.2).

$$x_1, x_2 = \frac{-b \pm \sqrt{b^2 - 4ac}}{2a} \tag{3.2}$$

A user-defined function can be created to solve such quadratic equations for finding the roots in MATLAB. An example of such a user-defined function is given below:

```
function [ x1, x2 ] = quad_roots( a,b,c )
x1 = (-b + sqrt(b^2 - 4 * a * c))/(2*a);
x2 = (-b - sqrt(b^2 - 4 * a * c))/(2*a);
end
```

Fig. 3.37 Code—Quadratic equation function

```
%Quadratic Equation: 2x^2 + 3x + 5=0
a = 2; b = 3; c = 5;
[x1, x2] = quad_roots(a,b,c)

x1 = -0.7500 + 1.3919i
x2 = -0.7500 - 1.3919i
```

Fig. 3.38 Output—Quadratic equation function

3.7.1 MATLAB Example 3.6: User-Defined Function for Solving Quadratic Equation

Find the roots of the following quadratic equation using MATLAB:

$$\text{Quadratic equation} : 2x^2 + 3x + 5 = 0$$

The MATLAB code is provided in Fig. 3.37 with its output in Fig. 3.38.

3.8 Conclusion

As MATLAB is a high-level programming language, it is imperative for the readers to learn about the conditional statements and loops that can be implemented in MATLAB. For the convenience of the readers, multiple examples are provided that are implemented in MATLAB. While writing large programs, user-defined functions play important roles in many aspects. In MATLAB, two types of user-defined functions are mostly used, which are demonstrated in this chapter with adequate examples. This chapter will help the readers to build up the knowledge to initiate writing programs and details about creating and using a user-defined function in a MATLAB program.

Table 3.2 Hexadecimal codes of various colors

Color	Hexadecimal code
Red	#FF0000
Green	#00FF00
Blue	#0000FF
Orange	#FFA500
Yellow	#FFFF00
Black	#000000
White	#FFFFFF

Exercise 3

1. What is the difference between a script and a live script in MATLAB?
2. Enlist the logical operators in MATLAB with examples.
3. Create a MATLAB program which will take a numerical user input. If the user input is within the 0–100 range, the program would display "Inside range"; otherwise, the program would display "Outside range." Moreover, if the input is greater than 25 and less than or equal to 50, the program would additionally display "First half," and if the number is more than 50 but less than or equal to 75, the program would additionally display "Second half."
4. Hexadecimal codes of colors are vastly used in computer science. The hexadecimal codes of certain colors are given in Table 3.2.
   Write a MATLAB code using switch-case so that the users can provide the given hexadecimal code as input to know the corresponding color. If the hexadecimal color is wrong or is not available in the table, the code should output an error message: "The color code you entered is wrong/not available."
5. For two Cartesian coordinates (x_1, y_1) and (x_2, y_2), the formula to determine the distance is $d = \sqrt{\left\{ (x_1 - x_2)^2 + (y_1 - y_2)^2 \right\}}$. Write a user-defined function "distance" to take two coordinates as input and return the distance as the output. Check the function for:

   (a) $(2, -1)$ and $(-2, -2)$
   (b) $(3, 5)$ and $(-1, -6)$

6. Create a user-defined function "usercal," which will take two numbers as input from the user, and another number to determine the mathematical operation to perform. If the third number is 1, 2, or 3, addition, subtraction, or multiplication will be performed. If any other third number is entered, an error message "Wrong operation entered" will be demonstrated.

Chapter 4
Complex Numbers

4.1 Introduction

Complex numbers have penetrated deep into the world of electrical engineering. It is impossible to imagine terms such as impedance, inductance, capacitance, reactive power, etc. without referencing complex quantities. As such, an electrical engineer must be fluent with the use and execution of complex quantities for solving engineering problems. In this chapter, the main concentration will be on complex numbers and their different representations in rectangular and polar forms. Later, the chapter also introduces different mathematical series, where complex numbers play significant roles. Another important concept named equilibrium point is covered in this chapter. At the end, some examples related to the field of engineering are demonstrated, where the previous concepts will be utilized.

4.2 Origin of Complex Numbers

The origin of complex numbers came from the concept of dealing with the solution of the square root of any negative numbers. While solving polynomial equations, the square root term of the negative number appears. During the sixteenth century, the square root of any negative number was a mystery for mathematicians. Therefore, solving such problems requires a unique solution. In 1530, Niccolò Tartaglia first used a formula in a competition to solve cubic equations, where he found a way to deal with the root of a negative number. The opponent of that competition was a mathematician named Girolamo Cardano. Cardano published the method applied by Tartaglia after some time of that competition in 1545, where he improvised the method to create a generalized formula to deal with the mysterious number—the root of a negative number, for solving all quadratic and cubic equations. This is considered as the discovery of the complex number for which he is regarded as its creator.

© The Author(s), under exclusive license to Springer Nature Switzerland AG 2022
E. Hossain, *MATLAB and Simulink Crash Course for Engineers*,
https://doi.org/10.1007/978-3-030-89762-8_4

Later, Rafael Bombelli derived general rules for arithmetic operations for the root of a negative number. The imaginary unit, i, was first introduced later by Mathematician Leonhard Euler in 1777 for the roots of the equation: $x^2 + 1 = 0$. The usage of complex numbers became popular after mathematicians were able to interpret it both geometrically and arithmetically.

The general form of any complex number is:

$$a + ib ,\tag{4.1}$$

where i refers to the imaginary unit, which is represented as $i = \sqrt{-1}$. One of the interesting characteristics of i is that the square of that number is a negative value, i.e., $i^2 = -1$, although the square of any number is considered as positive. The introduction of the complex number plays a significant role in mathematics, such as solving any polynomial equation.

4.3 Rectangular Form

The rectangular form of a complex number can be referred to as follows:

$$C = a + i * b\tag{4.2}$$

Here, i is the imaginary number; a and b are coefficients; C represents the complex number.

In this complex number, there are two parts—real and imaginary parts. In the above equation, a represents the real part, whereas b indicates the imaginary part of the complex value. In a rectangular coordinate system, the position of a complex number has been shown in Fig. 4.1 to understand the parameters.

Fig. 4.1 Rectangular co-ordinate system

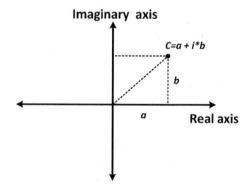

If the real and imaginary values are known, the complex number can be generated using the above-mentioned formula. In MATLAB, there is also an in-built function *complex(real, imaginary)*—which can be utilized to generate the complex number. In this complex function, there are two input parameters—real and imaginary values. An example is given below, where the complex number is generated using MATLAB in both ways:

4.3.1   MATLAB Example 4.1: Rectangular Form

The following code in Fig. 4.2 depicts the process of displaying complex numbers in the rectangular form.

4.4   Polar Form

The general format of the polar form of any complex number is:

$$| C | \angle\theta \tag{4.3}$$

Here, $|C|$ is the absolute value or magnitude of the complex number, $C = a + i * b$; and θ represents the angle.

The absolute value or the magnitude of a complex number is given by $| C | = \sqrt{a^2 + b^2}$ and the angle is given by $\theta = \frac{b}{a}$ rad $= \left(\frac{b}{a}\right) \times \frac{180^\circ}{\pi}$ degree. In a polar

```
% Real value, a=2
% Imaginary value, b=5
% Complex number, C=a+i*b
a=2; b=5;
C=a+i*b

 C = 2.0000 + 5.0000i
```

```
% Real value, a=2
% Imaginary value, b=5
% Complex number, C=a+i*b
a=2; b=5;
C=complex(a,b)

 C = 2.0000 + 5.0000i
```

```
% Complex number, C=a+i*b = 2+5i
a=2; b=5;
C=complex(a,b);
[C_angle, C_mag]=cart2pol(a,b);
C_angle=C_angle*(180/pi); %in degree
C_polar=[C_mag,C_angle]

 C_polar = 1-2
     5.3852    68.1986
```

Fig. 4.2 Code and output—Rectangular co-ordinate system

Fig. 4.3 Polar co-ordinate system

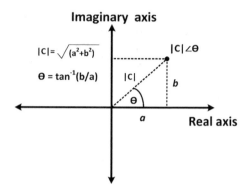

Polar Co-ordinate

```
% Complex number, C=a+i*b = 2+5i
a=2; b=5;
C_mag=sqrt(a^2+b^2);
C_angle=atan(b/a)*(180/pi); % in degree
C_polar=[C_mag,C_angle]

  C_polar = 1×2
        5.3852    68.1986
```

```
% Complex number, C=a+i*b = 2+5i
a=2; b=5;
C=complex(a,b);
C_mag=abs(C);
C_angle=angle(C)*(180/pi); % in degree
C_polar=[C_mag,C_angle]

  C_polar = 1×2
        5.3852    68.1986
```

Fig. 4.4 Code and output—Polar co-ordinate system

coordinate system, the position of a complex number has been shown in Fig. 4.3 to understand the parameters.

To convert a complex number from its rectangular form to polar form, we can utilize the above formula. MATLAB also provides some in-built functions to calculate both magnitude and angle directly by using *abs*() and *angle*(), respectively. Another way is to utilize the direct conversion of Cartesian to polar form by using *cart2pol*(). An example showing all these ways is given below:

4.4.1 MATLAB Example 4.2: Polar Form

The following code in Fig. 4.4 shows how to get the polar form of any complex number from the rectangular form.

4.5   Euler's Series

Euler's formula provides a relationship between trigonometric and exponential terms. It can be deduced from the exponential series by inserting the imaginary unit in there. Mathematician Euler first came up with this idea, and set forth the following formula, which is known as Euler's formula:

$$e^{ix} = \cos \cos x + i \sin \sin x, \tag{4.4}$$

where e is the base of the natural logarithm, x is any real number, and i is the imaginary unit.

In the Cartesian coordinate system, any complex number C can be written as $C = a + ib$, while in the polar form, it can be written as $|C| \angle \theta$, where $|C| = \sqrt{a^2 + b^2}$ and $\theta = \frac{b}{a}$ rad. As discussed earlier, it is also possible to convert a complex number from its polar form to Cartesian form by using the following relationships:

$$a = |C| \cos \cos \theta \tag{4.5}$$
$$b = |C| \sin \sin \theta \tag{4.6}$$

Therefore, the complex number in a Cartesian format becomes:

$$C = |C|(\cos \cos \theta + i \sin \sin \theta) \tag{4.7}$$

Euler's formula shows its beauty while it is implemented in the above-mentioned equation, and suddenly the complex number can be represented in an exponential format as below:

$$C = |C|(\cos \cos \theta + i \sin \sin \theta) = |C|e^{i\theta} \tag{4.8}$$

If the magnitude and the angle are known for a complex number in its polar coordinate system, it can be converted back into the exponential form using Euler's formula, which makes the multiplication and division of complex numbers easier to follow.

Example Consider two complex number, $C_1 = 10 \angle 45$ rad and $C_2 = 20 \angle 30$ rad. Determine:

i. $C_1 . C_2$
ii. C_1/C_2

Solution Using Euler's formula to convert the polar form into the exponential form:

$$C_1 = 10\angle 45 = 10e^{i45} \tag{4.9}$$

$$C_2 = 20\angle 30 = 20e^{i30} \tag{4.10}$$

i. $C_1 . C_2 = 10e^{i45} \times 20e^{i30} = 10 \times 20 \; e^{i(45 + 30)} = 200e^{i75} = 200 \angle 75$

ii. $\frac{C_1}{C_2} = \frac{10e^{i45}}{20e^{i30}} = 0.5e^{i(45-30)} = 0.5e^{i15} = 0.5\angle 15$

4.5.1 MATLAB Example 4.3: Euler's Formula

Two complex numbers are $C1 = 2 + 5i$ and $C2 = 5 + 10i$. Determine:

(a) $C1$ and $C2$ in exponential form
(b) $C1*C2$ and $C1/C2$ in exponential form

The MATLAB code is provided in Fig. 4.5 with its output in Fig. 4.6.

```
% C1= 2+5i
% C2= 5+10i
% Convert C1 and C2 in exponential form
% Determine M=C1*C2 and D=C1/C2 in exponential form

C1=complex(2,5);
C2=complex(5,10);
C1_mag=abs(C1);
C1_angle=angle(C1);
C2_mag=abs(C2);
C2_angle=angle(C2);
% C=a+bi=|C|(cos(theta)+i*sin(theta))=|C|exp(i*theta)
disp(['C1 in exponential form: ', num2str(C1_mag),'exp(i*',...
                               num2str(C1_angle),')']);
disp(['C2 in exponential form: ', num2str(C2_mag),'exp(i*',...
                               num2str(C2_angle),')']);
% C1*C2
M_mag=C1_mag*C2_mag;
M_angle=C1_angle+C2_angle;
disp(['C1*C2 in exponential form: ', num2str(M_mag),'exp(i*',...
                               num2str(M_angle),')']);
% C1/C2
D_mag=C1_mag/C2_mag;
D_angle=C1_angle-C2_angle;
disp(['C1/C2 in exponential form: ', num2str(D_mag),'exp(i*',...
                               num2str(D_angle),')']);
```

Fig. 4.5 Code—Euler's formula

```
Command Window
  C1 in exponential form: 5.3852exp(i*1.1903)
  C2 in exponential form: 11.1803exp(i*1.1071)
  C1*C2 in exponential form: 60.208exp(i*2.2974)
  C1/C2 in exponential form: 0.48166exp(i*0.083141)
fx >>
```

Fig. 4.6 Output—Euler's formula

Output

Application of Euler's Series for Solving Initial Value Problem
Euler's method can be used to produce a numerical solution for the initial value
problem. The general format of such a problem is as follows:

$$\frac{dy}{dx} = y' = f(x, y), \text{ where } y(x_0) = y_0 \tag{4.11}$$

The variable x may have a certain range, which can be defined as lower and upper
limit values. The procedure for generating solutions to such problem using Euler's
series can be arranged as follows:

1. Deciding number of steps (N) for iteration over the range of values x, based on
   which a step size (h) will be determined.
2. For each iteration, the values of y and x will be updated as follows:

$$y_{n+1} = y_n + h * f(x_n, y_n); \tag{4.12}$$
$$x_{n+1} = x_0 + n * h; \text{ where } n = 0, 1, 2, \ldots\ldots.N \tag{4.13}$$

4.5.2   MATLAB Example 4.4: Euler's Series for Solving Initial Value Problem

Problem
$$\frac{dy}{dx} = y' = f(x, y) = 2x^2 + y - 2; 0 \le x \le 5; y(0) = 0.1 \tag{4.14}$$

The MATLAB code is provided in Figs. 4.7 and 4.8 with its output in Figs. 4.9
and 4.10.

```
function  Sol=Euler_series(f,L,U,N,y1)
% Input parameters:
% f: A function that provides the differfential equation
% L: Lower limit of x
% U: Upper limit of x
% N: Number of steps
% y1: Initial value of y
% Output Parameters:
% Sol: [x,y]; x: abscissas; y: ordinate

% Consider the step size, h

h=(U-L)/N;
y=zeros(N+1,1);
x=zeros(N+1,1);
x(1)=L;
y(1)=y1;
for i=1:N
    y(i+1)=y(i)+h*f(x(i),y(i));
    x(i+1)=x(1)+ i*h;
end
plot(x,y,'*');
grid on;
xlabel('x');
ylabel('y');
title('Differential equation: dy/dx=2x^2+y-2;0<=x<=5;y(0)=0.1');
Sol=[x,y];
end
```

Fig. 4.7 Code—Creating the function for Euler's series for solving initial value problem

```
clc; clear all;
% Generate a function for the differential equation
% Differential equation: dy/dx=y'=2x^2+y-2
% Condition:   0 <= x <= 5;  y(0)=0.1
% N.B. MATLAB iterates from 1, not from 0.
% Hence, for the initial value y(0),we will consider it y(1).
% Number of step, N=18

f=@(x,y) 2*x^2+y-2;
L=input('Enter the lower limit of x:');
U=input('Enter the upper limit of x:');
N=input('Enter the number of step:');
y1=input('Enter initial value of y:');
Euler_series(f,L,U,N,y1);
```

Fig. 4.8 Code—Using the function for Euler's series for solving initial value problem

Fig. 4.9 Output—Euler series for solving initial value problem

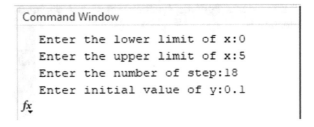

```
Command Window

    Enter the lower limit of x:0
    Enter the upper limit of x:5
    Enter the number of step:18
    Enter initial value of y:0.1
fx
```

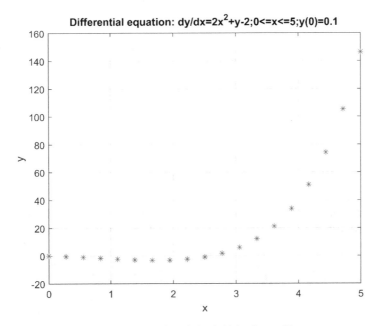

Fig. 4.10 Graphical output—Euler series for solving initial value problem

Output

4.6 Fourier Series

Fourier series is a series through which any periodic function can be represented as a summation of sine and cosine functions. It provides an opportunity to represent any periodic function to a trigonometric representation.

Consider a periodic function $f(t)$, with a period T. The Fourier series of this function is:

$$f(t) = \frac{a_0}{2} + \sum_{k=1}^{\infty} a_k \cos(k\omega t) + \sum_{k=1}^{\infty} b_k \sin(k\omega t); k = 0, 1, 2, \ldots\ldots \quad (4.15)$$

where:

$$a_0 = \frac{2}{T} \int_{-T/2}^{T/2} f(t) \, dt \qquad (4.16)$$

$$a_k = \frac{2}{T} \int_{-T/2}^{T/2} f(t) * \cos(k\omega t) \, dt \qquad (4.17)$$

$$b_k = \frac{2}{T} \int_{-T/2}^{T/2} f(t) * \sin(k\omega t) \, dt \qquad (4.18)$$

4.6.1   MATLAB Example 4.5: Fourier Series

The Fourier series of a square function with $\omega = \pi$ and $T = 2$ for $k = 5$ terms can be determined using the MATLAB code provided in Fig. 4.11 with its output in Fig. 4.12.

```
% Input: Square wave function
% T=2; Magnitude = 1; Omega=pi
t=-7:0.01:7;
x=1-square(pi*(t+1));
plot(t,x,'LineWidth',1.5)
grid on;
hold on;

% Fourier Series
syms k t
omega=pi;
T=2;
k=1:5;
a_0=(2/T)*int(2,t,0,1);
a_k=(2/T)*int(2*cos(k*omega*t),t,0,1);
b_k=(2/T)*int(2*sin(k*omega*t),t,0,1);
f=(a_0/2)+sum(a_k.*cos(k*omega*t))+sum(b_k.*sin(k*omega*t));
ezplot(f,[0,7])
grid on;
```

Fig. 4.11 Code—Fourier series

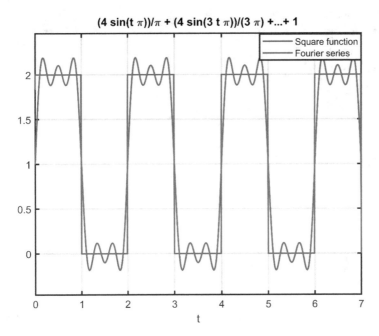

Fig. 4.12 Graphical output—Fourier series

Output

Discrete Fourier Transform
Discrete Fourier Transform (DFT) is almost similar to continuous Fourier transform, except that in DFT, a finite sequence of the input is considered as an input. In this method, any continuous signal is sampled for a finite length to make the input a finite discrete sequence.

Consider a sequence $x[k]$, for which the DFT can be determined using Eq. (4.19):

$$X[n] = \sum_{k=0}^{N-1} e^{\frac{-j2\pi nk}{N}} . x[k], \text{ where } n = 0, 1, 2, \ldots, n-1 \qquad (4.19)$$

Here, the length of the sequence is N.

It is also essential to study inverse DFT, which can be accomplished by using the formula as shown in Eq. (4.20):

$$x[k] = \sum_{n=0}^{N-1} e^{\frac{+j2\pi nk}{N}} . X[n], \text{ where } k = 0, 1, 2, \ldots, n-1 \qquad (4.20)$$

Here, the output of DFT is considered as input of the inverse DFT function, and we can reproduce the original sequence, $x[k]$, by implementing inverse DFT.

```
clc; clear all;
disp('Input sequence: ');
x=[1 4 5 7]
F=fft(x);
disp('Fourier transform of x: ');
F
inv_F=ifft(F);
disp('Inverse Fourier transform of F: ');
inv_F
```

Fig. 4.13 Code—Discrete and inverse discrete Fourier transform

- **In MATLAB, the DFT of an input vector can be determined by using a built-in function:** *fft*().
- **Inverse DFT can be achieved by using a built-in function:** *ifft*().

4.6.2 MATLAB Example 4.6: DFT and Inverse DFT

Consider an input sequence, $x = [1\ 4\ 5\ 7]$. Determine:

 (i) Fourier transform of the input vector
(ii) Inverse Fourier transform of the output of (i)

The MATLAB code is provided in Fig. 4.13 with its output in Fig. 4.14.

Output

4.7 Taylor Series

Taylor series was first discovered by Mathematician Brook Taylor, who came up with a general formula to represent any function as the summation of infinite terms that incorporates the derivatives of the function at a single point. The sum of this infinite series provides a finite value, which is equal to the original value of the function near that single point. This series is called the Taylor series.

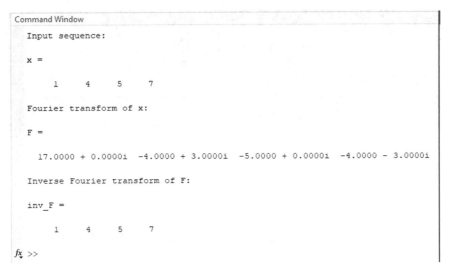

Fig. 4.14 Output—Discrete and inverse discrete Fourier transform

The Taylor series of a real function $f(x)$ about a point $x = a$ can be represented as follows:

$$f(x) = f(a) + (x-a)\frac{f'(a)}{1!} + (x-a)^2\frac{f''(a)}{2!} + (x-a)^3\frac{f'''(a)}{3!} + \ldots \quad (4.21)$$

In general format, this series can be represented as follows:

$$f(x) = \sum_{n=0}^{\infty}(x-a)^n \frac{f^n(a)}{n!} \quad (4.22)$$

Here, f^n represents the nth derivative of the function $f(x)$.

When the Taylor series is determined at a point $x = 0$, i.e., $a = 0$, it becomes the Maclaurin series. The general format of the Maclaurin series is shown in Eq. (4.23):

$$f(x) = \sum_{n=0}^{\infty}(x)^n \frac{f^n(0)}{n!} \quad (4.23)$$

The higher order we can go of a Taylor series, the more accurate the result will be for a particular function at a certain point. MATLAB provides a built-in function for determining the Taylor series of a certain function at a specific point, which is *taylor(f, var)*.

MATLAB function for Taylor series for a function *f* **at a point** *var* = 0 **up
to fifth order (default) is** *taylor*(*f*, *var*).
MATLAB function for Taylor series for a function *f* **at a point** *var* = *a* **up
to fifth order (default) is** *taylor*(*f*, *var* , *a*).
MATLAB function for Taylor series for a function *f* **at a point** *var* = *a* **up
to** *n* **order (default) is** *taylor*(*f*, *var* , *a*, *'order'*, *n*).

4.7.1   MATLAB Example 4.7: Taylor Series

The MATLAB code to expand the Taylor series is provided in Fig. 4.15 with its
output in Figs. 4.16 and 4.17.

Output

When $a = 0$, the Taylor series becomes Maclaurin series and can be determined
using the same *taylor*() function as shown above.

```
clc; clear all;
% Taylor series expansion
% Function: f(x)=2*sin(x) at a point x=a=0.5;
% Taylor series expansion up to 4th and 10th order

syms x;
f = 2*sin(x);
a = 0.5;
T_4 = taylor(f,x,a,'order',4);
T_10 = taylor(f,x,a,'order',10);
disp('Taylor series expansion of 2*sin(x) at a=0.5 up to 4th order:');
T_4
fplot(T_4,'b','Linewidth',1.5);
hold on;
fplot(T_10,'g','Linewidth',1.5);
hold on;
fplot(f,'r','Linewidth',1.5);
hold off;
xlim([-4 4]);
ylim([-4 2]);
grid on
legend('Taylor series up to 4th order',...
       'Taylor series up to 10th order',...
       'Original function: 2*sin(x)','Location','Best')
title('Taylor series of 2*sin(x) at a = 0.5');
```

Fig. 4.15 Code—Taylor series

```
Command Window
  Taylor series expansion of 2*sin(x) at a=0.5 up to 4th order:

  T_4 =

  2*sin(1/2) - sin(1/2)*(x - 1/2)^2 + 2*cos(1/2)*(x - 1/2) - (cos(1/2)*(x - 1/2)^3)/3

fx >>
```

Fig. 4.16 Output—Taylor series

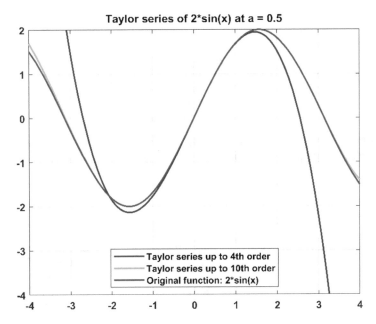

Fig. 4.17 Graphical output—Taylor series

4.8   Equilibrium Point

Mathematically, the equilibrium point can be defined as the constant solution of a differential equation. It represents the steady-state points of a dynamic system. Consider a differential equation as follows:

$$\frac{dy}{dt} = f(t, y) \tag{4.24}$$

For such a differential equation, the equilibrium points will be the solutions of the following equation:

$$\frac{dy}{dt} = f(t, y) = 0 \tag{4.25}$$

Here, the values of y will be the equilibrium points, for satisfying the above condition. It can be useful to determine the stability of a system. In linearization, the equilibrium point also plays an important role.

4.8.1   MATLAB Example 4.8: Equilibrium Points

Consider the following two differential equations:

$$\frac{dy}{dt} = 4x^2 - xy \tag{4.26}$$

$$\frac{dy}{dx} = 2y - x^2 \tag{4.27}$$

Determine the equilibrium points using MATLAB.
The MATLAB code is provided in Fig. 4.18 with its output in Fig. 4.19.

Output

```
clc; clear all;
% Two differential equation
% dy/dt = 4x^2-xy
% dx/dt = 2y-x^2
% Determine the equilibrium points of the system
% For determining equilibrium points consider,
% dy/dt = 0 and dx/dt = 0
% The solutions x and y will be the equilibrium points of the system

clc; clear all;
syms x y
[solx,soly] = solve(4*x^2-x*y == 0, 2*y-x^2 == 0);
disp('Equilibrium points:')
E_point1=[solx(1) soly(1)]
E_point2=[solx(2) soly(2)]
```

Fig. 4.18   Code—Equilibrium points of a differential equation

Fig. 4.19 Output—
Equilibrium points of a
differential equation

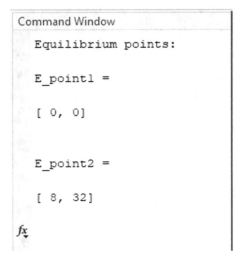

4.9 Energy Calculation

In an electrical system, the power can be represented using Eq. (4.28):

$$S = P + jQ, \tag{4.28}$$

where, S, P, and Q are apparent, real, and reactive power, respectively, and j is the imaginary unit. Therefore, the electrical power can be considered as a complex number, which is multiplied with scalar value time to calculate the electrical energy of a system. Hence, the electrical energy can be calculated using Eq. (4.29):

$$E = |S| \times t = |(P + jQ)| \times t, \tag{4.29}$$

where E is the electrical energy and t represents the time of energy usage. The units of S, P, and Q are VA, W, and VAR.

4.9.1 MATLAB Example 4.9: Energy Calculation

An electrical system consumes 10 W real power and 20 VAR reactive power for 24 h. Using MATLAB, determine:

(i) Apparent power in rectangular form
(ii) Apparent power in polar form
(iii) Electrical energy

The MATLAB code is provided in Fig. 4.20 with its output in Fig. 4.21.

Output

```
clc; clear all;
% Real power,P = 10 W
% Reactive power,Q = 20 VAR
% Time, t = 24 hours = 86400 sec
% Aparent power,S is a complex number.
% Electrical energy, E=P*t

P=10; Q=20; t=86400;
% Rectangular form
S_rec=complex(P,Q);
disp('Apparent power in rectangular form:');
S_rec
% Polar form
S_mag=abs(S_rec);
S_angle=angle(S_rec)*(180/pi); % Unit: Degree
S_polar=[S_mag,S_angle];
disp('Apparent power in polar form- [Magnitude   Angle(Degree)]:');
S_polar
% Electrical energy
E=P*t;
disp(['Electrical energy: ',num2str(E),' Joule']);
```

Fig. 4.20   Code—Power and energy calculation

```
Command Window

  Apparent power in rectangular form:

  S_rec =

     10.0000 +20.0000i

  Apparent power in polar form- [Magnitude   Angle(Degree)]:

  S_polar =

      22.3607    63.4349

  Electrical energy: 864000 Joule
fx >>
```

Fig. 4.21   Output—Power and energy calculation

4.10 Impedance Calculation

Impedance is a parameter in an electrical network that is used to depict the overall opposition of the flow of current. Impedance consists of two parts—resistance and reactance. Impedance is also a complex number that can be represented as in Eq. (4.30):

$$Z = R + jX \tag{4.30}$$

Here, Z is the impedance, R is the resistance, and X represents the reactance of an electrical circuit.

Another definition of impedance is that it is the ratio of the voltage to the current of a circuit. Therefore, impedance can also be represented by Eq. (4.31):

$$Z = \frac{V}{I} \tag{4.31}$$

Here, V and I represent the voltage and the current of a circuit, respectively.

4.10.1 MATLAB Example 4.10: Impedance Calculation

In a series ac electrical circuit, consider a voltage source of $100 \angle 60°$ V. There is an impedance connected in series with the voltage source. The current flowing through the circuit is $3 \angle 30°$ A. Determine the impedance in both rectangular and polar forms.

The MATLAB code is provided in Fig. 4.22 with its output in Fig. 4.23.

Output

Consider a series AC circuit with a voltage source of $100 \angle 90°$ V. In series, there is a resistor of 5 ohms, a capacitor of 2 μF, and an inductor of 15 mH. Determine the impedance of the circuit in both rectangular and polar forms.

The MATLAB code is provided in Fig. 4.24 with its output in Fig. 4.25.

Output

```
% Voltage,V = 100∠60
% Current, I = 3∠30
% Impedance, Z = V/I
V_mag=100;
V_angle=60;
I_mag=3;
I_angle=30;
% Rectangular form
[Vx,Vy]=pol2cart(V_angle,V_mag);
V_rec=Vx+i*Vy;
[Ix,Iy]=pol2cart(I_angle,I_mag);
I_rec=Ix+i*Iy;
Z_rec=V_rec/I_rec;
disp('Impedance in rectangular form:');
Z_rec
% Polar form
Z_mag=abs(Z_rec);
Z_angle=angle(Z_rec)*(180/pi); % Unit: Degree
Z_polar=[Z_mag,Z_angle];
disp('Impedance in polar form- [Magnitude  Angle(Degree)]:');
Z_polar
```

Fig. 4.22 Code—Impedance Calculation 1

```
Command Window

  Impedance in rectangular form:

  Z_rec =

     5.1417 -32.9344i

  Impedance in polar form- [Magnitude   Angle(Degree)]:

  Z_polar =

     33.3333   -81.1266
```

Fig. 4.23 Output—Impedance Calculation 1

```
% Resistance, R=5 ohms
% Capacitance, C=2 micro F
% Inductance, L=15 mH
% Frequency, f=60Hz
% Impedance, Z=R+jX=R+j(X_L-X_C)
% Here,
% X_L=Inductive Reactance=omega*L=2*pi*f*L
% X_C=Capactive Reactance=1/(omega*C)=-1/(2*pi*f*C)

R=5; L=15*10^(-3); C=2*10^(-6); f=60;
X_L=2*pi*f*L;
X_C=-1/(2*pi*f*C);
% Rectangular form
Z_rec=R+i*(X_L-X_C);
disp('Impedance in rectangular form:');
Z_rec
% Polar form
Z_mag=abs(Z_rec);
Z_angle=angle(Z_rec)*(180/pi); % Unit: Degree
Z_polar=[Z_mag,Z_angle];
disp('Impedance in polar form- [Magnitude  Angle(Degree)]:');
Z_polar
```

Fig. 4.24 Code—Impedance Calculation 2

```
Command Window

  Impedance in rectangular form:

  Z_rec =

     5.0000e+00 + 1.3319e+03i

  Impedance in polar form- [Magnitude  Angle(Degree)]:

  Z_polar =

     1.0e+03 *

       1.3320     0.0898
```

Fig. 4.25 Output—Impedance Calculation 2

4.11   Conclusion

From this chapter, the readers will be able to learn about complex numbers and their importance in the engineering domain. In MATLAB, how to define a complex number and their different forms of representations are shown in this chapter. In addition, some of the important series, such as Euler's series, Taylor's series, Fourier series, inverse Fourier series, that are widely used in the engineering domain are explained and implemented in MATLAB to solve different mathematical problems. The chapter concludes with a demonstration of how to use this concept in real engineering applications such as energy calculation and impedance calculation.

Exercise 4

1. Write down some of the applications of complex numbers in engineering context.
2. Define the following functions with examples:

   (a) cart2pol()
   (b) abs()
   (c) taylor()
   (d) ifft()
   (e) ezplot()

3. Take a user input for a complex number in rectangular form.

   (a) Convert the complex number into polar form and save it in a variable m.
   (b) Determine the magnitude and angle of the complex number.
   (c) Consider another complex number $n = 5 - i$. Determine $m * n$ in exponential form.

4. Given a differential equation with initial values as follows:

   (a) $\frac{dy}{dx} = y' = f(x, y) = 5x^4 + x^2 - x + 2y - 14; 7 \leq x \leq 20; y(0) = 0.2$
   (b) $\frac{dy}{dx} = y' = f(x, y) = x^2 + 5x - y; -2 \leq x \leq 2; y(0) = 0.05$.

   Develop a function for Euler's series, such that the function takes the equation, lower and upper limit, step number, and the initial value to generate the values of x and y. Plot the 'x versus y' graph.
5. Consider an input sequence $x =$ linspace($-2, 2, 10$). Determine the Fourier transform of x and the inverse Fourier transform of the output.
6. Perform the Taylor series expansion of the function $f = 2 \cos(x) + 3 \sin(x)$ at a point 0.6, for the order up to fourth and tenth.
7. In a series ac electrical circuit, consider a voltage source of $220\angle 30°$ V. There is a resistor of 10 ohms, an inductor of 20×10^{-3} H, and a capacitor of 4 μF attached to the circuit in series. The current flowing through the circuit is $5\angle 45°$ A. The

systems connected to it take 20 W real power and 35 VAR reactive power for a single day.

(a) Determine the impedance in both rectangular and polar forms.
(b) Calculate the apparent power in rectangular and polar form.
(c) Determine the electrical energy.

Chapter 5
Visualization

5.1 Introduction

Visualization is one of the best features of MATLAB. In this chapter, different visualization techniques of MATLAB are elaborately explained and demonstrated. Similar to other programming languages, MATLAB covers a wide variety of visualization techniques, such as line plot, bar plot, area plot, surface plot, pie plot, heat map, radar plot, etc. In this chapter, all of these visualization methods are produced in MATLAB by using data relevant to engineering fields. In MATLAB, three-dimensional plotting is also available, which is also covered in this chapter. In the end, the process of exporting high-quality figures from MATLAB will be demonstrated with step-by-step directions.

5.2 Line Plot

In visualization, a line plot is the most basic and important form of graphics. In MATLAB, a 2D line plot can be drawn by utilizing the *plot* command, which offers several features, and user customization options. A plot command incorporating some common features is given below with necessary illustrations:

> **Line plot command for MATLAB:**
> *plot(xvalue, yvalue, 'Line Color Line Style Marker', 'Linewidth', n)*

xvalue: x-axis values
   yvalue: y-axis values
   The above-mentioned plot command will plot a 2D line of *xvalue* vs *yvalue*. Therefore, *plot(xvalue, yvalue)* command is the simplest version of this command

© The Author(s), under exclusive license to Springer Nature Switzerland AG 2022
E. Hossain, *MATLAB and Simulink Crash Course for Engineers*,
https://doi.org/10.1007/978-3-030-89762-8_5

Table 5.1 Line colors in
MATLAB

Color	Commands
Blue	'b'
Red	'r'
Green	'g'
Black	'k'
Yellow	'y'
White	'w'
Magenta	'm'
Cyan	'c'

Table 5.2 Line styles in
MATLAB

Line style	Symbols
Solid line	-
Dashed line	--
Dash-dotted line	-.
Dotted line	:

through which a 2D line plot can be drawn using MATLAB. For all the other features, MATLAB will use default values or styles.

However, it is also important to know about the other features to learn how to make a plot self-customized. Hence, descriptions of the other features of the *plot* command are given below:

Line Color: The color of the plotted line can be specified, or MATLAB will use the default color. Therefore, it is an optional feature. MATLAB supports a wide range of colors. For the most common colors, there are some specific commands. A table incorporating commands of most common colors is given in Table 5.1.

Apart from the above-mentioned colors, MATLAB also supports all combinations of red, green, and blue (RGB) colors, which are represented by a single vector in MATLAB. By changing the values of the vector, all kinds of variations in colors can be made possible. An RGB color vector in MATLAB contains pixel values of red, green, and blue that range from (0,255) individually. Therefore, a vector of [1 120 230] defines a specific color, and thus, a wide variety of colors can be formed and utilized in MATLAB.

Line Style: Line style defines the nature of the line that will be used to plot a 2D line plot. The line can be solid, or dashed, or any other form. MATLAB provides the opportunity to customize such styles by introducing some specific symbols to define them. A table incorporating different line styles is provided in Table 5.2.

It is to be noted that MATLAB uses solid lines as the default line style.

Marker: Marker is also an optional feature provided by MATLAB. The feature allows marking specific points in a line plot by using a customized marker. Some of the line markers that are offered by MATLAB are given in Table 5.3 with corresponding symbols to use in the code.

Linewidth: The width of the plotted line can be weighted by choosing a numerical value. By default, MATLAB will choose linewidth as 1. By increasing the number, the width can be customized. The standard procedure to add linewidth feature in the

Table 5.3 Line Markers in MATLAB

Marker	Symbol
Circle	O
Asterisk	*
Point	.
Cross	X
Plus	+
Diamond	d
Square	s
Hexagram	h
Pentagram	p
Triangle (pointed upward)	^
Triangle (pointed downward)	v
Triangle (pointed right)	>
Triangle (pointed left)	<

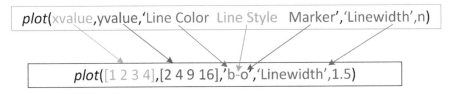

Fig. 5.1 Plot command syntax and example

Table 5.4 Axis labels and titles in MATLAB plots

Features	Significance
xlabel()	To label the x-axis to enable a better understanding of the x-axis data, this function is used, which takes strings as input
ylabel()	To label the y-axis to enable a better understanding of the y-axis data, this function is used, which takes strings as input
title()	To provide a title of the plotted figure, this function is used, which also takes strings as input

plot command is to incorporate—'*Linewidth'*, *n*—command within the plot command, where *n* represents a numerical value referring to the level of the weight of the linewidth.

A plot command for input *x* and *y*, with other features, can be written as shown in Fig. 5.1.

Basic Features There are some basic features of visualizations that are not only restricted to be used in a line plot but also for other forms of visualizations. To beautify and also to add relevant information to a figure, the features mentioned in Table 5.4 are essential and used widely in MATLAB.

5.2.1 MATLAB Example 5.1: Line Plot

A line plot of global temperature from 2010 to 2020 is shown below using the MATLAB code in Fig. 5.2. The data is collected from NOAA National Centers for Environmental Information [1]. In the output presented in Fig. 5.3, the x-axis represents the years, while the y-axis represents the temperature in degree Celsius.

Subplot Subplot is essential for visualization when comparisons or decisions need to be made from the side-by-side display. In a subplot, multiple figures can be plotted separately in one frame side by side. For subplotting multiple figures, *subplot*

```
% Line plot
% Data: Global temperature in degree celcius (2010-2020)
year=2010:1:2020;
temp=[14.46 14.55 14.48 14.67 14.82 15.16 14.83 ...
        14.88 14.89 15.05 14.78];
plot(year,temp,'b-o','linewidth',1.5);
xlabel('Years');
ylabel('Temperature (Degree celcius)');
title('Line plot of global temperature (2010-2020)')
grid on;
```

Fig. 5.2 Code—Line plot

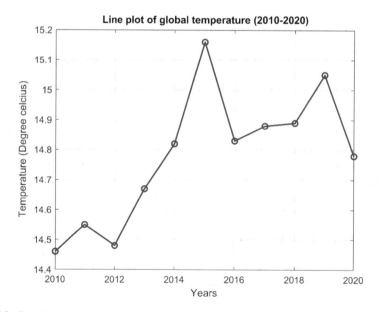

Fig. 5.3 Graphical output—Line plot

command is used in MATLAB, where a dimension size is defined in the input to create a frame. For example, *subplot*(2, 3, 1) signifies the total figure window will create a subspace of two by three blocks, i.e., there will be six blocks available in the entire figure window arranged in two rows and three columns. The last numerical number 1 represents the placement—first block—where the next plot command will generate a sub-figure.

5.2.2 MATLAB Example 5.2: Subplot

The line plots of global temperatures over the years 2010–2020 can be plotted using the MATLAB code in Fig. 5.4, and the output is shown as subplots in Fig. 5.5.

Here, a *subplot*(1, 2, 1) is used before the first plot command, which signifies there will be two sub-figures in one row and two columns and the first sub-figure will be plotted in the first place. For plotting the second sub-figure in the second block (first row and second column), the *subplot*(1, 2, 2) command is used right after the plot command of the second plot.

Double-Axis Plot In a line plot, sometimes we need to plot two line plots in the same figure with the double-axis feature. For creating double-axis feature, *yyaxis left*

```
% Subplot
% Data of global temperature from 2010:2020
% Temperature unit: degree celcius and farenhite
year=2010:1:2020;
temp_C=[14.46 14.55 14.48 14.67 14.82 15.16 14.83 ...
        14.88 14.89 15.05 14.78];
temp_F=[58.028 58.19 58.064 58.406 58.676 59.288 ...
        58.694 58.78 58.802 59.09 58.604];
subplot (1,2,1);
plot(year,temp_C,'b-o','linewidth',1.5);
xlabel('Years');
ylabel('Temperature (degree celcius)');
title('Global temperature (2010-2020)')
grid on;
subplot (1,2,2);
plot(year,temp_F,'k-o','linewidth',1.5);
xlabel('Years');
ylabel('Temperature (farenhite)');
title('Global temperature (2010-2020)')
grid on;
```

Fig. 5.4 Code—Subplot

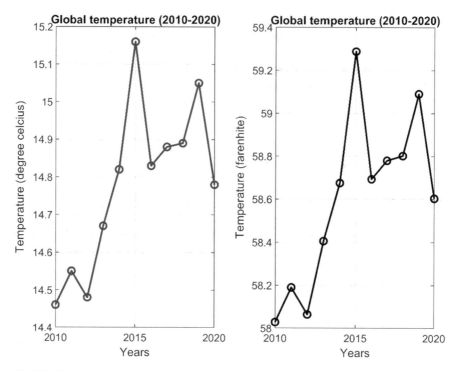

Fig. 5.5 Graphical output—Subplot

command is used right before the first plot command, which defines the left y-axis. Similarly, *yyaxis right* command is used to define the right y-axis. Between the two plot commands, *hold on* command is used to keep both the plots. *hold on* is usually used to hold the previous plotted figure until the *hold off* command. If two *plot* commands are used consecutively without *hold on* and *hold off* command, the first figure will be replaced by the second one. A *hold on* command placed between these two *plot* commands allows MATLAB to keep both of these plots in one figure without replacing anyone. Using *hold on* even multiple plots can be incorporated in one figure until we finish it by *hold off* command.

5.2.3   MATLAB Example 5.3: Double-Axis Plot

The code in Fig. 5.6 is for the price comparison of two energy commodities—steel and copper—from 2000 to 2020 (Data source [2]). The output is presented in Fig. 5.7. The left y-axis represents the prices of copper, whereas the right y-axis signifies the prices of steel over the years.

```
%% Double-axis plot
% Data: Price of Copper and Steel (2000-2020)
year=2000:5:2020;
copper_price=[1813 3679 7535 5631 5786];
steel_price=[296 633 716 543 491];
colororder({'k','b'})
yyaxis left
plot(year,copper_price,'k-o','linewidth',1.5);
xlim([2000 2020]);
ylim([0 8000]);
xlabel('Year');
ylabel('Copper Price ($/Tonne)');
hold on
yyaxis right
plot(year,steel_price,'b-o','linewidth',1.5);
ylim([200 1000]);
ylabel('Steel Price ($/Tonne)');
legend({'Copper Price','Steel Price'},'Location','Northwest');
title('Price of Copper and Steel (2000-2020)')
grid on
```

Fig. 5.6 Code—Double axis plot

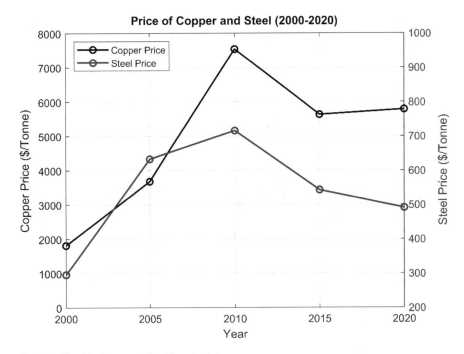

Fig. 5.7 Graphical output—Double axis plot

5.3   Bar Plot

Bar plot is an important visualization that facilitates better comparative analysis. The bar plot command for MATLAB is listed below:

> **Bar plot command for MATLAB:**
> *bar(xvalue, yvalue,'Bar color')*

Here, the *Bar color* can be defined similarly as mentioned in the Line Color part in Sect. 5.2.

5.3.1   MATLAB Example 5.4: Bar Plot

A bar plot showing the global CO_2 emissions in Giga metric ton (Gt) over the years 2010–2020 is illustrated below using the MATLAB code in Fig. 5.8, whose output is given in Fig. 5.9.

It is to be noted that the data of the above plot is obtained from the International Energy Agency (IEA) [3].

Horizontal Bar Plot

MATLAB offers some variations in bar plot visualization, such as horizontal bar plots. For plotting horizontal bar, the command line for MATLAB will be as follows:

> **Horizontal bar plot command for MATLAB:**
> *barh(xvalue, yvalue,'Bar color')*

```
% Bar plot
% Data: Global CO2 emission (2010-2020)
year=2010:1:2020;
CO2=[30.5824 31.4595 31.806 32.3707 32.3886 32.3655 ...
     32.3747 32.8374 33.5133 36.4568 34.0752];
bar(year,CO2,'b');
ylim([20 40]);
xlabel('Years');
ylabel('CO2 emission (Gt)')
title('Global CO2 Emission (2010-2020)')
grid on;
```

Fig. 5.8 Code—Bar plot

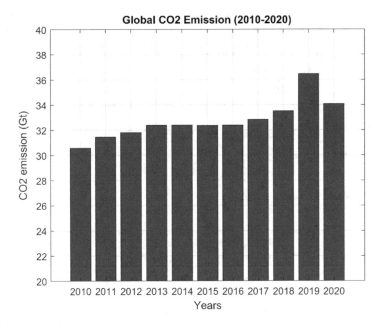

Fig. 5.9 Graphical output—Bar plot

In all bar plots, the "Bar color" defines the color of all the bars in a plot. However, MATLAB also provides the opportunity to customize the colors of each bar individually. An example for customizing colors of each bar is given in the following example, which is applicable not only for horizontal bar plots but also for vertical bar plots.

5.3.2   MATLAB Example 5.5: Horizontal Bar Plot

The data of electricity consumption by different household entities as end-users in the USA (2018) is utilized to plot a horizontal bar. The unit considered for electricity consumption per household is kWh/household. The data has been collected from Ref. [2]. The code is given in Fig. 5.10 with the output in Fig. 5.11.

Output

From the output (Fig. 5.11), it is clear that the highest consumption/household is obtained from the air conditioning system, whereas the least is from refrigeration. To color each bar individually, we have assigned the *barh* plot to a variable C, through which we have accessed each bar, and assigned an RGB vector representing separate colors for each of them.

```
% Horizontal bar plot
% Data: Electricity consumption by household entities in USA
X = categorical({'Refrigeration','Water Heating','Lighting','Air Conditioning','Other'});
X = reordercats(X,{'Refrigeration','Water Heating','Lighting','Air Conditioning','Other'});
Y = [879 1056 1628 2545 2127];
C=barh(X,Y);
C.FaceColor = 'flat';
C.CData(1,:)=[0 1 1];
C.CData(2,:)=[0 0 1];
C.CData(3,:)=[0 0.4470 0.7410];
C.CData(4,:)=[0 1 0];
C.CData(5,:)=[0.4660 0.6740 0.1880];
xlabel('KWh / Household');
title('Electricity consumption by household entities in USA')
grid on
```

Fig. 5.10 Code—Horizontal bar plot

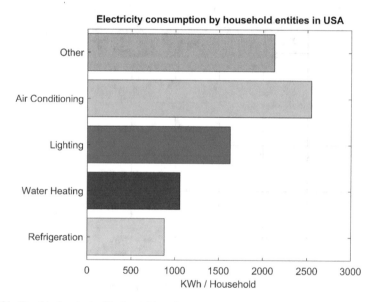

Fig. 5.11 Graphical output—Horizontal bar plot

5.4 Area Plot

Area plot is another interesting and essential visualization technique. The command for area plot is MATLAB is demonstrated below:

Area plot command for MATLAB:
area(input _ matrix)

```
% Area plot
% Data: Typical load curve over 24 hours in MW
clc; clear;
a=xlsread('Area_plot_2.xlsx');
y=a(:,2:4)
area(y);
xlim([0 24]);
xlabel('Hours');
ylabel('Load (MW)');
title('Load curve in MW')
legend('Base load','Shoulder load','Peak load','Location','northwest');
```

Fig. 5.12 Code—Area plot

The dimension of the area matrix may vary according to the dataset. In the area plot, each of the features stacked over each other sequentially over the y-axis. An example is given below to facilitate the understanding.

5.4.1   MATLAB Example 5.6: Area Plot

In a power system, the load curve illustrates the variation of loads within a certain time period. When the time period is 24 h, we consider it as a daily load curve. The loads can be of different types, such as base load, shoulder load, and peak load. Base load is the constant load that does not change over time. Shoulder load may change during certain time intervals by ramping up or ramping down and stay constant for the rest of the time. On the other hand, peak load only occurs for a short period of time in a day. It may happen twice, for example, during early morning or late afternoon.

An ideal scenario of load curve over 24 h in terms of base load, shoulder load, and peak load is shown in the following MATLAB code (Fig. 5.12) using an area plot.

Output
The data of the input matrix, y, used for the area plot is shown below (Fig. 5.13) to understand the plotted figure more conveniently. The first, second, and third columns of the y matrix represent the data of base load, shoulder load, and peak load, respectively.

From Fig. 5.14, the different load patterns over a day can be observed to find a pattern. The curves on different intervals in a day can be utilized to plan on dispatch opportunities and decide on energy market price as well.

5.5   Surface Plot

Surface plot provides a 3D illustration that facilitates better visualization interpretation. The MATLAB command for the surface plot is given below:

Fig. 5.13 Output—Area plot

```
Command Window

y =

        40      0      0
        40      0      0
        40     10      0
        40     20      0
        40     30      0
        40     30      0
        40     30      0
        40     30     10
        40     30     10
        40     30     10
        40     30      0
        40     20      0
        40     20      0
        40     30      0
        40     30     10
        40     30     15
        40     30     15
        40     30     15
        40     30     10
        40     30      0
        40     20      0
        40     10      0
        40      0      0
```

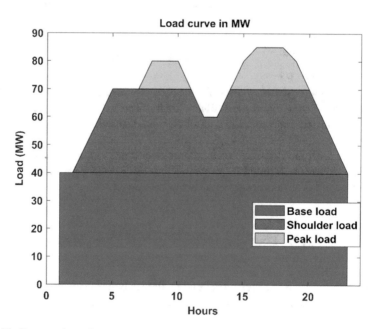

Fig. 5.14 Output—Area plot

> **Surface plot command for MATLAB:**
> $surf(x, y, z)$

Here, x, y, and z are the input matrices, having the same dimension. The three inputs utilize a 3D coordinate system to create a surface plot.

5.5.1   MATLAB Example 5.7: Surface Plot

The data of the day-ahead locational marginal price (LMP) of electricity, collected from California ISO Open Access Same-Time Information System (OASIS), is used to plot a surface plot [4]. In this plot, only 1-week data started from January 1 to January 7, 2019 is utilized for a 24-hour time horizon to understand the interval of a particular day at which the LMP is maximum, or minimum. The surface plot can be useful to decide on arbitrage or regulation opportunities during a particular interval on a specific day. Moreover, the surface plot can also determine when an energy storage system needs to be charged and when it can be discharged. The code is given in Fig. 5.15.

Output
The x-axis represents 7 days of the week, and the y-axis refers to the 24-h time horizon. The x- and y-axis data are used to make a mesh grid, which creates a coordinate array of dimensions 24×7. Later, the LMP matrix of dimension 7×24 indicates the z-axis value in the surface coordinate. The LMP data is shown in Fig. 5.16 to realize the surface plot generated afterward.

```
% Surface plot
% Data: Electricity day-ahead market price (2019)
% LMP: Local Marginal Price ($/MW)
% Day: Jan-01-2019 to Jan-07-2019
% Hour: 24 hours
clc; clear;
LMP=xlsread('surface2.xlsx')
Day=1:7;
Hour=1:24;
[DAY,HOUR]=meshgrid(Day,Hour);
surf(DAY,HOUR,LMP')
colorbar
xlabel('Days (Jan 1 - Jan 7)');
ylabel('Hours');
zlabel('Price ($/MW)');
title('Electricity day-ahead market price (2019)');
```

Fig. 5.15 Output—Area plot

```
Command Window
>> LMP

LMP =

    Columns 1 through 13

    44.9967   56.2480   55.3468   51.2607   45.9139   44.3831   44.1278   42.1309   41.2310   40.0213   38.7816   38.4742   39.5576
    44.8305   53.3640   61.9953   58.2229   52.9929   50.4424   47.1042   44.8625   43.3866   42.3716   42.2208   42.4111   43.4045
    51.5220   71.3100   71.9753   62.5451   58.7773   53.8963   46.6466   42.7969   41.0990   39.7071   40.4144   39.6398   41.3517
    49.8493   69.3141   71.0427   60.6604   57.1362   50.6729   44.8080   41.3024   40.1341   38.8131   38.8424   39.5107   40.0722
    50.0476   68.4420   67.1451   55.1748   52.6856   47.9470   43.6028   39.8518   37.8449   37.5594   36.5538   36.5499   36.5499
    41.8361   51.5499   52.8814   47.9464   45.0682   40.6657   38.8953   36.5517   34.5361   34.5215   32.6502   33.0992   34.5179
    38.8869   46.7549   48.8842   44.2088   42.0300   38.5365   34.5758   33.6571   34.5573   33.2878   32.2634   33.4595   35.1086

    Columns 14 through 24

    42.7888   42.8066   40.7658   35.7666   29.0960   22.7794   18.9996   16.0533   14.8840   21.8590   36.8867
    45.0377   57.9937   55.3043   42.5977   39.0772   36.4407   34.5510   31.9231   31.7445   34.6209   42.1169
    44.0753   56.8770   50.8418   41.5099   35.9709   33.7787   31.9940   29.0733   30.5193   33.3969   40.6107
    43.5883   55.3317   48.5251   40.6936   34.6420   33.1419   32.2070   30.7925   31.0946   32.3366   37.5603
    38.1673   39.3710   39.6269   39.6748   39.9456   38.8293   37.1539   36.0187   37.5305   38.5080   38.5241
    34.9405   36.5421   34.8724   34.4245   33.0124   29.6523   25.1282   23.6025   26.3636   29.6678   34.5341
    39.3838   48.2464   56.0296   43.8929   43.4156   40.1062   38.0461   34.6880   34.6625   34.3164   39.7262
```

Fig. 5.16 Output—Surface plot

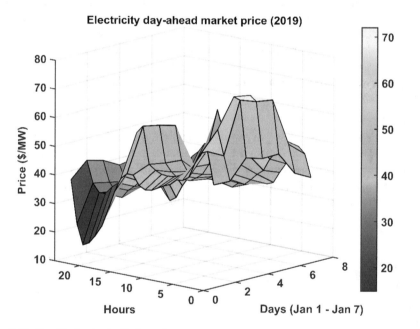

Fig. 5.17 Graphical output—Surface plot

Here, in this plot in Fig. 5.17, the yellow color signifies the highest locational marginal price on a certain day and certain interval. The deep blue on the other hand represents the period when the price is the lowest. The colors are chosen by default; no code has been used to select these colors in the plot. From an energy market perspective, when the price is higher, the stakeholder will be more interested to dispatch energy storage systems for the highest profit. Conversely, during lower price conditions, the energy storage system will be charged for future use. The surface plot, thus, can be very useful in regard to attaining higher profit from the energy market.

5.6   Pie Plot

Pie plot is a widely used visualization method. In MATLAB, a pie plot can be made using the following command:

> **Pie plot command for MATLAB:**
> *pie(x, explode)*

Here, x is the input vector, and explode is an optional feature of the function. The explode feature is used whenever single or multiple pieces of pie need to explode from their original position to signifying their impact. An example with both *explode* and without *explode* feature is shown in the following example.

5.6.1   MATLAB Example 5.8: Pie Plot

The data of electricity consumption by different sectors in the USA (2018) is demonstrated using a pie plot in MATLAB using the code in Fig. 5.18. A subplot is used to demonstrate two pie plots side by side with and without explode feature.

In the first subplot of Fig. 5.19, all the pieces are stuck together without the "explode" feature. In the second subplot, we can see that all the pieces explode from their rigid position. In the code, explode represents a vector, where the value "0" indicates not to explode, while the value "1" indicates to explode. As all the values in the explode vector in the above-mentioned code are "1's," all of the pieces are exploded in the second subplot.

```
% Pie plot
% Data: Electricity consumption by different sectors in USA (2018)
% x: Percentage of Consumption in four different sectors in USA
x = [35.4,25.9,2,38.5];
subplot(1,2,1);
pie(x);
title('Without explode feature')
subplot(1,2,2);
explode=[1,1,1,1];
pie(x,explode);
title('With explode feature')
labels = {'Commercial','Industrial','Transportation','Residential'};
legend(labels,'Location','best');
```

Fig. 5.18 Code—Pie plot

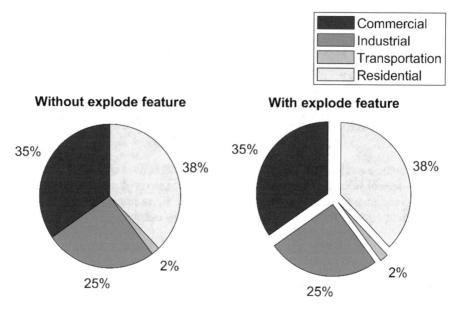

Fig. 5.19 Graphical output—Pie plot

5.7   Heat Map

A heat map is another visualization format for decision-making while considering three sets of parameters. The MATLAB command for generating a heat map is shown below:

Heat map plot command for MATLAB:
heatmap(x, y, z)

Here, x, y, and z are the three input parameters. An illustration of a heat map is shown in the following example to understand its significance.

5.7.1   MATLAB Example 5.9: Heat Map

The data of electricity day-ahead market price, more precisely, locational marginal price (LMP), for 1 week (January 1, 2019 to January 7, 2019) over the time horizon starting from 10 AM to 3 PM is utilized to generate a heat map using the MATLAB code in Fig. 5.20. The data is available on the California ISO OASIS site [4].

```
% Heatmap
% Data: Electricity day-ahead market price (2019)
% LMP: Local Marginal Price ($/MW)
% Date: Jan-01-2019 to Jan-07-2019
% Time: 10 AM to 03 PM
clc; clear;
LMP = [38.8 38.5 39.6 42.8 42.8 40.8;...
       42.2 42.4 43.4 45.0 58.0 55.3;...
       40.4 39.6 41.4 44.0 56.9 50.9;...
       38.8 39.5 40.0 43.6 55.3 48.5;...
       36.6 36.5 36.5 38.2 39.4 39.6;...
       32.7 33.1 34.5 35.0 36.5 34.9;...
       32.3 33.5 35.1 39.4 48.2 56.0];
Time = {'10 AM','11 AM','12 PM','01 PM','02 PM','03 PM'};
Date = {'Jan-01','Jan-02','Jan-03','Jan-04','Jan-05','Jan-06','Jan-07'};
H = heatmap(Time,Date,LMP);
H.Title = 'Electricity day-ahead market price ($/MW)';
H.XLabel = 'Time';
H.YLabel = 'Date';
```

Fig. 5.20   Code—Heat map

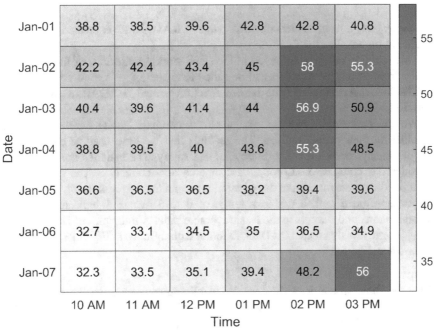

Fig. 5.21   Output—Heat map

Output

In the heat map in Fig. 5.21, the x-axis represents the time periods and the y-axis signifies the date. The individual blocks represent the value of LMP for a specific date and period. The color bar is used to differentiate among the different LMPs. The mild blue indicates the lowest LMP, while the deep blue regards the highest LMP.

By demonstrating each block in terms of different color intensity makes the heat map very useful to realize where the price is comparatively higher and lower to make better decisions.

5.8   Radar Plot

Radar plot is one of the most intriguing and sophisticated visualization techniques. In a radar plot, the center can be defined as zero. The coordinate starts to span outside from the center with equal distribution. There is no in-built function in MATLAB for the radar plot. However, due to its importance in the engineering domain, we have shown a user-defined function for allowing the scope of radar plot in MATLAB.

5.8.1   MATLAB Example 5.10: Radar Plot

For radar plot, a user-defined function named "RADAR.m" is created, which has five input parameters.

This function needs to be saved in the same working directory of the executing M-file from which the function will be called to execute.

The MATLAB code of user-defined function named "RADAR.m" is given in Fig. 5.22.

We are interested in plotting a radar plot to compare among different batteries considering their certain entities. The features that we have used to compare are power density, energy density, life cycle, and safety. The batteries that have been considered to compare are Li-ion, liquid supercapacitor, and NaS. The values of the mentioned features for each of these batteries are obtained from Ref. [5]. All of these values have been ranked within a range of 0 to 5, where 5 indicates the most favorable rank and 0 refers to the least favorable rank. A radar plot has been plotted using the previously created RADAR function with the code in Fig. 5.23. The output is shown in Fig. 5.24.

Output

From the above figure, it can be depicted that liquid supercapacitor battery provides the highest life cycle and power density opportunity compared to other batteries. On the other hand, Li-ion is highly favorable while considering safety and energy density. Therefore, the above radar plot provides an opportunity to decide on choosing a specific battery based on intended features.

```
function f=RADAR(I,Feature,Legend,line_color,Title)
%INPUT: Data, I: Input matrix; size row by col,
% row: number of examples; col: Features for each examples
% Feature: Labels of each examples
% Legend: a string array of legends,e.g. {'leg1','leg2'}
% line_color: a string vector of line colors, e.g. ['r','g']
% Title: A string representing the title, e.g. {'Title'}
row=size(I,1);
col=size(I,2);
Feature_num=size(Feature,2);
I=[I I(:,1)];
theta= (2*pi/col)*[1:col+1] + (pi/col);
R=ones(1,size(I,1));
[x,y]=pol2cart(theta,I);
P=plot(y',x','LineWidth',1.5);
legend(Legend,'Location','eastoutside');
title(Title);
for i=1:row
    set(P(i),'Color',line_color(i))
end
axis_max=max(max(I))*1.1;
axis([-axis_max axis_max -axis_max axis_max]);
axis equal
axis off
if Feature_num>0
    R_axis=linspace(0,max(max(I)),Feature_num);
    for k=1:Feature_num
        text(R_axis(k)*sin(pi/col-0.3),R_axis(k)*cos(pi/col-0.3),num2str(R_axis(k),2),...
                                    'FontSize',10)
    end
    [R,R_axis]=meshgrid(ones(1,col),R_axis);
    R_axis=[R_axis R_axis(:,1)];
    theta_axis=2*pi/col*[1:col+1]+pi/col;
    R=ones(1,size(R_axis,1));
    [y_axis,x_axis]=pol2cart(theta_axis,R_axis);
    hold on
    B=plot(x_axis,y_axis,':k');
    for i = 1:length(B)
        set(get(get(B(i),'Annotation'),'LegendInformation'),'IconDisplayStyle','off');
    end
    C=plot(x_axis',y_axis',':k');
    for i = 1:length(C)
        set(get(get(C(i),'Annotation'),'LegendInformation'),'IconDisplayStyle','off');
    end
end
if length(Feature)>=col
    theta_feature=2*pi/col*[1:col]+pi/col;
    R_feature=axis_max;
    [y_feature,x_feature]=pol2cart(theta_feature,R_feature);
    for k=1:col
        if ~sum(strcmpi({'' },Feature(k)))
            text(x_feature(k), y_feature(k),cell2mat(Feature(k)), 'FontSize',...
                                    12,'HorizontalAlignment','center')
        end
    end
end
```

Fig. 5.22   Code—Creation of RADAR function

```
clc;clear;
% Execution of Radar.m function
% Data: Four features of three different types battery
% Battery types: Li-ion, Liquid super capacitors, NaS
% Features: Power density, Energy density, Life cycle, Safety

% Input
I=[2 5 2 4;5 2.5 5 2;1 2 1.5 3];
Feature={'Power density','Energy density','Life cycle','Safety'};
Legend={'Li-ion','Liquid super capacitor','NaS'};
line_color=['r','g','b'];
Title={'Comparison of different battery types'};
% Function call
RADAR(I,Feature,Legend,line_color,Title)
```

Fig. 5.23 Code—Radar plot

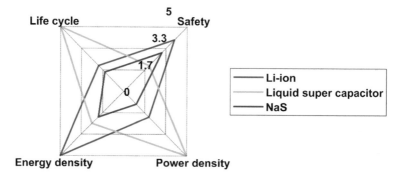

Fig. 5.24 Graphical output—Radar plot

5.9   3D Plot

3D plot enhances the visualization, by providing the opportunity to plot a figure which can be rotated in a three-dimensional space. We can realize any plot far better in a 3D view instead of 2D. MATLAB offers such an opportunity for some of the plots, such as pie plot, bar plot, etc. To make a 3D pie plot, the MATLAB command remains almost the same except for including "3" at the later of the name of the function. For example, the 3D pie plot command of MATLAB is as follows:

3D pie plot command for MATLAB:
pie3(x, explode.)

```
%% 3D pie plot
% Data: Electricity consumption by different sectors in USA (2018)
clc;clear;
x = [35.4,25.9,2,38.5];
explode=[0,0,1,0];
pie3(x,explode);
title('Electricity consumption by different sectors in USA (2018)')
labels = {'Commercial','Industrial','Transportation','Residential'};
legend(labels,'Location','best');
```

Fig. 5.25   Code—3D pie plot

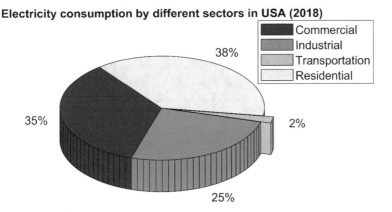

Fig. 5.26   Graphical output—3D pie plot

5.9.1   MATLAB Example 5.11: 3D Pie Plots

The following MATLAB code in Fig. 5.25 demonstrates the plotting of a 3D pie plot with its output in Fig. 5.26. The plot depicts the share of different sectors in the electricity consumption in the USA in 2018.

5.10   Exporting High-Quality Figure

In MATLAB, it is possible to customize the quality of the figure while exporting by choosing the value of dpi. It is expected by most of the journals to maintain a minimum quality of the figures. The recommended dpi level for a good quality figure is generally 300 dpi. From MATLAB, while exporting a figure, the dpi level can be customized manually. In default mood, the dpi level in MATLAB is below 300 dpi. Therefore, it is necessary to know how to increase the dpi level while exporting figure from MATLAB, which is described below:

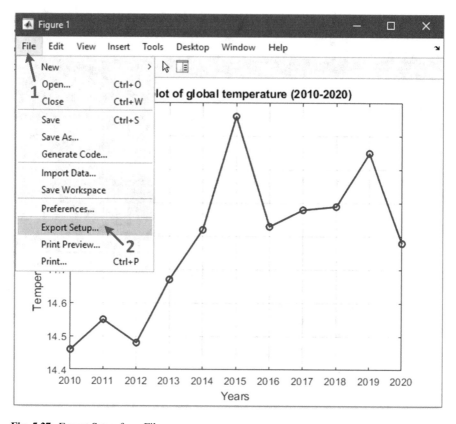

Fig. 5.27   Export Setup from File

Step 1   After producing a figure in MATLAB, the first step is to select File > Export Setup. A window as in Fig. 5.27 will appear.

Step2   After selecting "Export Setup," the following window (Fig. 5.28) will appear:

From this window, first, select the "Rendering" option. Later, click on the dropdown option named "Resolution (dpi)." From this dropdown box, the user may choose their desired dpi level. In Fig. 5.28, 600 dpi is selected as the resolution.

Step 3   Click on the "Apply to Figure" option, and then select the "Export" button. Save the figure in the desired format by browsing the desired saving location.

By following the above-mentioned steps, high-quality figures can be exported from MATLAB.

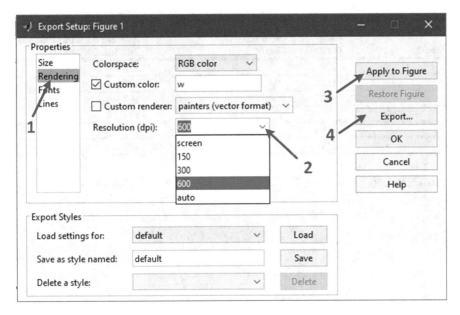

Fig. 5.28 Export Setup Window

5.11 Conclusion

The readers will be able to learn different visualization techniques that can be implemented in MATLAB from this chapter. This chapter demonstrated line plot, bar plot, area plot, surface plot, pie plot, heat map, and 3D plot. For all of these visualizations, MATLAB has built-in functions with many features to self-customize. All of these functions and the customization techniques are shown and implemented in this chapter. Radar plot, another important visualization in the engineering domain, is also illustrated in this chapter. For producing a radar plot, MATLAB does not have any built-in function. However, in this chapter, a user-defined function is created for producing radar plots in MATLAB. The chapter concludes with the demonstration of an essential topic that is the exportation of high-quality images produced in MATLAB.

Exercise 5

1. Name different types of plots that can be drawn in MATLAB.
2. What is the difference between:

   (a) Plot and subplot
   (b) Hold on and hold off

Table 5.5 Average prices of products A and B in 6 months

Months	Average price of product A Price (in dollars)	Average price of product B Price (in dollars)
1	129	178
2	155	198
3	145	183
4	131	174
5	160	181
6	151	193

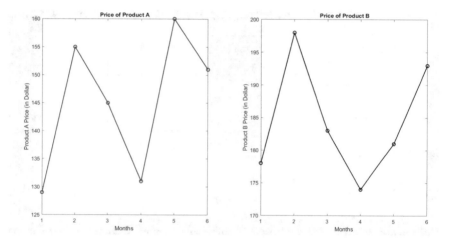

Fig. 5.29 Expected graphical output for Question 3a

    (c) Bar and barh
    (d) Pie and pie3

3. Two products have their average price varied over different months of the year as shown in Table 5.5. Demonstrate the graph with a line plot:

    (a) With subplots
    (b) With double axis, where the left axis shows the price of product A, and the right axis shows the price of product B

The graphs should resemble Figs. 5.29 and 5.30.

    (a)
    (b)

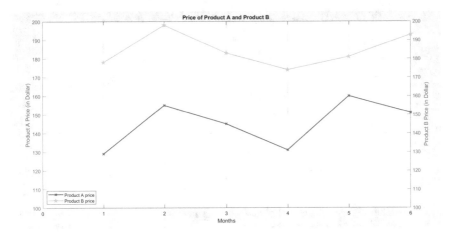

Fig. 5.30 Expected graphical output for Question 3b

Table 5.6 Number of members in each team

Team	Members
HR	5
Software team	15
Electrical team	22
Mechanical team	17
Management team	8
Marketing team	10

4. A new startup has team members as mentioned in Table 5.6. Using MATLAB, represent the data in a (a) horizontal bar chart and (b) vertical bar chart.

   The graphs should resemble Figs. 5.31 and 5.32.
   (a)
   (b)

5. In an apparatus purchase for university laboratories, the number of equipment is estimated as shown in Table 5.7.

   Using MATLAB, represent the mentioned data in a:

   (a) 2D pie chart (with and without explode feature of the smallest pie using subplot)
   (b) 3D pie chart (with and without explode feature of the largest pie using subplot).

   The graphs should resemble Figs. 5.33 and 5.34.
   (a)

   (b)

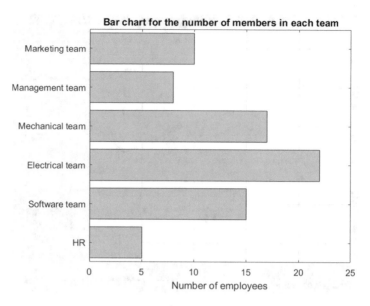

Fig. 5.31 Expected graphical output for Question 3b

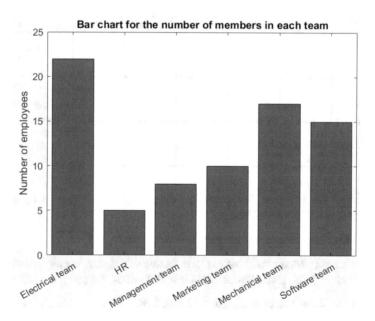

Fig. 5.32 Expected graphical output for Question 4b

Table 5.7 Number of each type of equipment

Equipment	Number
Electrical machines	7
Communication trainer kit	5
Electronics trainer kit	6
PLCs	15
Assorted IC boxes	10

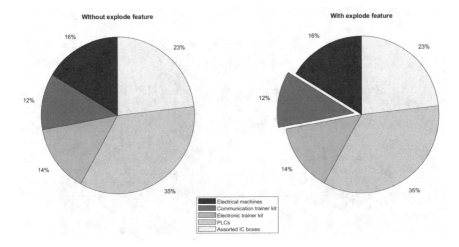

Fig. 5.5 Expected graphical output for Question 5a

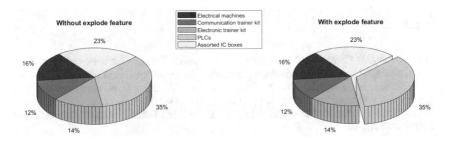

Fig. 5.34 Expected graphical output for Question 5b

6. The temperature for a particular location is recorded in Table 5.8.

Draw the heat map from the above data using MATLAB. The graph should appear as in Fig. 5.35.

Table 5.8 The temperature, in °C, at different times in different dates

Date/time	12 AM	4 AM	8 AM	12 PM	4 PM	8 PM
1 march	31.5	29.3	30.1	33.2	32.5	31.7
2 march	29.8	28.4	29.0	30.3	30.1	28.8
3 march	27.9	28.1	29.2	30.0	29.5	28.1
4 march	30.7	29.6	30.2	31.6	32.9	32.7
5 march	31.2	30.5	30.4	30.9	31.8	31.4
6 march	31.1	30.6	31.6	32.5	33.7	32.2

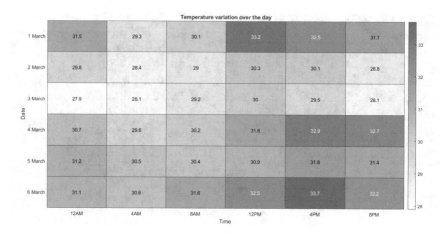

Fig. 5.35 Expected graphical output for Question 6

References

1. NOAA. National Centers for Environmental information, Climate at a Glance: Global Time Series, published April 2021, retrieved on April 14, 2021 from https://www.ncdc.noaa.gov/cag/.
2. Sakib N, Hossain E, Ahamed SI. A qualitative study on the United States internet of energy: a step towards computational sustainability. IEEE Access. 2020;8:69003–37.
3. https://www.iea.org/data-and-statistics?country=WORLD&fuel=CO2%20emissions& indicator=TotCO2.
4. http://oasis.caiso.com/mrioasis/logon.do.
5. http://www.flashchargebatteries.com/solution/.

Chapter 6
Solving Equations

6.1 Introduction

In the engineering domain, solving equations is very crucial, which may take significant time and brainwork if solved manually. In MATLAB, all of these equations can be solved very easily with a few lines of code, without spending too much time and effort. In this chapter, the basics of linear algebra are explained, and techniques to solve different types of algebraic and differential equations are presented with several examples using MATLAB.

6.2 Linear Algebra

Rank The rank of a matrix can be defined as the maximum number of linearly independent rows or columns. Here, a row or a column vector can be regarded as linearly independent if:

(a) The vector is not a scalar multiple of other vectors
(b) The vector is not a resultant of the combinations of other vectors

If the dimension of a matrix is *row* × *col*, the rank of that matrix will be either the maximum number of linearly independent rows or columns. It can be distinguished more precisely as follows:

(a) If *row* > *col*, the rank of the matrix≡ maximum number of linearly independent columns.
(b) If *row* < *col*, the rank of the matrix≡ maximum number of linearly independent rows.

The rank of a matrix can be determined using MATLAB, by using *rank* function as follows:

> **MATLAB command for determining the rank of a matrix, A:**
> *rank(A)*

6.2.1 MATLAB Example 6.1: Rank

Determine the ranks of the following two matrices A and B:

(i) $A = [1\ 2\ 4; 2\ 4\ 8\ ]$
(ii) $B = [1\ 1\ 2; 5\ 2\ 7; 0\ 4\ 4; 2\ 6\ 8\ ]$

```
% Determining rank
% Input matrix, A
clc; clear;
A=[1 2 4;2 4 8];
disp('The rank of the matrix A:')
rank(A)
% Input matrix, B
B=[1 1 2;5 2 7;0 4 4;2 6 8];
disp('The rank of the matrix B:')
rank(B)
```

Fig. 6.1 Code—Determination of rank of a matrix

Fig. 6.2 Output—
Determination of rank of a
matrix

```
Command Window

  The rank of the matrix A:

  ans =

      1

  The rank of the matrix B:

  ans =

      2
```

The MATLAB code for this example is given in Fig. 6.1 with its output in Fig. 6.2.

Output

Eigenvector and Eigenvalue

Consider a square matrix X of dimension $n \times n$. The eigenvector, v, of the square matrix A signifies a linear transformation that follows the following condition in Eq. (6.1):

$$Xv = \lambda v \tag{6.1}$$

Here, λ is a scalar value, which can also be regarded as eigenvalue; and $v \in R^n$ is a nonzero vector.

For determining the eigenvalues, the following characteristic equation can be solved for λ:

$$|XI - \lambda| = 0, \tag{6.2}$$

where I is the identity matrix having the same dimension as X. In MATLAB, using $eig()$ function eigenvalues can be determined easily.

> **Finding eigenvalues of a matrix, X, in MATLAB:**
> $eig(X)$

It is to be noted that the input matrix X always needs to be a square matrix.

6.2.2   MATLAB Example 6.2: Eigenvalue

Find the eigenvalues of the following matrix, X:

$$X = [1\ 2\ 0; 0\ 5\ 0; 1\ 3\ 1] \tag{6.3}$$

The MATLAB code for this example is given in Fig. 6.3 with its output in Fig. 6.4.

Fig. 6.3 Code—
Determination of
eigenvalues of a matrix

```
% Eigenvalue
% Input matrix, X
clc; clear;
X=[1 2 0;0 5 0;1 3 1];
disp('The eigenvalues of X are:')
eig(X)
```

6.2.3   MATLAB Example 6.3: Eigenvector

Find the eigenvector of the following matrix, X:

$$X = [1\ 2\ 0; 0\ 5\ 0; 1\ 3\ 1] \qquad (6.4)$$

The MATLAB code for this example is given in Fig. 6.5 with its output in Fig. 6.6.

Here, the MATLAB command $[vector, lambda] = eig(X)$ provides two outputs—the eigenvector (*vector*) and a diagonal vector containing the eigenvalues (*lambda*). Using the above code, both eigenvalues and eigenvectors can be determined.

Fig. 6.4 Output—Determination of eigenvalues of a matrix

```
Command Window

   The eigenvalues of X are:

   ans =

        1
        1
        5
```

```
% Eigenvalue
% Input matrix, X
% [vector, lambda]=eig(X)
% Here, vector is the eigenvector
% lambda is a diagonal vector containing the eigenvalues
clc; clear;
X=[1 2 0;0 5 0;1 3 1];
[vector,lambda]=eig(X);
disp('The eigenvalues of X:')
lamda=sum(lambda)
disp('The eigenvector of X:')
vector
```

Fig. 6.5 Code—Determination of eigenvectors of a matrix

Fig. 6.6 Output—
Determination of
eigenvectors of a matrix

```
Command Window

  The eigenvalues of X:

  lamda =

        1      1      5

  The eigenvector of X:

  vector =

              0     0.0000    0.3522
              0          0    0.7044
         1.0000    -1.0000    0.6163
```

6.3   Quadratic Equations

MATLAB has built-in functions to solve any quadratic equations by using the *solve* command. The "solve" command also becomes useful to solve multiple equations with multiple variables. The entities of the *solve* function of MATLAB are given below:

> **Solving Equations in MATLAB:**
> *solve(equation, variable)*

Here, the variables can be single or multiple, and the output of the solve function can also be single or multiple. Some examples solving different types of equations using the *solve* function are given below.

6.3.1   MATLAB Example 6.4: Solving Quadratic Equation

Consider a quadratic equation: $2x^2 + 4x + 5 = 0$. Determine the values of x using MATLAB.

The MATLAB code for this example is given in Fig. 6.7 with its output in Fig. 6.8.

Fig. 6.7 Code—Solving
quadratic equation

```
% Solving quadratic equation
% 2x^2+4x+5=0
% Determine the values of x
clc;clear;
syms x
x_val=solve(2*x^2+4*x+5==0,x);
disp('The solutions are:');
x_val
```

Fig. 6.8 Output—Solving
quadratic equation

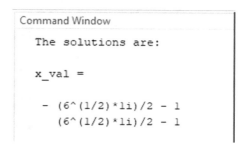

```
Command Window
    The solutions are:

    x_val =

      - (6^(1/2)*1i)/2 - 1
        (6^(1/2)*1i)/2 - 1
```

Output

Here, the highest degree of x is 2. Therefore, the solution will have two values of x. The output in Fig. 6.8 confirms that theory.

6.3.2   MATLAB Example 6.5: "Solve" Function

Consider the following two equations:

$$2x^2 + 4xy + 5 = 0 \tag{6.5}$$

$$3y^2 + 5xy - 2 = 0 \tag{6.6}$$

Determine the values of x and y using MATLAB.
The MATLAB code for this example is given in Fig. 6.9 with its output in Fig. 6.10.

In the above example, two quadratic equations with two variables have been solved using MATLAB. Thus, the *solve* function provides a wide opportunity to solve different types of equations. It is to be noted that the *syms* is used to define the unknown variables at the start of implementing *solve* function in each case.

```
%% Two quadratic equations
% 2x^2+4xy+5=0
% 3y^2+5xy-2=0
% Determine the values of x and y
clc;clear;
syms x y
[x_val,y_val]=solve(2*x^2+4*x*y+5==0,3*y^2+5*x*y-2==0);
disp('The solutions are:');
disp('x =');
disp(x_val);
disp('y =');
disp(y_val);
```

Fig. 6.9 Code—solve function

```
Command Window
 The solutions are:
 x =
  (153*(47/28 - (5*849^(1/2))/84)^(1/2))/20 - (21*(47/28 - (5*849^(1/2))/84)^(3/2))/10
  (153*((5*849^(1/2))/84 + 47/28)^(1/2))/20 - (21*((5*849^(1/2))/84 + 47/28)^(3/2))/10
  (21*(47/28 - (5*849^(1/2))/84)^(3/2))/10 - (153*(47/28 - (5*849^(1/2))/84)^(1/2))/20
  (21*((5*849^(1/2))/84 + 47/28)^(3/2))/10 - (153*((5*849^(1/2))/84 + 47/28)^(1/2))/20

 y =
  -(47/28 - (5*849^(1/2))/84)^(1/2)
  -((5*849^(1/2))/84 + 47/28)^(1/2)
   (47/28 - (5*849^(1/2))/84)^(1/2)
   ((5*849^(1/2))/84 + 47/28)^(1/2)
```

Fig. 6.10 Output—solve function

6.4 Differential Equations

6.4.1 Ordinary Differential Equations

Using MATLAB, ordinary differential equations can be solved by *dsolve* function. It not only can be used for first-order differential equations but also covers higher degrees of differential equations. The input parameters of *dsolve* function are given below:

Solving Equations:
dsolve(equation, condition)

dsolve function has two parameters—*equation*, which represents the differential equation that needs to be solved, and *condition*, which is an optional feature. If the problem has initial conditions, the function will provide an exact solution based on those conditions. Otherwise, the function will provide a generalized result with an unknown constant *C1*. Hence, *dsolve* function can be used to solve initial value problems as well.

Some examples covering first-order to third-order differential equation solves are shown in the following examples:

6.4.2 MATLAB Example 6.6: First-Order Differential Equation

Consider the following first-order differential equation:

$$\frac{dy}{dx} = 2x + y \tag{6.7}$$

(i) Solve the differential equation.
(ii) If initial condition $y(0) = 1$, find the solution of the differential equation.

The MATLAB code for this example is given in Fig. 6.11 with its output in Fig. 6.12.

Output

```
% 1st order differential equation
% dy/dx=2*x+y;
% Solve the differential equation
clc;clear;
syms y(x)
diff_eq= diff(y,x)==2*x+y;
disp('Solution without initial condition:')
Sol_y(x)=dsolve(diff_eq)
% If the initial condition y(0)=1
condition=y(0)==1;
disp('Solution with initial condition:')
Sol_y(x)=dsolve(diff_eq,condition)
```

Fig. 6.11 Code—First-order differential equation

Fig. 6.12 Output—First-order differential equation

```
Command Window

    Solution without initial condition:

    Sol_y(x) =

    C1*exp(x) - 2*x - 2

    Solution with initial condition:

    Sol_y(x) =

    3*exp(x) - 2*x - 2
```

Fig. 6.12 Output—First-order differential equation

```
% 2nd order differential equation
% (dy/dx)^2= 2*x^2+ 3*dy/dx-5;
% Initial conditions: y(0)=1, y'(0)=1
% Solve the differential equation
clc;clear;
syms y(x)
diff_eqn=diff(y,x,2)==2*x^2+3*diff(y,x)-5;
condition1=y(0)==1;
dy=diff(y,x);
condition2=dy(0)==1;
condition=[condition1 condition2];
Sol_y(x)=dsolve(diff_eqn,condition)
```

Fig. 6.13 Code—Second-order differential equation

6.4.3 MATLAB Example 6.7: Second-Order Differential Equation

Solve the following second-order differential equation:

$$\frac{d^2y}{dx^2} = 2x^2 + 3\frac{dy}{dx} - 5; \quad y(0) = 1, y'(0) = 1 \tag{6.8}$$

The MATLAB code for this example is given in Fig. 6.13 with its output in Fig. 6.14.

Output

6.4.4 MATLAB Example 6.8: Third-Order Differential Equation

Solve the following third-order differential equation:

$$\frac{d^3y}{dx^3} = 3x^2 + 3\frac{d^2y}{dx^2} - 2\frac{dy}{dx} + 1; \quad y(0) = 1, y'(0) = 0, y''(0) = 1 \qquad (6.9)$$

The MATLAB code for this example is given in Fig. 6.15 with its output in Fig. 6.16.

```
Command Window

  Sol_y(x)  =

  (41*x)/27 - (14*exp(3*x))/81 - (2*x^2)/9 - (2*x^3)/9 + 95/81
```

Fig. 6.14 Output—Second-order differential equation

```
% 3rd order differential equation
% (dy/dx)^3= 3*x^2+3*(dy/dx)^2-2*dy/dx+1;
% Initial conditions: y(0)=1, y'(0)=0, y''(0)=1;
% Solve the differential equation
clc;clear;
syms y(x)
diff_eqn=diff(y,x,3)==3*x^2+3*diff(y,x,2)-2*diff(y,x)+1;
dy=diff(y,x);
d2y=diff(y,x,2);
condition1=y(0)==1;
condition2=dy(0)==0;
condition3=d2y(0)==1;
condition=[condition1 condition2 condition3];
Sol_y(x)=dsolve(diff_eqn,condition)
```

Fig. 6.15 Code—Third-order differential equation

```
Command Window

  Sol_y(x) =

  (23*x)/4 + (9*exp(2*x))/8 - 8*exp(x) + (9*x^2)/4 + x^3/2 + 63/8
```

Fig. 6.16 Output—Third-order differential equation

Output

6.4.5   Partial Differential Equations

In the partial derivative of a function of multiple variables, the differentiation is performed with respect to one variable while considering the rest of the variables as constant. To solve a partial differential equation, the first step is to solve the partial derivative terms using *diff()* command. The distinction between the ordinary and the partial derivative will be made while defining the symbols of the variable at the start of MATLAB code. In ordinary differentiation, we define the output variable as a function of other variables. However, in partial derivatives, we treat only one as an active variable, while the rest are considered constants. An example is given below for further understanding.

6.4.6   MATLAB Example 6.9: Partial Differential Equation

Consider the following partial differential equation:

$$\frac{\partial}{\partial x}\left(2x^2 + y - 5\right) - x^2 = 0 \qquad (6.10)$$

Solve the partial differential equation for *x*.

The MATLAB code for this example is given in Fig. 6.17 with its output in Fig. 6.18.

Output

Here, after solving each partial derivative terms, we have used *solve()* function to finally solve the partial differential equation for *x*.

```
% Partial differential equation
% del(F)/del(x) - x^2 = 0
% Here, F = 2*x^2+y-5
% Solve the partial differential equation for x
clc;clear;
syms x y
F=2*x^2+y-5;
P_diff=diff(F,x);
disp('Solution:')
Sol_x=solve(P_diff-x^2==0,x)
```

Fig. 6.17 Code—Partial differential equation

Fig. 6.18 Output—Partial
differential equation

```
Command Window

  Solution:


  Sol_x =


   0

   4
```

6.5   Integral Equations

In MATLAB, integration can be solved using *int()* function. Integrals can be of two types: definite integral and indefinite integrals. In an indefinite integral, the limits are not defined; whether in a definite integral, the limits are defined for the variable with respect to which the integration will be performed. To solve any integral equation, the very basic step is to first determine the values of the integral terms. The MATLAB command for determining integration of a function is as follows:

> **Integration of a function f with respect to x with a limit of [a, b] in MATLAB:**
> $int(f, x, a, b)$

After determining the integral terms, by using *solve()*, integral equations can be solved.

6.5.1   *MATLAB Example 6.10: Single Variable Integral Equation*

Consider the following single variable integral equation:

$$\int 2x^2 .dx - 3x = 0 \qquad\qquad (6.11)$$

(i) Solve the above integral equation for x.
(ii) If the limit of x is [0 2], find the solution for x.

```
% Integral equation
% Integration[2*x^2].dx - 3x = 0
% Without limit
clc;clear;
syms x
I1=int(2*x^2,x);
disp('The solution without limit:')
x_sol=solve(I1-3*x==0,x)
% With limit of [0 2]
I2=int(2*x^2,x,0,2);
disp('The solution with limit:')
x_sol=solve(I2-3*x==0,x)
```

Fig. 6.19 Code—Single variable integral equation

Fig. 6.20 Output—Single variable integral equation

```
Command Window

    The solution without limit:

    x_sol =

                    0
        -(3*2^(1/2))/2
         (3*2^(1/2))/2

    The solution with limit:

    x_sol =

    16/9
```

The MATLAB code for this example is given in Fig. 6.19 with its output in Fig. 6.20.

Output

6.5.2 MATLAB Example 6.11: Multivariable Integral Equation

Consider the following multivariable integral equation:

$$\int (x^2 + e^y).dx = 0 \tag{6.12}$$

(i) Solve the above integral equation for y.
(ii) If the limit of x is [0 1], find the solution for y.

The MATLAB code for this example is given in Fig. 6.21 with its output in Fig. 6.22.

Output

```
% Integral equation
% Integration[x^2+exp(y))].dx=0
% Without limit
clc;clear;
syms x y
I1=int(x^2+exp(y),x);
disp('The solution without limit:')
y_sol=solve(I1==0,y)
% With limit of [0 1]
I2=int(x^2+exp(y),x,0,1);
disp('The solution with limit:')
y_sol=solve(I2==0,y)
```

Fig. 6.21 Code—Multi-variable integral equation

Fig. 6.22 Output—Multi-
variable integral equation

> Command Window
>
>   The solution without limit:
>
>   y_sol =
>
>   log(-x^2/3)
>
>   The solution with limit:
>
>   y_sol =
>
>   - log(3) + pi*1i

6.6   Conclusion

Linear algebra is an important field in the engineering domain. In this chapter, some essential concepts of linear algebra, such as rank, eigenvalues, and eigenvectors, are discussed with MATLAB examples. The key part of this chapter is the detailed MATLAB implementations to solve different types of equations. The readers will be able to learn about different MATLAB functions and their implementations to solve quadratic equations, differential equations, and integral equations. The possible variations of each of these equations are brought in the discussion of this chapter, to facilitate the readers with a detailed understanding of the topic.

Exercise 6

1. What are eigenvalues and eigenvectors? How these are mathematically determined?
2. Mention the application of *solve()* and *dsolve()* on MATLAB with examples.
3. Given   two   matrices (i) $M = \begin{bmatrix} -4 & 5 \\ 8 & -11 \end{bmatrix}$

   and (ii) $N = \begin{bmatrix} 0.33 & 1 & 3.3 \\ 0.5 & 0.45 & -5.12 \\ 2 & -2 & 0 \end{bmatrix}$:

   (a) Determine the rank of M and N.
   (b) Determine the inverse of M and N.
   (c) Determine the eigenvalues and eigenvector of M and N.

4. Solve the following algebraic equations using MATLAB:

(a) $x^2 + 5x + 9 = 0$
(b) $101x^2 + 36x + 255 = 4$
(c) $2.60x^2 + 5.34x - 7 = 7.44$
(d) $9x^2 + 3xy - 2 = -3;\ 4x^2 + 7xy + 5/2 = 0$
(e) $16x^2 + xy - 3 = 9x^2 - 11xy + 2 = 7$

5. Solve the following differential equations using MATLAB. Show the solution with the initial condition where applicable. Use *vpa()* to summarize large expression up to two significant digits:

(a) $\frac{dy}{dx} = 3x + 2y;\ y(0) = 2$
(b) $\frac{dy}{dx} = -7x + 4y + 2;\ y(0) = 1$
(c) $\frac{d^2y}{dx^2} = 5x^2 + 9\frac{dy}{dx} + 2y;\ \ y(0) = 2, y'(0) = 1$
(d) $\frac{d^2y}{dx^2} = -3x^2 - \frac{dy}{dx} + 2;\ \ y(0) = 1, y'(0) = 1$
(e) $\frac{d^3y}{dx^3} = 5x^2 + 11\frac{d^2y}{dx^2} + \frac{dy}{dx} + 8;\ \ y(0) = 2, y'(0) = 1, y''(0) = -1$
(f) $\frac{d^3y}{dx^3} = -6x^2 + \frac{d^2y}{dx^2} - 23;\ \ y(0) = 1, y'(0) = 0, y''(0) = 1$
(g) $\frac{\partial}{\partial x}(x^2 + 4y + 3) + 7x^2 = 3$
(h) $\frac{\partial}{\partial x}(3x^2 - 11y) - 2x^2 = 0$

6. Consider the following integral equations:

(a) $\int (\log(x))^2.\ dx - 2x$
(b) $\int (2^x - e^y).\ dx.$

For each of the above:

(i) Solve the above integral equations for *y*.
(ii) If the limit of x is [0 2], find the solution for *y*.

Chapter 7
Numerical Methods in MATLAB

7.1 Introduction

Numerical methods are important to solve mathematical problems that contain continuous variables, and not possible to solve explicitly. It is an important area that involves both mathematics and computer science. In the field of engineering, numerical methods can be applied in numerous applications. With the recent advent of digital computers, the implication of numerical analysis has become a part and parcel of engineering applications. MATLAB can provide the platform to utilize different numerical methods in various applications. In this chapter, some of the important numerical methods are described and implemented using MATLAB, such as the Gauss-Seidel method, the Newton-Raphson method, and the Runge-Kutta method.

7.2 Gauss-Seidel Method

Gauss-Seidel method is an iterative method through which a set of equations can be solved for determining unknown variables. Carl Friedrich Gauss first developed the Gauss iteration method, which was later improved by Philipp Ludwig Seidel. Due to both of their contributions, the improved algorithm was named after both of them, which is called the Gauss-Seidel method.

Consider the following set of equations, where x, y, and z are the unknown variables:

$$a_1x + b_1y + c_1z = d_1 \tag{7.1}$$

$$a_2x + b_2y + c_2z = d_2 \tag{7.2}$$

E. Hossain, *MATLAB and Simulink Crash Course for Engineers*,
https://doi.org/10.1007/978-3-030-89762-8_7

$$a_3 x + b_3 y + c_3 z = d_3 \tag{7.3}$$

Here, a_1, a_2, a_3 are the coefficients of x; b_1, b_2, b_3 are the coefficients of y; c_1, c_2, c_3 represents the coefficients of z; and d_1, d_2, d_3 are the constants of the above equations.

In the Gauss-Seidel method, the aim is to solve these equations for x, y, and z, respectively. Hence, these equations can be rewritten as follows:

$$x = \frac{1}{a_1}(d_1 - b_1 y - c_1 z) \tag{7.4}$$

$$y = \frac{1}{b_2}(d_1 - a_2 x - c_2 z) \tag{7.5}$$

$$z = \frac{1}{c_3}(d_1 - a_3 x - b_3 y) \tag{7.6}$$

The next step is to assign the initial values of x, y, and z into the above Eqs. (7.4)–(7.6) to determine the values of x^1, y^1, z^1, which indicates the first approximation of the values of x, y, and z, respectively after the first iteration. In the previous Gauss iteration method, all the initial values are used during the first iteration; and later the results of the first iterations are used for the successive iteration. To understand it more clearly, let us assume the initial values of x, y, and z are x_0, y_0, and z_0, respectively.

In the Gauss iteration method, the first approximations of the values are determined as follows:

$$x^1 = \frac{1}{a_1}(d_1 - b_1 y_0 - c_1 z_0) \tag{7.7}$$

$$y^1 = \frac{1}{b_2}(d_1 - a_2 x_0 - c_2 z_0) \tag{7.8}$$

$$z^1 = \frac{1}{c_3}(d_1 - a_3 x_0 - b_3 y_0) \tag{7.9}$$

To determine the values of x^2, y^2, z^2 in the second iteration, the determined values x^1, y^1, z^1 will be used in place of x_0, y_0, and z_0, respectively. This algorithm continues until convergence.

Later, with an improvisation made by Seidel, the Gauss-Seidel method appears, where instead of using all the values of the previous iteration for the calculation of the next iterated values, this method suggests the usage of the most recent updated values all the time to converge faster than before.

In the Gauss-Seidel method, the first approximations are calculated as follows:

$$x^1 = \frac{1}{a_1}(d_1 - b_1 y_0 - c_1 z_0) \qquad (7.10)$$

$$y^1 = \frac{1}{b_2}(d_1 - a_2 x^1 - c_2 z_0) \qquad (7.11)$$

$$z^1 = \frac{1}{c_3}(d_1 - a_3 x^1 - b_3 y^1) \qquad (7.12)$$

Here, in the first iteration, for determining y^1, the latest updated value x^1 is used instead of x_0. The same continues for other values and other iterations as well. Due to this improvisation, the number of iterations greatly reduces, and the algorithm converges more quickly.

After repeating these steps for several iterations, we can determine the values of x, y, and z more accurately. Generally, the greater the number of iterations, the results will start to become more accurate. Therefore, it is a good question to ask—what is the standard number of iterations before we can stop the procedure? There is no specific number to answer this question; however, in this iterative method, the approximate values become almost constant after some iterations. When this occurs, we can conclude that the convergence has happened, and it is probably the best place to stop the iteration. There are also some other methods such as calculating the tolerance and making a decision based on our expectation of tolerance for a certain problem to set up stopping criteria.

The tolerance is calculated by using the following formula:

$$\text{Tol\_}x^{i+1} = \frac{|x^{i+1} - x^i|}{x^i} \qquad (7.13)$$

Here, $\text{Tol\_}x^{i+1}$ indicates the tolerance for x^{i+1} in the $(i+1)^{th}$ iteration. Similarly, the tolerance for both y^i and z^i can be calculated. A threshold tolerance can be defined based on which the stopping decision will be made. For example, if the tolerance values for all x, y, and z fall below 0.0001, we can make a decision that the convergence has been achieved at that iteration; and further iterations can be curtailed.

MATLAB is one of the most suitable platforms to perform this iterative analysis. Using MATLAB, it is easier to perform a higher number of iterations more easily to reach the convergence. An example is given below where a set of equations are solved for determining the unknown variables using the Gauss-Seidel method.

7.2.1 MATLAB Example 7.1: Gauss-Seidel Method

Consider the following set of equations to determine the values of x, y,and z using the Gauss-Seidel method:

$$80x - 10y + 2z = 85 \tag{7.14}$$

$$5x + 50y + 12z = 112 \tag{7.15}$$

$$4x + 9y + 30z = 68 \tag{7.16}$$

Consider the tolerance for x, y, and z to be less than 0.00001.

Solution The first step is to rewrite Eqs. (7.14)–(7.16) as follows:

$$x = \frac{1}{80}(85 - 10y + 2z) \tag{7.17}$$

```
% Gauss-seidel method
% Set of eqautions:
% F1(x,y,z)= 80x+10y-2z==85
% F2(x,y,z)= 5x+50y+12z==112
% F3(x,y,z)= 4x+9y+30z==68
% Stopping criteria: Tolerance for (x,y,z)< 0.0000 1
clc;clear
fx=@(x,y,z) (1/80).*(85-10*y+2*z);
fy=@(x,y,z) (1/50).*(112-5*x-12*z);
fz=@(x,y,z) (1/30).*(68-4*x-9*y);
xo=0; yo=0; zo=0;
N=100;
for j=1:N
    x=fx(xo,yo,zo);
    y=fy(x,yo,zo);
    z=fz(x,y,zo);
    tol_x=abs(x-xo)/xo;
    tol_y=abs(y-yo)/yo;
    tol_z=abs(z-zo)/zo;
    fprintf('x:%.5f Tol_x: %.5f y: %.5f Tol_y: %.5f z: %.5f Tol_z: %.5f \n',...
                        x,tol_x,y,tol_y,z,tol_z);
    xo=x; yo=y; zo=z;
    % Stopping criteria
    if (tol_x<0.00001 && tol_y<0.00001 && tol_z<0.00001)
        break;
    end
end
fprintf('The solution after %dth iteration:\n',j);
fprintf('x: %f  y: %f  z: %.5f \n',x,y,z);
```

Fig. 7.1 Code—Gauss-Seidel method

```
Command Window
  x:1.06250 Tol_x: Inf y: 2.13375 Tol_y: Inf z: 1.48487 Tol_z: Inf
  x:0.83290 Tol_x: 0.21609 y: 1.80034 Tol_y: 0.15626 z: 1.61551 Tol_z: 0.08798
  x:0.87785 Tol_x: 0.05396 y: 1.76449 Tol_y: 0.01991 z: 1.62027 Tol_z: 0.00295
  x:0.88245 Tol_x: 0.00524 y: 1.76289 Tol_y: 0.00091 z: 1.62014 Tol_z: 0.00008
  x:0.88264 Tol_x: 0.00022 y: 1.76290 Tol_y: 0.00001 z: 1.62011 Tol_z: 0.00002
  x:0.88264 Tol_x: 0.00000 y: 1.76291 Tol_y: 0.00000 z: 1.62011 Tol_z: 0.00000
  The solution after 6th iteration:
  x: 0.882640   y: 1.762910   z: 1.62011
```

Fig. 7.2 Output—Gauss-Seidel method

$$y = \frac{1}{50}(112 - 5x - 12z) \tag{7.18}$$

$$z = \frac{1}{30}(68 - 4x - 9y) \tag{7.19}$$

These equations will be our input to the MATLAB program. The initial values of x, y,and z will be considered as zero.

The MATLAB code for this example is given in Fig. 7.1 with its output in Fig. 7.2.

Here, we can observe that at the sixth iteration, the tolerances for x, y, and z are all below the threshold—0.00001. Hence, the iteration is stopped after the sixth iteration, and the values obtained at this iteration are our desired outcomes.

7.3 Newton-Raphson Method

Newton-Raphson (N-R) method is one of the most effective methods to approximate the root of a nonlinear function, which is differentiable. It uses the concept of the tangent to produce an approximation of the root, and it only needs one initial value, which needs to be close to the value of the actual root of that function. The closer will be the initial value, the accuracy of approximation will enhance accordingly. Consider a function $f(x)$ and we are interested in approximating its root using the N-R method. If the initial guess of x is x_o, the procedure to approximate the root through iteration is to follow the steps below:

The first step is to determine the first approximate value of the root by using the following formula:

$$x^1 = x_0 - \frac{f(x_0)}{f'(x_0)} \tag{7.20}$$

where x^1 indicates the first approximated root determined after the first iteration and $f'(x)$ is the first derivative of the input function $f(x)$. In the above equation, the ratio of the original function and its derivative at the point $x = x_0$ is used. In the next step, x^1 replaces x_0, and the second iteration is performed to determine the second approximate root. This step is repeated until convergence is attained. Therefore, the general formula for the approximation of root using the N-R method can be written as follows:

$$x^{n+1} = x^n - \frac{f(x)}{f'(x)} \qquad (7.21)$$

Here, i represents the number of iterations. Sometimes the initial value can be a small range instead of a single value. In that case, the good strategy is to take the mean value of that range as our initial guess and start from there. For the stopping criteria of iteration, we can follow the same procedure mentioned previously in the Gauss-Seidel method. Here, as we are only interested in approximate the root, we only need to calculate the tolerance of x in each iteration using the following formula as mentioned earlier as well:

$$\text{Tol}\_x^{i+1} = \frac{\left| x^{i+1} - x^i \right|}{x^i} \qquad (7.22)$$

7.3.1 MATLAB Example 7.2: Newton-Raphson Method

Consider the function, $f(x) = 2x + \sin(x) - 2$, which has a root within the range of [0,2]. Using the N-R method, approximate the root of the function in MATLAB. Also, make sure that the tolerance for the value of root is less than 0.0001.

The MATLAB code for this example is given in Fig. 7.3 with its output in Fig. 7.4.

Here, after the fourth iteration, the tolerance has fallen below 0.00001. Therefore, the approximation of the root value of our function is obtained in the fourth iteration, which is 0.68404.

7.4 Runge-Kutta Method

Runge-Kutta method is an iterative method to solve or approximate ordinary differential equations. Out of the Runge-Kutta family, "RK4" or the fourth-order Runge-Kutta method is the most widely used method. Sometimes it is also regarded as the classical Runge-Kutta method.

```
% Newton raphson method
% Find the root of 2*x+sin(x)-2
% Stopping criteria: Tolerance < 0.00001
clc;clear;
F=@(x) 2*x+sin(x)-2;
syms x
% Derivative
dF(x)=diff(F(x));
a=0;
b=2;
xo=mean([a b]);
N=100;
 for i=1:N
     x=xo-(F(xo)/dF(xo));
     tol=abs(x-xo)/xo;
     fprintf('x: %.5f  Tolerance: %.5f \n',x,tol);
     xo=x;
     %Stopping criteria
     if (tol<0.00001)
         break;
     end
 end
fprintf('Root of the equation after %dth iteration: %.5f\n',i,x);
```

Fig. 7.3 Code—Newton-Raphson method

```
Command Window

 x: 0.66875   Tolerance: 0.33125
 x: 0.68401   Tolerance: 0.02282
 x: 0.68404   Tolerance: 0.00004
 x: 0.68404   Tolerance: 0.00000
 Root of the equation after 4th iteration: 0.68404
```

Fig. 7.4 Output—Newton-Raphson method

Consider a differential equation $y' = f(x, y)$, with the initial value of $y(0) = y_0$. For approximating the solution of this differential equation for the value of y using the Runge-Kutta method, the following formula needs to be followed:

$$x^{i+1} = x^i + h \tag{7.23}$$

$$y^{i+1} = y^i + \frac{1}{6}[k_1 + 2k_2 + 2k_3 + k_4], \tag{7.24}$$

where:

$$k_1 = hf\left(x^i, y^i\right) \tag{7.25}$$

$$k_2 = hf\left(x^i + \frac{h}{2}, y^i + \frac{k_1}{2}\right) \tag{7.26}$$

$$k_3 = hf\left(x^i + \frac{h}{2}, y^i + \frac{k_2}{2}\right) \tag{7.27}$$

$$k_4 = hf\left(x^i + h, y^i + k_3\right) \tag{7.28}$$

Here, the value of x is updated with a certain time step h in every ith iteration. For each of that, the updated value of x, the value of y is approximated in every iteration as mentioned in the equation. In the above equation k_1, k_2, k_3, k_4 are the weights of the fourth-order Runge-Kutta method, which are updated in every iteration following the formula mentioned above. For a certain range of the x value, the solution y is approximated in each iteration step. An example of the implementation of the Runge-Kutta method in MATLAB is given below to understand the method more clearly.

7.4.1   MATLAB Example 7.3: Runge-Kutta Method

Consider the following differential equation:

$$\frac{dy}{dx} = (x + 2y)\cos(y); 0 \le x \le 2, y(0) = 5 \tag{7.29}$$

Solve the equation for the value of y using the Runge-Kutta method for the step size of 0.2 in MATLAB.

The MATLAB code for this example is given in Fig. 7.5 with its output in Fig. 7.6.

Here, the solution of the differential equation is plotted for the range of $x[0, 2]$ using the step size 0.2. The obtained final value of y for $x = 2$ is 7.49116.

7.5   Conclusion

In this chapter, some of the important numerical methods are explained, along with the implementation in MATLAB. From this chapter, the readers will be able to learn about three numerical methods—Gauss-Seidel method, Newton-Raphson method, and Runge-Kutta method—that have been described in detail. The algorithms for

```
% Runge-kutta method
% Differential equation dy/dx= (x+2*y)*cos(y)
% Conditions: 0<=x<=2; y(0)=5; Step size,h=0.2
% Solve for y
clc;clear;
F=@(x,y) (x+2*y)*cos(y);
h = 0.2;
x0 = 0;
y0 = 5;
xn = 2;
N=length(x0:h:xn);
for j=1:N-1
    k1=h*F(x0,y0);
    k2=h*F(x0+0.5*h,y0+0.5*k1);
    k3=h*F(x0+0.5*h,y0+0.5*k2);
    k4=h*F(x0+h,y0+k3);
    y(j)=y0+(1/6)*(k1+2*k2+2*k3+k4);
    x0=x0+h;
    y0=y(j);
end
x=0.2:h:xn;
plot(x,y,'o-b','LineWidth',1.5);
xlabel('x');
ylabel('y');
title('Runge-kutta method')
grid on;
fprintf('The final solution for x = 2 is: %.5f\n',y(j));
```

```
The final solution for x = 2 is: 7.49116
```

Fig. 7.5 Code—Runge-Kutta method

each of these algorithms are demonstrated both theoretically and practically to grasp the concept. Examples of mathematical problems and their solutions utilizing these three numerical methods are shown in this chapter by using MATLAB programming language. Therefore, the contents of this chapter will immensely help the readers to build up a basic understanding of numerical methods and their practical applications to solve different mathematical problems using MATLAB.

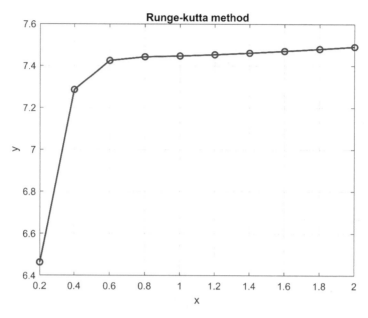

Fig. 7.6 Graphical output—Runge-Kutta method

Exercise 7

1. Write down the basic steps of the following:

   (a) Gauss-Seidel method
   (b) Newton-Raphson method
   (c) Runge-Kutta method

2. State the major differences in the calculation of the three methods mentioned in Question 1.
3. Given a set of equation as follows:

$$20x - 2y - z = 122$$

$$4x - 60y + 18z = 76$$

$$2x - 15y + 35z = 50$$

   (a) Solve the equation using Gauss-Seidel method in MATLAB. Consider the tolerance for x, y, and z to be less than 0.00001.
   (b) What happens when you decrease the tolerance to 0.0001 and 0.001? Do the values vary from the values determined in question (a)? Why do you think this occurs?

(c) What happens when you increase the tolerance to 0.000001 and 0.0000001? Do the values vary from the values determined in question (a)? Why do you think this occurs?

4. Solve the following equation using Newton-Raphson method, which has a root within the range of $[0, 2]$. Consider a tolerance for the value of root less than 0.0001:

(a) $3x + 2\cos(x) - 5$
(b) $x^5 - x - 2$

5. Use the classical fourth-order Runge-Kutta method to solve the following differential equations for the step size of 0.2, for $0 \le x \le 2$, and with an initial condition of $y(0) = 5$ in MATLAB:

(a) $\frac{dy}{dx} = -4x^3 - 6x^2 - 10x + 2$
(b) $\frac{dy}{dx} = xsin(y) + y\cos(x)$

Chapter 8
Electrical Circuit Analysis

8.1 Introduction

Electrical circuit analysis can be carried out by implementing different circuit theories and formulas by utilizing MATLAB programming language. The electrical circuit analyses can be divided initially into two main parts—DC and AC circuit analysis. Both in DC and AC circuit analysis, some of the important theorems and methods relevant to solving individual circuit problems are explained with necessary MATLAB illustrations. Apart from that, two important electrical components—operational amplifier and transistor—are covered in this chapter with their relevant circuit problems.

8.2 DC Circuit Analysis

In DC circuit analysis, MATLAB can be useful for circuit solutions and demonstrating different circuit laws. In this section, some of the important laws and circuit problem solutions will be discussed utilizing MATLAB.

8.2.1 Ohm's Law

Ohm's law demonstrates the relationship among voltage, current, and resistance as shown in Eq. (8.1) with the consideration of constant temperature:

$$V = IR, \tag{8.1}$$

where V is the voltage, I is the current, and R is the resistance of a DC circuit.

© The Author(s), under exclusive license to Springer Nature Switzerland AG 2022
E. Hossain, *MATLAB and Simulink Crash Course for Engineers*,
https://doi.org/10.1007/978-3-030-89762-8_8

This formula demonstrates a proportional relationship between voltage and current. In MATLAB, this relationship can be demonstrated graphically to realize this proportionality.

8.2.1.1   MATLAB Example 8.1: Ohm's Law

Consider a DC series circuit whose voltage across a resistance changes within the range of 1 to 10 V, if the resistance is 5 ohms. Plot a graph showing the changes of current in accordance with the changes of voltage.

The MATLAB code for this example is given in Fig. 8.1 with its output in Fig. 8.2.

Output

8.2.2   Equivalent Resistance

In an electrical circuit analysis, equivalent resistance calculation is of supreme importance. The equivalent resistance signifies the overall resistance of a circuit, where the multiple resistance can be connected in series, parallel, or a combination of both.

If multiple resistances are connected in series, the equivalent resistance is the summation of all of them. Therefore, for a series-connected resistive circuit, the equivalent resistance will be as shown in Eq. (8.2):

```
% Ohm's Law: V=IR
% Voltage, V=[1:10]
% Resistance, R=5 ohms
% Plot voltage vs current
V=1:10; R=5;
I=V/R;
plot(V,I,'o-b','Linewidth',1.2);
xlabel('Voltage, Volt');
ylabel('Current, Amp');
title('Ohms Law');
grid on;
```

Fig. 8.1 Code—Ohm's law

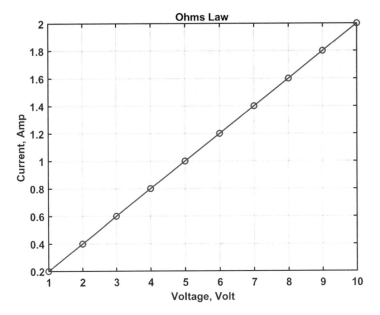

Fig. 8.2 Output—Ohm's law

Fig. 8.3 A series-parallel resistive circuit

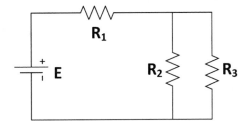

$$R_{eq} = R_1 + R_2, \tag{8.2}$$

where R_1 and R_2 are the two resistances connected in series. For parallel or shunt connected circuit, the equivalent resistance can be represented by Eq. (8.3):

$$R_{eq} = \frac{1}{R_1} + \frac{1}{R_2} = \frac{R_1 \times R_2}{R_1 + R_2} \tag{8.3}$$

An example is provided below, where both the combination of series and parallel resistances can be found in a single circuit.

In the circuit in Fig. 8.3, R_2 and R_3 are connected in parallel, with which R_1 is in series. Therefore, the equivalent resistance of the circuit can be written as follows:

```
% Equivalent resistance
% R1 + (R2||R3)
R1=10; R2=5; R3=4;
Equivalent_R= R1 + ((R2*R3)/(R2+R3));
fprintf('Equivalent resistance: %f',Equivalent_R);
```

Fig. 8.4 Code—Equivalent resistance

Fig. 8.5 Output—
Equivalent resistance

> Command Window
>
>   Equivalent resistance: 12.222222

$$R_{eq} = R_1 + \left(R_2 \| R_3\right) = R_1 + \frac{R_2 \times R_3}{R_2 + R_3} \tag{8.4}$$

8.2.2.1   MATLAB Example 8.2: Equivalent Resistance

Determine the equivalent resistance for the circuit shown in Fig. 8.3. Consider the value of $R_1 = 10$; $R_2 = 5$; $R_3 = 4$ ohms.

The MATLAB code for this example is given in Fig. 8.4 with its output in Fig. 8.5.

8.2.3   Delta-Wye Conversion

In some scenarios, the resistances of a circuit may not be connected either in series or parallel. During that scenario, delta-wye conversion may become useful.

Consider the following circuit in Fig. 8.6, where R_1, R_2, and R_3 are in a delta configuration, and R_4, R_5, and R_3 are in another delta configuration. To determine the equivalent resistance, one of the easiest ways is to convert the delta configuration into a wye configuration to make the calculation easy. Before we solve the following circuit for equivalent resistance, let us have a look at the formula to convert any delta configuration to a wye configuration and vice versa.

In Fig. 8.7, the delta configuration and the wye configuration of resistances have been shown with a diagram. In the delta configuration, the resistances are connected in such a way that it looks like the delta symbol Δ. Conversely, in the wye configuration, the resistances create an appearance of "Y." This configuration is

Fig. 8.6 A delta connected
resistive circuit

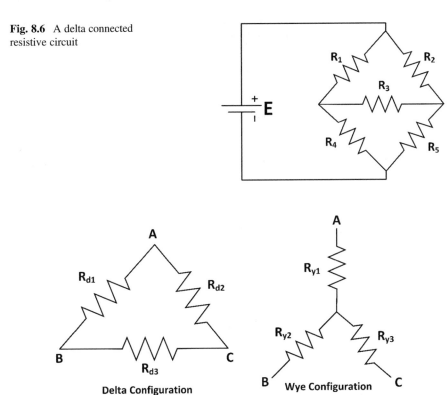

Fig. 8.7 Delta and wye configuration in electrical circuits

termed star configuration as well. The conversion among delta to wye and wye to
delta are both possible and can be defined with a generalized formula.

Delta to Wye Conversion

In Fig. 8.7, R_{d1}, R_{d2}, and R_{d3} are the resistances that are connected in a delta
configuration. This delta configuration can be transformed into an equivalent wye
configuration, where the new resistances will be R_{y1}, R_{y2}, and R_{y3}. Both of the
configurations have three common nodes A, B, and C. The formulas for converting
each of the delta resistances to its corresponding wye resistances are enlisted as
Eqs. (8.5)–(8.7):

$$R_{y1} = \frac{R_{d1}\,R_{d2}}{R_{d1} + R_{d2} + R_{d3}} \tag{8.5}$$

$$R_{y2} = \frac{R_{d1}\,R_{d3}}{R_{d1} + R_{d2} + R_{d3}} \tag{8.6}$$

$$R_{y3} = \frac{R_{d2}\,R_{d3}}{R_{d1} + R_{d2} + R_{d3}} \tag{8.7}$$

8.2.3.1   MATLAB Example 8.3: Delta to Wye Conversion

Consider a delta configured circuit as shown in Fig. 8.7, having the resistances $R_{d1} = 10\ \Omega$, $R_{d2} = 5\ \Omega$, and $R_{d3} = 20\ \Omega$. Determine the equivalent wye resistances R_{y1}, R_{y2}, and R_{y3} using MATLAB.

The MATLAB code for this example is given in Fig. 8.8 with its output in Fig. 8.9.

Output

Wye to Delta Conversion

To convert a wye configured resistances R_{y1}, R_{y2}, and R_{y3} to a delta configured equivalent resistances R_{d1}, R_{d2}, and R_{d3}; the following relationships shown in Eqs. (8.8)–(8.10) can be utilized:

```
% Delta to wye conversion
% Delta configured resistances:
% Rd1=10 ohms, Rd2= 5 ohms, Rd3= 20 ohms
% Equivaelent wye configured resistances:
% Ry1, Ry2, Ry3
clc;clear;
Rd1=10; Rd2=5; Rd3=20;
Ry1= (Rd1*Rd2)/(Rd1+Rd2+Rd3);
Ry2= (Rd1*Rd3)/(Rd1+Rd2+Rd3);
Ry3= (Rd2*Rd3)/(Rd1+Rd2+Rd3);
fprintf('Equivalent wye configured resistances:\n');
fprintf('Ry1= %f   Ry2= %f    Ry3= %f\n',Ry1,Ry2,Ry3);
```

Fig. 8.8 Code—Delta to wye conversion

```
Command Window

   Equivalent wye configured resistances:
      Ry1= 1.428571   Ry2= 5.714286   Ry3= 2.857143
```

Fig. 8.9 Output—Delta to wye conversion

$$R_{d1} = \frac{R_{y1} \cdot R_{y2} + R_{y2} \cdot R_{y3} + R_{y3} \cdot R_{y1}}{R_{y3}} \tag{8.8}$$

$$R_{d2} = \frac{R_{y1} \cdot R_{y2} + R_{y2} \cdot R_{y3} + R_{y3} \cdot R_{y1}}{R_{y2}} \tag{8.9}$$

$$R_{d3} = \frac{R_{y1} \cdot R_{y2} + R_{y2} \cdot R_{y3} + R_{y3} \cdot R_{y1}}{R_{y1}} \tag{8.10}$$

8.2.3.2 MATLAB Example 8.4: Delta to Wye Conversion

Consider a delta configured circuit as shown in Fig. 8.7, having the resistances $R_{d1} = 10\ \Omega$, $R_{d2} = 5\ \Omega$, and $R_{d3} = 20\ \Omega$. Determine the equivalent wye resistances R_{y1}, R_{y2}, and R_{y3} using MATLAB.

The MATLAB code for this example is given in Fig. 8.10 with its output in Fig. 8.11.

```
% Wye to delta conversion
% Delta configured resistances:
% Ry1=10 ohms, Ry2= 5 ohms, Ry3= 20 ohms
% Equivaelent wye configured resistances:
% Rd1, Rd2, Rd3
clc;clear;
Ry1=10; Ry2=5; Ry3=20;
Rd1= (Ry1*Ry2+Ry2*Ry3+Ry3*Ry1)/Ry3;
Rd2= (Ry1*Ry2+Ry2*Ry3+Ry3*Ry1)/Ry2;
Rd3= (Ry1*Ry2+Ry2*Ry3+Ry3*Ry1)/Ry1;
fprintf('Equivalent delta configured resistances:\n');
fprintf('Rd1= %.3f    Rd2= %.3f    Rd3= %.3f\n',Rd1,Rd2,Rd3);
```

Fig. 8.10 Code—Wye to delta conversion

Command Window

Equivalent delta configured resistances:
Rd1= 17.500   Rd2= 70.000   Rd3= 35.000

Fig. 8.11 Code—Wye to delta conversion

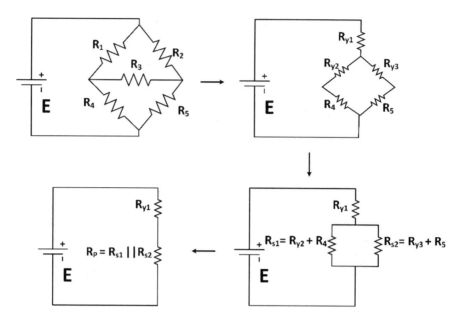

Fig. 8.12 Simplifying the circuit by determining the equivalent resistance using delta-wye conversion

Output

8.2.3.3 MATLAB Example 8.5: Equivalent Resistance with Delta-Wye Conversion

Consider Fig. 8.6 to determine its equivalent resistance by utilizing the delta-wye conversion method. The values of the resistances are $R_1 = 2\,\Omega$; $R_2 = 4\,\Omega$; $R_3 = 6\,\Omega$; $R_4 = 3\,\Omega$; and $R_5 = 2\,\Omega$.

For clarification, in Fig. 8.12, the procedures for determining equivalent resistance are shown graphically, which is implemented in the MATLAB code below. The MATLAB code for this example is given in Fig. 8.13 with its output in Fig. 8.14.

Output

8.2.4 Kirchhoff's Laws

Gustav Robert Kirchhoff first proposed two very fundamental laws of electrical circuits known as Kirchhoff's Current Law (KCL) and Kirchhoff's Voltage Law (KVL).

```
% Equivalent resistance with Delta-wye conversion
% R1, R2, R3: Delta configuration
% First step: Conversion into wye configuration
% Hence,find Ry1, Ry2, Ry2
% Second step: Find Rs1 and Rs2
% Third step: Find Rp
% Fourth step: Find overall equivalent resistance, Req
clc;clear;
R1=2; R2=4; R3=6; R4=3; R5=2;
Ry1= (R1*R2)/(R1+R2+R3);
Ry2= (R1*R3)/(R1+R2+R3);
Ry3= (R2*R3)/(R1+R2+R3);
Rs1= Ry2+R4;
Rs2= Ry3+R5;
Rp= (Rs1*Rs2)/(Rs1+Rs2);
Req= Ry1+Rp;
fprintf('The equivalent resistance: %.3f ohms\n',Req);
```

Fig. 8.13 Code—Equivalent resistance using delta-wye conversion

```
Command Window

     The equivalent resistance: 2.667 ohms
```

Fig. 8.14 Output—Equivalent resistance using delta-wye conversion

Kirchhoff's Current Law (KCL) According to KCL, the sum of all the currents entering a specific node is always zero. It can also be interpreted in the following way:

The summation of all the currents entering a node is equal to the summation of all the currents drawing out from that node.

Kirchhoff's Voltage Law (KVL) According to KVL, the summation of all the voltages in a closed loop is always zero.

These two formulas can be utilized to solve electrical circuits for determining various parameters such as voltage, current, etc.

Consider the circuit in Fig. 8.15, where there are two loops, whose currents are considered as I_{L1} and I_{L2}. The resistances of the circuit are $R_1 = 2\ \Omega$, $R_2 = 4\ \Omega$, and $R_3 = 4\ \Omega$. The voltage source, $E = 10\ V$. At node B, the current I_1 is entering, which gets divided into two parts—I_2 and I_3. We are interested to determine the current I_1, I_2, and I_3, and the voltage across the resistance R_3, which is termed as V_{R3}.

In the first ABCD loop, we can apply KVL, and the equation will be as follows:

$$E - V_{R1} - V_{R2} = 0$$

or,

$$E = I_{L1}R_1 + (I_{L1}R_2 - I_{L2}R_2) = 2I_{L1} + 4I_{L1} - 4I_{L2} = 6I_{L1} - 4I_{L2}$$

Hence,

$$6I_{L1} - 4I_{L2} = 10 \ (1)$$

Here, V_{R1} and V_{R2} are the voltages across resistance R_1 and R_2, respectively. For the second loop BEFC, applying KVL:

$$V_{R2} + V_{R3} = 0$$

or,

$$(I_{L2}R_2 - I_{L1}R_2) + I_{L2}R_3 = 0$$

Hence,

$$-4I_{L1} + 8I_{L2} = 0 \ (2)$$

By solving (1) and (2), the determined values of I_{L1} and I_{L2} are $I_{L1} = 2.5 \ A$ and $I_{L1} = 1.25 \ A$.

From the figure, we can write the following equations:

$$I_1 = I_{L1} = 2.5 \ A$$

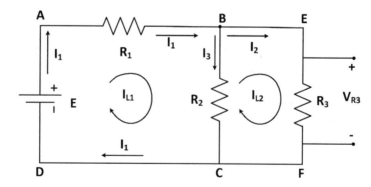

Fig. 8.15 A resistive electrical circuit with two loops

```
% Circuit problem
% R1= 2 ohms; R2=R3= 4 ohms
% Voltage source, E =10 V
% Determine loop current IL1 and IL2
% Determine current, I1, I2, and I3
% Determine voltage across resistance R3: VR3
% Determine voltage across resistance R2: VR2
% Determine voltage across resistance R1: VR1
clc;clear;
R1=2; R2=4; R3=4;
syms IL1 IL2
eqn1= 6*IL1-4*IL2==10;
eqn2= -4*IL1+8*IL2==0;
[IL1,IL2]=solve(eqn1,eqn2);
fprintf('The ABCD loop current, IL1: %.3f A\n',IL1);
fprintf('The BEFC loop current, IL2: %.3f A\n',IL2);

I1=IL1; I2=IL2;
I3= I1-I2;

fprintf('The currents in the circuit:\n');
fprintf('I1= %.3f A   I2= %.3f A   I3= %.3f A\n',I1,I2,I3);

VR1=I1*R1;
VR2=I3*R2;
VR3=I2*R3;

fprintf('The voltage across R1, VR1= %.3f V\n',VR1);
fprintf('The voltage across R2, VR2= %.3f V\n',VR2);
fprintf('The voltage across R3, VR3= %.3f V\n',VR3);
```

Fig. 8.16   Code—Using KVL and KCL in a circuit

Fig. 8.17   Output—Using
KVL and KCL in a circuit

Command Window

```
    The ABCD loop current, IL1: 2.500 A
    The BEFC loop current, IL2: 1.250 A
    The currents in the circuit:
    I1= 2.500 A   I2= 1.250 A   I3= 1.250 A
    The voltage across R1, VR1= 5.000 V
    The voltage across R2, VR2= 5.000 V
    The voltage across R3, VR3= 5.000 V
```

$$I_2 = I_{L2} = 1.25\,A$$

At node B, applying KCL, we can write the following equation:

$$I_1 = I_2 + I_3$$

Hence,

$$I_3 = I_1 - I_2 = 1.25\,A$$

The voltage across the resistance R_3 can be determined by applying Ohm's law as follows:

$$V_{R3} = I_2 \times R_3 = 1.25 \times 4 = 5\,V$$

The MATLAB implementation of the above solution of the circuit is given below:

8.2.4.1   MATLAB Example 8.6: Circuit Problem

Using loop analysis, solve the circuit shown in Fig. 8.15, considering $R_1 = 2\,\Omega$, $R_2 = 4\,\Omega$, $R_3 = 4\,\Omega$, and $E = 10\,V$, for determining:

(a) The loop currents in ABCD and BEFC loops
(b) The currents I_1, I_2, and I_3
(c) The voltages across resistances R_1, R_2, and R_3

The MATLAB code for this example is given in Fig. 8.16 with its output in Fig. 8.17.

Output

8.2.5   Voltage Divider and Current Divider Laws

Voltage Divider Rule In a series circuit, the voltages are divided across all the series-connected resistances. Consider the circuit in Fig. 8.18, where three resistances R_1, R_2, and R_3 are connected in series. Using the voltage divider rule, the

Fig. 8.18 An electrical circuit with a voltage source and series resistance

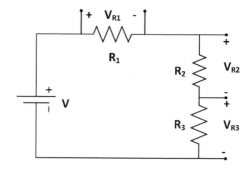

voltages across each resistance can be determined using the following formulas in Eqs. (8.11)–(8.13):

$$V_{R1} = \frac{R_1}{R_1 + R_2 + R_3} \times V \tag{8.11}$$

$$V_{R2} = \frac{R_2}{R_1 + R_2 + R_3} \times V \tag{8.12}$$

$$V_{R3} = \frac{R_3}{R_1 + R_2 + R_3} \times V \tag{8.13}$$

It can be noted that the total voltage V is equal to the summation of V_{R1}, V_{R2}, and V_{R3}.

8.2.5.1   MATLAB Example 8.7: Voltage Divider

Consider the circuit shown in Fig. 8.18, where $R_1 = 2\ \Omega$, $R_2 = 4\ \Omega$, $R_3 = 8\ \Omega$, and $E = 24\ V$. Determine the voltage V_{R2} and V_{R3} using voltage divider rule.

The MATLAB code for this example is given in Fig. 8.19 with its output in Fig. 8.20.

Output

```
% Voltage divider
% R1= 2 ohms; R2= 4 ohms; R3=8 ohms; E= 24 V
% Determine the voltage across the resistances R2 and R3
clc;clear;
R1=2; R2=4; R3=8; E=24;
VR2=(R2/(R1+R2+R3))*E;
VR3=(R3/(R1+R2+R3))*E;
fprintf('Voltage across the resistance R2: %.3f V\n',VR2);
fprintf('Voltage across the resistance R3: %.3f V\n',VR3);
```

Fig. 8.19 Code—Voltage divider rule

```
Command Window

   Voltage across the resistance R2:  6.857 V
   Voltage across the resistance R3: 13.714 V
```

Fig. 8.20 Output—Voltage divider rule

Fig. 8.21 An electrical circuit with a current source and parallel resistance

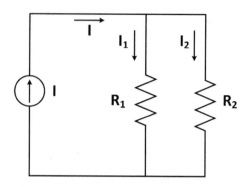

Current Divider Rule In a parallel circuit, the current is divided among all the parallel resistive paths based on different resistances in each path. Consider the circuit in Fig. 8.21, where two resistances R_1 and R_2 are connected in parallel. Using the current divider rule, the current through each of these resistances can be determined using the following formulas:

$$I_1 = \frac{R_1}{R_1 + R_2} \times I \tag{8.14}$$

$$I_2 = \frac{R_2}{R_1 + R_2} \times I \tag{8.15}$$

It is also to be noted that the total current, I, will be the summation of I_1 and I_2.

8.2.5.2 MATLAB Example 8.8: Current Divider

Consider the circuit in Fig. 8.21, where $R_1 = 2\,\Omega$, $R_2 = 4\,\Omega$, and $I = 16\,A$. Determine the currents I_1 and I_2 flowing through the resistances R_1 and R_2, respectively, using the current divider rule. The MATLAB code for this example is given in Fig. 8.22 with its output in Fig. 8.23.

Output

8.2.6 Thevenin's Theorem

According to Thevenin's theorem, any linear circuit can be represented with an equivalent series circuit, incorporating an open-circuit voltage in the terminal, V_{th}, and an input equivalent resistance, which is determined considering all the voltage

```
% Current divider
% R1= 2 ohms; R2= 4 ohms; I= 16 A
% Determine the current through the resistances R1 and R2
clc;clear;
R1=2; R2=4; I=16;
I1=(R2/(R1+R2))*I;
I2=(R1/(R1+R2))*I;
fprintf('Current through the resistance R1, I1: %.3f A\n',I1);
fprintf('Current through the resistance R2, I2: %.3f A\n',I2);
```

Fig. 8.22 Code—Current divider rule

```
Command Window

    Current through the resistance R1, I1: 10.667 A
    Current through the resistance R2, I2: 5.333 A
```

Fig. 8.23 Output—Current divider rule

sources replaced by short circuit, and the current sources replaced by open circuit. A simple illustration is provided in Fig. 8.24.

Here, Fig. 8.24a is the original circuit. For determining its Thevenin's circuit, Thevenin's voltage V_{th} and Thevenin's equivalent resistance R_{th} need to be determined. In Fig. 8.24b, the circuit representation for determining V_{th} is shown, from where it can be observed that the voltage across the resistance R_2 is equal to V_{th}. By applying the voltage divider rule, the value of V_{th} can be determined as follows:

$$V_{th} = \frac{R_2}{R_1 + R_2} \times V \tag{8.16}$$

The equivalent resistance R_{th} can be determined from Fig. 8.24c, where the voltage source is replaced by a short circuit. If there was a current source, it would be replaced by an open circuit. From the figure, Thevenin's equivalent resistance can be calculated as follows:

$$R_{th} = \left(R_1 \| R_2 \right) + R_3 = \frac{R_1 \times R_2}{R_1 + R_2} + R_3 \tag{8.17}$$

The final equivalent of Thevenin's circuit is demonstrated in Fig. 8.24d.

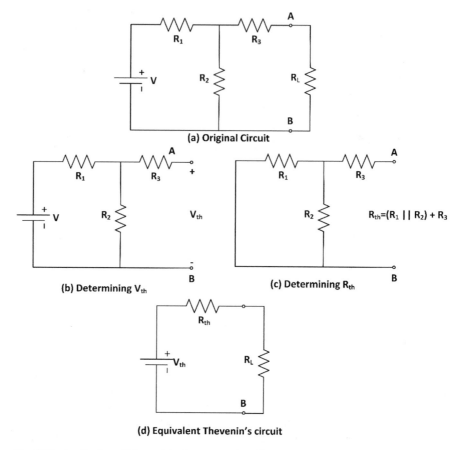

(a) Original Circuit

(b) Determining V_{th}

(c) Determining R_{th}

(d) Equivalent Thevenin's circuit

Fig. 8.24 Application of Thevenin's theorem to simplify an electrical circuit

8.2.6.1   MATLAB Example 8.9: Thevenin's Theorem

Consider the circuit shown in Fig. 8.24a, where the parameters are $R_1 = 4\ \Omega$, $R_2 = 2\ \Omega$, $R_3 = 3\ \Omega$, $R_L = 5\ \Omega$, and $V = 10\ V$. Using Thevenin's theorem, find the following parameters:

(a) Thevenin's voltage, V_{th}
(b) Thevenin's equivalent resistance, R_{th}
(c) The current flowing through the load resistance, R_L

The MATLAB code for this example is given in Fig. 8.25 with its output in Fig. 8.26.

Output

```
% Thevenin's theorem
% R1= 4 ohms; R2= 2 ohms; R3= 3 ohms; RL= 5 ohms; V= 10 V;
% Determine: Thevenin's voltage, Vth
% Determine: Thevenin's equivalent resistance, Rth
% Determine: Load current, IRL
clc;clear;
R1=4; R2=2; R3=3; RL=5; V=10;
Vth= ((R2)/(R1+R2))*V;
Rth= ((R1*R2)/(R1+R2))+R3;
fprintf('Thevenin voltage: %.3f V\n',Vth);
fprintf('Thevenin equivalent resistance: %.3f ohms\n',Rth);
IRL=Vth/(Rth+RL);
fprintf('Load current: %.3f A\n',IRL);
```

Fig. 8.25 Code—Thevenin's theorem

```
Command Window

   Thevenin voltage: 3.333 V
   Thevenin equivalent resistance: 4.333 ohms
   Load current: 0.357 A
```

Fig. 8.26 Output—Thevenin's theorem

8.2.7 Maximum Power Transfer Theorem

According to the maximum power transfer theorem, the maximum power can be achieved from a circuit if its load resistance matches with the Thevenin's equivalent resistance.

Consider the following Thevenin's equivalent circuit, where the load resistance is R_L and the Thevenin's voltage and equivalent resistance are V_{th} and R_{th}. According to this theorem, maximum power will be achieved when $R_L = R_{th}$; and the maximum power will be calculated using Eq. (8.18):

$$P_{max} = I^2 R_L = \left(\frac{V_{th}}{R_{th} + R_L}\right)^2 \cdot R_L \qquad (8.18)$$

8.2.7.1 MATLAB Example 8.10: Maximum Power Transfer Theorem

Consider the circuit shown in Fig. 8.27, where the parameters are $R_{th} = 5\ \Omega$ and $V_{th} = 10\ V$. Vary the load resistance starting from 1 to 12 ohms, and determine the output power for all scenarios to prove the maximum power transfer theorem;

Fig. 8.27 Thevenin's
equivalent circuit

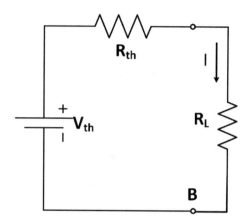

```
% Maximum power transfer theorem
% Rth= 5 ohms; Vth= 10 V
clc;clear;
Rth=5; Vth=10;
RL= 1:1:26;
for i=1:1:26
    I(i)=Vth/(Rth+RL(i));
    Power(i)=I(i)^2*RL(i);
end
plot(RL,Power,'o-b','LineWidth',1.2);
xlabel('Load resistance,R_L (Ohms)');
ylabel('Output power, P (W)');
title('Maximum power transfer theorem');
grid on;
% Maximum power, when RL=Rth
RL=5;
P_max=(Vth/(Rth+RL))^2*RL;
fprintf('Maximum output power= %.3f\n',P_max);
```

Fig. 8.28 Code—Maximum power transfer theorem

Fig. 8.29 Output—
Maximum power transfer
theorem

Command Window

Maximum output power= 5.000

also determine the maximum output power. The MATLAB code for this example is
given in Fig. 8.28 with its output in Figs. 8.29 and 8.30.

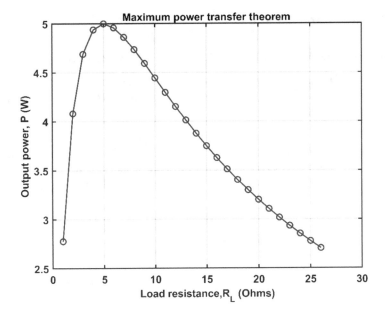

Fig. 8.30 Graphical output—Maximum power transfer theorem

Output

Here, from the figure, it can be observed that the highest maximum power can be attained only when the load resistance is equal to Thevenin's equivalent resistance.

8.3   AC Circuit Analysis

An AC circuit can be referred to those circuits having sinusoid inputs as voltage or current source. A sinusoid can be either a sine or cosine signal. The general representation of an AC voltage and current source can be defined as follows:

$$v(t) = V_M \sin(\omega t) \tag{8.19}$$

$$i(t) = I_M \sin(\omega t) \tag{8.20}$$

Here, V_M and I_M are the magnitude of voltage and current signal, respectively. Both sources are a function of time; therefore, after a certain time, the voltage and current may become both positive and negative. ω indicates the angular frequency in radian.

8.3.1   Some Terminologies

Some terminologies on the aspect of AC voltage are defined below:

Peak Value   The maximum value of a sinusoid from the zero level is regarded as the peak value. V_p represents the peak voltage, which is the maximum positive value of the voltage.

RMS Value   RMS value implies the root mean square value of the amplitude of a sinusoid signal. It can also be regarded as the effective value of an AC circuit. The RMS value of voltage plays an important role while calculating different terms of an AC circuit. The RMS value of an AC voltage can be calculated using the following formula:

$$\text{RMS Voltage, } V_{\text{RMS}} = \frac{1}{\sqrt{2}} \times V_p \qquad (8.21)$$

Average Value   The average value indicates the area under the sinusoid signal. The average value of an AC voltage signal can also be calculated from its peak value by using the following formula:

$$\text{Average Voltage, } V_{\text{avg}} = \frac{2}{\pi} \times V_p \qquad (8.22)$$

Instantaneous Value   Instantaneous value represents the exact value of a sinusoid at a specific time. As AC voltage is a function of time, by providing a specific time, AC voltage at that time can be determined.

$$\text{Instantaneous Voltage, } V_{\text{inst}}(t) = V_p \sin\left(2\pi ft\right) \qquad (8.23)$$

Here, t is the time at which the instantaneous voltage can be calculated. f indicates the frequency of the input voltage.

8.3.1.1   MATLAB Example 8.11: AC Circuit Terminologies

The input voltage of an AC circuit is $v(t) = 2 \sin\left(2\pi ft\right)$, where $f = 60$ Hz. Answer the following questions using MATLAB:

(a) Plot the input voltage in MATLAB for $t = 0{:}0.1$.
(b) Find the values of peak voltage, peak-to-peak voltage, RMS voltage, and average voltage.
(c) Find the instantaneous voltage at $t = 0.02$.

The MATLAB code for this example is given in Fig. 8.31 with its output in Figs. 8.32 and 8.33.

```
% v(t)=10 sin(2*pi*f*t)
% f= 60 Hz; t= 0:0.1 sec
% Determine: Peak voltage, Vp
% Determine: Peak to peak voltage, Vpp
% Determine: RMS voltage, V_rms
% Determine: Average voltage, V_avg
% Determine: Instantaneous voltage at T=0.02 sec, v_inst
clc;clear;
f = 60;
t = 0:0.0001:0.1;
v = 2*sin(2*pi*f*t);
plot(t,v,'LineWidth',1.5);
xlabel('Time (sec)');
ylabel('Voltage (volt)');
ylim([-2.5 2.5]);
grid on;
Vp=max(abs(v));
fprintf('Peak voltage: %.3f\n',Vp);
Vpp=2*Vp;
fprintf('Peak to peak voltage: %.3f\n',Vpp);
V_rms=(1/sqrt(2))*Vp;
fprintf('RMS voltage: %.3f\n',V_rms);
V_avg=(2/pi)*Vp;
fprintf('Average voltage: %.3f\n',V_avg);
T=0.02;
V_inst=2*sin(2*pi*f*T);
fprintf('Instantaneous voltage at T=0.02 sec: %.3f\n',V_inst);
```

Fig. 8.31 Code—Determination of AC circuit voltage parameters

```
Command Window

  Peak voltage: 2.000
  Peak to peak voltage: 4.000
  RMS voltage: 1.414
  Average voltage: 1.273
  Instantaneous voltage at T=0.02 sec: 1.902
```

Fig. 8.32 Output—Determination of AC circuit voltage parameters

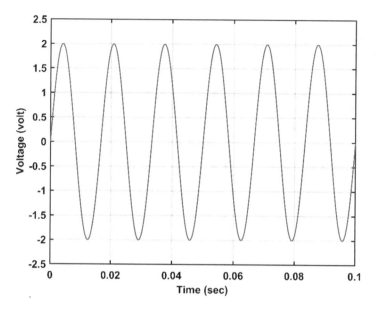

Fig. 8.33 Graphical output—Determination of AC circuit voltage parameters

Output

8.3.2 Impedance

Impedance signifies the total impediment of the flow of charge in an AC circuit. The impedance can be divided into two parts—resistance and reactance. The resistance is a zero-frequency component; on the other hand, reactance is dependent on frequency. The general expression for overall impedance can be written as follows:

$$\text{Impedance}, Z = R + jX \tag{8.24}$$

Here the real part is the resistance, and the imaginary part is the reactance. Again, the reactance can be categorized into two parts—inductive reactance, (X_L), and capacitive reactance, (X_C). With the inclusion of both of these reactances with resistance in series, the formula of impedance can be rewritten as follows:

$$Z = R + j(X_L - X_C) = R + j\left(\omega L - \frac{1}{\omega C}\right) \qquad (8.25)$$

Therefore,

$$|Z|\angle\theta = \sqrt{R^2 + \left(\omega L - \frac{1}{\omega C}\right)^2} \angle \frac{\left(\omega L - \frac{1}{\omega C}\right)}{R} \qquad (8.26)$$

Here, L is the inductance, C is the capacitance, and ω represents the angular frequency. From the formula, it can also be observed that $X_L = \omega L$ and $X_C = \frac{1}{\omega C}$. An inductive circuit is called a lagging circuit, as the current through the reactance lags the voltage across it. Conversely, a capacitive circuit has a leading current with respect to voltage. A pure inductive circuit lags the current by $-90°$, whereas a purely capacitive circuit leads the current by $+90°$. If the circuit has both inductive and capacitive reactance, the nature of the circuit can be determined as follows:

if $(X_L - X_C) > 0$
Inductive reactance;
Power factor lagging;
elseif $(X_L - X_C) < 0$
Capacitive reactance;
Power factor leading;
else
Resistive
Power factor Unity;
end

Based on the above discussion, an impedance triangle can be drawn (Fig. 8.34), which shows the relationship between resistance and reactance more clearly. In an impedance triangle, the horizontal line indicates the resistance, as it is a zero-frequency component. The reactance indicates the perpendicular line, as they shift the voltage or current by $+90°$ or $-90°$. An upward perpendicular line refers to an overall inductive reactance, while a downward perpendicular line depicts an overall

Fig. 8.34 The impedance triangle

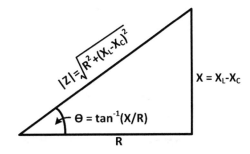

capacitive reactance. The hypotenuse of the triangle demonstrates the magnitude of the overall impedance, and the angle between the hypotenuse and the horizontal line refers to the phase angle of impedance, which is usually regarded as θ. The cosine of this angle introduces a crucial term in the AC circuit, which is called power factor.

8.3.2.1   MATLAB Example 8.12: Impedance

Consider a series RLC circuit with $R = 10$ ohm, $L = 0.02$ H, and $C = 0.05$ F. If the frequency is 60 Hz, determine:

```
% Impedance
% R= 10 ohms; L= 0.02 H; C= 0.05 F; f= 60 Hz
% Determine Impendance: Z
% Power facor: PF
clc;clear;
R=10; L=0.02; C=0.05;
f=60;
XL=2*pi*f*L;
XC=1/(2*pi*f*C);
disp('Impedance:')
Z=R+j*(XL-XC)
Imp_magnitude=abs(Z);
Phase_angle=angle(Z)*(180/pi);
disp('In polar form:');
fprintf('|Z|= %.3f ohms;   Phase angle= %.3f degree\n',...
    Imp_magnitude,Phase_angle);
PF=cos(Phase_angle);
fprintf('Power factor= %.3f\n',PF);
```

Fig. 8.35 Code—Impedance in an RLC circuit

```
Command Window

  Impedance:

  Z =

     10.0000 + 7.4868i

  In polar form:
  |Z|= 12.492 ohms;   Phase angle= 36.821 degree
  Power factor= 0.639
```

Fig. 8.36 Output—Impedance in an RLC circuit

(a) Impedance of the circuit
(b) Power factor

The MATLAB code for this example is given in Fig. 8.35 with its output in Fig. 8.36.

Output

8.3.3 Power Triangle

In an AC circuit, the power can also be divided into two components—real power (P) and reactive power (Q). The real power is the resistive power, which does not depend on frequency. On the other hand, reactive power is the frequency-dependent part. The vector summation of these two components is called the apparent power (S). These three components can be represented using the power triangle (Fig. 8.37), where the horizontal line represents the real power, P; the perpendicular line indicates the reactive power, Q; and the hypotenuse is the apparent power, S, which can be represented using the following formula:

$$S = P + jQ \tag{8.27}$$

Hence,

$$|S| \angle \theta = \sqrt{P^2 + Q^2} \angle \frac{Q}{P} \tag{8.28}$$

For the inductive circuit, the reactive power will be indicated by upward perpendicular line; and for the capacitive circuit, the reactive power will be represented by a downward perpendicular line. The angle between the horizontal line and hypotenuse is called the power angle (θ), cosine of which is also referred to as the power factor.

Fig. 8.37 The power triangle

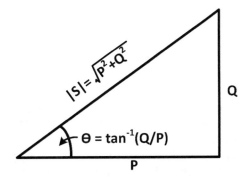

Therefore, both in the impedance and power triangle, this angle corresponds to the same entity.

$$Q = \sqrt{Q_L^2 + Q_C^2} \qquad (8.29)$$

if $(Q_L - Q_C) > 0$
Inductive reactance;
Power factor lagging;
elseif $(Q_L - Q_C) < 0$
Capacitive reactance;
Power factor leading;
else
Resistive
Power factor Unity;
end

8.3.3.1   MATLAB Example 8.13: Power Triangle

Consider a series RLC circuit with $P = 10$ W and $Q = 5$ Var. Determine:

(a) Apparent power, S
(b) Power factor, PF

The MATLAB code for this example is given in Fig. 8.38 with its output in Fig. 8.39.

Output

8.3.4   Three-Phase AC Circuit Analysis

In a three-phase circuit, the configuration can either be in wye or delta configuration, based on which the properties change. To understand the relationship among different parameters in both wye and delta connection, we first need to categorize the circuit into two parts based on the nature of the load, which are balanced load and unbalanced load. In an unbalanced load, all the loads are not equally distributed among three phases, whereas in a balanced load condition, all the loads are

```
% Power traingle
% Find: Apparent power, S
% Find Power factor, PF
clc;clear;
% Example 1: Real power,P= 10 W; Reactive power,Q=5 Var
fprintf('Example 1: Positive reactive power\n');
fprintf('------------------------------------\n');
P=10; Q=5;
disp('Apparent power:')
S=P+j*Q
S_mag=abs(S);
S_angle=angle(S)*(180/pi);
fprintf('Apparent power in polar form:\n');
fprintf('|S|= %.3f VA    Power angle= %.3f degree\n',S_mag,S_angle);
PF=cos(S_angle);
if Q>0
    fprintf('Power factor= %.3f; Lagging\n',PF);
elseif Q<0
    fprintf('Power factor= %.3f; Leading\n',PF);
else
     fprintf('Power factor= %.3f; Unity\n',PF);
end
fprintf('\n');
% Example 2: Real power,P= 10 W; Reactive power,Q=-5 Var
P=10; Q=-5;
fprintf('Example 2: Negative reactive power\n');
fprintf('------------------------------------\n');
disp('Apparent power:')
S=P+j*Q
S_mag=abs(S);
S_angle=angle(S)*(180/pi);
fprintf('Apparent power in polar form:\n');
fprintf('|S|= %.3f VA     Power angle= %.3f degree\n',S_mag,S_angle);
PF=cos(S_angle);
if Q>0
    fprintf('Power factor= %.3f; Lagging\n',PF);
elseif Q<0
    fprintf('Power factor= %.3f; Leading\n',PF);
else
     fprintf('Power factor= %.3f; Unity\n',PF);
end
fprintf('\n');
% Example 3: Real power,P= 10 W; Reactive power,Q=0 Var
P=10; Q=0;
fprintf('Example 3: Zero reactive power\n');
fprintf('------------------------------------\n');
disp('Apparent power:')
S=P+j*Q
S_mag=abs(S);
S_angle=angle(S)*(180/pi);
fprintf('Apparent power in polar form:\n');
fprintf('|S|= %.3f VA     Power angle= %.3f degree\n',S_mag,S_angle);
PF=cos(S_angle);
if Q>0
    fprintf('Power factor= %.3f; Lagging\n',PF);
elseif Q<0
    fprintf('Power factor= %.3f; Leading\n',PF);
else
     fprintf('Power factor= %.3f; Unity\n',PF);
end
```

Fig. 8.38 Code—Determination of the parameters of the power triangle

```
Command Window

   Example 1: Positive reactive power
   ------------------------------------
   Apparent power:

   S =

     10.0000 + 5.0000i

   Apparent power in polar form:
   |S|= 11.180 VA     Power angle= 26.565 degree
   Power factor= 0.138; Lagging

   Example 2: Negative reactive power
   ------------------------------------
   Apparent power:

   S =

     10.0000 - 5.0000i

   Apparent power in polar form:
   |S|= 11.180 VA     Power angle= -26.565 degree
   Power factor= 0.138; Leading

   Example 3: Zero reactive power
   ------------------------------------
   Apparent power:

   S =

       10

   Apparent power in polar form:
   |S|= 10.000 VA     Power angle= 0.000 degree
   Power factor= 1.000; Unity
```

Fig. 8.39 Output—Determination of the parameters of the power triangle

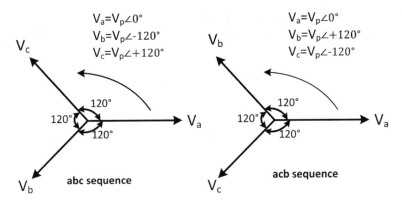

Fig. 8.40 abc and acb phase sequence in three-phase systems

Fig. 8.41 A delta connected unbalanced system

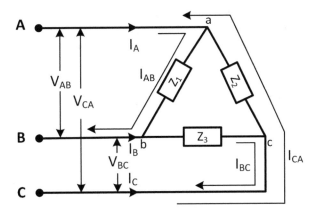

distributed evenly among the phases. In a balanced system, all the voltages, or currents, in three phases are in $120°$ phase difference from each other. Based on the nature of the order of the sequence of phase voltages, two-phase sequences are available—"abc" sequence and "acb" sequence. The phase sequence implies the order of sequence based on which the individual phase voltage or current reaches their peak values. A simple representation of these two types of phase sequences is given in Fig. 8.40.

8.3.4.1 Delta-Connected Unbalanced Load

A delta-connected unbalanced load is shown in Fig. 8.41, where the impedances are of different values in each phase to make the system unbalanced.

Table 8.1 Parameters in a delta-connected unbalanced load system

Line-to-line voltage	Phase voltage	Line current	Phase current
V_{AB}	V_a	I_A	I_{AB}
V_{BC}	V_b	I_B	I_{BC}
V_{CA}	V_c	I_C	I_{CA}

Table 8.2 Relationship between the parameters in a delta-connected unbalanced load system

Line-to-line voltage	Phase current	Line current
$V_{AB} = V_a - V_b$	$I_{AB} = \frac{V_{AB}}{Z_1}$	$I_A = I_{AB} - I_{CA}$
$V_{BC} = V_b - V_c$	$I_{BC} = \frac{V_{BC}}{Z_2}$	$I_B = I_{BC} - I_{AB}$
$V_{CA} = V_c - V_a$	$I_{CA} = \frac{V_{CA}}{Z_3}$	$I_C = I_{CA} - I_{BC}$

The relevant parameters for a delta-connected unbalanced load system are enlisted in Table 8.1 in accordance with Fig. 8.41.

For a delta-connected unbalanced system, the parameters can be related with each other by the formulas as mentioned in Table 8.2.

8.3.4.2 MATLAB Example 8.14: Delta-Connected Unbalanced Load

Consider a system shown in Fig. 8.41 with the following parameters:

$$V_{AB} = 120\angle 0° \ V$$

$$V_{BC} = 110\angle 120° \ V$$

$$V_{CA} = 150\angle 240° \ V$$

$$Z_1 = 10\angle 10°$$

$$Z_2 = 15\angle -25°$$

$$Z_3 = 20\angle -10°$$

Determine:

(a) Phase currents I_{AB}, I_{BC}, and I_{CA}
(b) Line currents I_A, I_B, and I_C.

The MATLAB code for this example is given in Fig. 8.42 with its output in Fig. 8.43.

Output

```
% Delta connected unbalanced load
% Line to line voltages:
% V_AB=120 V angle 0 deg; V_BC=110 V angle 120 deg; V_CA=150 V angle 240 deg
% Impedances:
% Z1=10 Ohms angle 10 degl Z2=15 Ohms angle -25 degl Z3=20 Ohms angle -10 deg;
% Find: Phase currents I_AB, I_BC, I_CA
% Find: Line currents: I_A, I_B, I_C
clc, clear;
% Line to Line voltages
V_AB=120*cos(0)+i*120*sin(0);
V_BC=110*cos(120*(pi/180))+i*110*sin(120*(pi/180));
V_CA=150*cos(240*(pi/180))+i*150*sin(240*(pi/180));
% Impedances
Z1=10*cos(10*(pi/180))+i*10*sin(10*(pi/180));
Z2=15*cos(-25*(pi/180))+i*15*sin(-25*(pi/180));
Z3=20*cos(-10*(pi/180))+i*20*sin(-10*(pi/180));
% Phase currents
I_AB=V_AB/Z1;
I_BC=V_BC/Z2;
I_CA=V_CA/Z3;
I_AB_mag=abs(I_AB);
I_AB_ang=angle(I_AB)*180/pi;
I_BC_mag=abs(I_BC);
I_BC_ang=angle(I_BC)*180/pi;
I_CA_mag=abs(I_CA);
I_CA_ang=angle(I_CA)*180/pi;
fprintf('Phase currents:\n');
fprintf('I_AB= %.3f A      Angle=%.3f degree\n',I_AB_mag,I_AB_ang);
fprintf('I_BC= %.3f A      Angle=%.3f degree\n',I_BC_mag,I_BC_ang);
fprintf('I_CA= %.3f A      Angle=%.3f degree\n',I_CA_mag,I_CA_ang);
% Line currents
I_A=I_AB-I_CA;
I_B=I_BC-I_AB;
I_C=I_CA-I_BC;
I_A_mag=abs(I_A);
I_A_ang=angle(I_A)*180/pi;
I_B_mag=abs(I_B);
I_B_ang=angle(I_B)*180/pi;
I_C_mag=abs(I_C);
I_C_ang=angle(I_C)*180/pi;
fprintf('Line currents:\n');
fprintf('I_A= %.3f A      Angle=%.3f degree\n',I_A_mag,I_A_ang);
fprintf('I_B= %.3f A      Angle=%.3f degree\n',I_B_mag,I_B_ang);
fprintf('I_C= %.3f A      Angle=%.3f degree\n',I_C_mag,I_C_ang);
```

Fig. 8.42 Code—Delta connected unbalanced load

Command Window

```
Phase currents:
I_AB= 12.000 A          Angle=-10.000 degree
I_BC= 7.333 A           Angle=145.000 degree
I_CA= 7.500 A           Angle=-110.000 degree
Line currents:
I_A= 15.215 A           Angle=19.041 degree
I_B= 18.902 A           Angle=160.563 degree
I_C= 11.769 A           Angle=-72.994 degree
```

Fig. 8.43 Output—Delta connected unbalanced load

Table 8.3 Parameters in a delta-connected balanced load system

Line-to-line voltage	Phase voltage	Line current	Phase current
V_{AB}	V_a	I_A	I_{AB}
V_{BC}	V_b	I_B	I_{BC}
V_{CA}	V_c	I_C	I_{CA}

8.3.4.3   Delta-Connected Balanced Load

A delta-connected balanced load is shown in Fig. 8.44, where the impedances are equally distributed in each phase to make the system balanced. In a balanced delta system, the magnitudes of the input line to line to voltages and phase voltages are equal.

The relevant parameters of a delta-connected balanced system are enlisted in Table 8.3.

Fig. 8.44 A delta connected unbalanced system

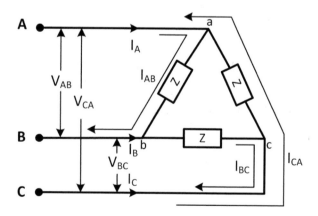

Table 8.4 Relationship between the parameters in a delta-connected balanced load system

Line-to-line voltage	Phase current	Line current
$V_{AB} = V_a - V_b$ $V_{BC} = V_b - V_c$ $V_{CA} = V_c - V_a$ $\lvert V_{AB}\rvert = \lvert V_{BC}\rvert = \lvert V_{CA}\rvert = V_L$ **Therefore, considering acb sequence,** $V_{AB} = V_L \angle 0°$ $V_{BC} = V_L \angle +120°$ $V_{CA} = V_L \angle -120°$	$I_{AB} = \dfrac{V_{AB}}{Z}$ $I_{BC} = \dfrac{V_{BC}}{Z}$ $I_{CA} = \dfrac{V_{CA}}{Z}$ $\lvert I_{AB}\rvert = \lvert I_{BC}\rvert = \lvert I_{CA}\rvert = I_P$ **Therefore, considering acb sequence,** $I_{AB} = I_P \angle 0°$ $I_{BC} = I_P \angle +120°$ $I_{CA} = I_P \angle -120°$	$I_A = I_{AB} - I_{CA}$ $= I_P \angle 0° - I_P \angle -120°$ $= I_P(1 \angle 0° - 1 \angle -120°)$ $= \sqrt{3} I_P \angle 30°$ $I_B = I_{BC} - I_{AB}$ $= I_P \angle +120° - I_P \angle 0°$ $= I_P(1 \angle 120° - 1 \angle 0°)$ $= \sqrt{3} I_P \angle 150°$ $I_C = I_{CA} - I_{BC}$ $= I_P \angle -120° - I_P \angle +120$ $= I_P(1 \angle 120° - 1 \angle 0°)$ $= \sqrt{3} I_P \angle -90°$ **Therefore,** $\lvert I_A\rvert = \lvert I_B\rvert = \lvert I_C\rvert = \sqrt{3} I_P = I_L$

```
% Delta connected balanced load
% Line to line voltages:
% V_AB=120 V angle 0 deg; V_BC=120 V angle 120 deg; V_CA=120 V angle 240 deg
% Impedances:
% Z=10 Ohms angle 10 deg
% Find: Phase currents I_AB, I_BC, I_CA
% Find: Line currents I_A, I_B, I_C
clc; clear;
% Line to line voltages
V_AB=120*cos(0)+i*120*sin(0);
V_BC=120*cos(120*(pi/180))+i*120*sin(120*(pi/180));
V_CA=120*cos(240*(pi/180))+i*120*sin(240*(pi/180));
% Impedances
Z=10*cos(10*(pi/180))+i*10*sin(10*(pi/180));
%Phase currents
I_AB=V_AB/Z;
I_BC=V_BC/Z;
I_CA=V_CA/Z;
Ip=abs(I_AB);
I_AB_ang=angle(I_AB)*180/pi;
I_BC_ang=angle(I_BC)*180/pi;
I_CA_ang=angle(I_CA)*180/pi;
fprintf('Phase Currents:\n');
fprintf('I_AB= %.3f A        Angle=%.3f degree\n',Ip,I_AB_ang);
fprintf('I_BC= %.3f A        Angle=%.3f degree\n',Ip,I_BC_ang);
fprintf('I_CA= %.3f A        Angle=%.3f degree\n',Ip,I_CA_ang);
% Line currents
I_A=I_AB-I_CA;
I_B=I_BC-I_AB;
I_C=I_CA-I_BC;
IL=abs(I_A);
I_A_ang=angle(I_A)*180/pi;
I_B_ang=angle(I_B)*180/pi;
I_C_ang=angle(I_C)*180/pi;
fprintf('Line currents:\n');
fprintf('I_A= %.3f A        Angle=%.3f degree\n',IL,I_A_ang);
fprintf('I_B= %.3f A        Angle=%.3f degree\n',IL,I_B_ang);
fprintf('I_C= %.3f A        Angle=%.3f degree\n',IL,I_C_ang);
```

Fig. 8.45 Code—Delta connected balanced load

The relationship among the parameters for a delta-connected balanced load system can be summarized as shown in Table 8.4.

8.3.4.4 MATLAB Example 8.15: Delta-Connected Balanced Load

Consider a system shown in Fig. 8.44 with the following parameters:

```
Command Window
    Phase currents:
    I_AB= 12.000 A          Angle= -10.000 degree
    I_BC= 12.000 A          Angle= 110.000 degree
    I_CA= 12.000 A          Angle= -130.000 degree
    Line currents:
    I_A= 20.785 A           Angle= 20.000 degree
    I_B= 20.785 A           Angle= 140.000 degree
    I_C= 20.785 A           Angle= -100.000 degree
```

Fig. 8.46 Output—Delta connected balanced load

$$V_{AB} = 120\angle 0^\circ \ V$$

$$V_{BC} = 110\angle 120^\circ \ V$$

$$V_{CA} = 150\angle 240^\circ \ V$$

$$Z = 10\angle 10^\circ$$

Determine:

(a) Phase currents I_{AB}, I_{BC}, and I_{CA}
(b) Line currents I_A, I_B, and I_C

The MATLAB code for this example is given in Fig. 8.45 with its output in Fig. 8.46.

Output

From the output, we can observe that the magnitude of the line current in each phase is equivalent to the square root of three multiplied with the phase current in each phase.

8.3.4.5   Wye-Connected Four-Wire Unbalanced Load

A wye-connected four-wire unbalanced load is shown in Fig. 8.47, where the impedances are different in each phase to make the system unbalanced. In addition, the common point is connected to a neutral, which explains the reason for naming it a four-wire system.

The relevant parameters of a wye-connected four-wire unbalanced system are enlisted in Table 8.5.

Fig. 8.47 A wye connected
four wire unbalanced system

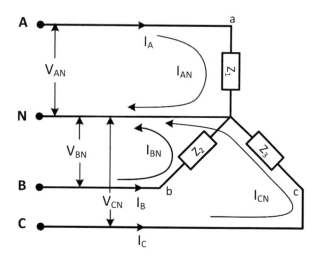

Table 8.5 Parameters in a wye-connected four-wire unbalanced load system

Line-to-line voltage/line voltage	Line to neutral voltage/phase voltage	Line current	Phase current
V_{AB}	$V_a = V_{AN}$	I_A	I_{AN}
V_{BC}	$V_b = V_{BN}$	I_B	I_{BN}
V_{CA}	$V_c = V_{CN}$	I_C	I_{CN}

Table 8.6 Relationship between the parameters in a wye-connected four-wire unbalanced load system

Line-to-line voltage	Phase current	Line current
$V_{AB} = V_{AN} - V_{BN}$	$I_{AN} = \frac{V_{AN}}{Z_1}$	$I_A = I_{AN}$
$V_{BC} = V_{BN} - V_{CN}$	$I_{BN} = \frac{V_{BN}}{Z_2}$	$I_B = I_{BN}$
$V_{CA} = V_{CN} - V_{AN}$	$I_{CN} = \frac{V_{CN}}{Z_3}$	$I_C = I_{CN}$

The relationship among the parameters for a wye-connected four wire unbalanced load system can be summarized as shown in Table 8.6.

8.3.4.6 MATLAB Example 8.16: Wye-Connected Four-Wire Unbalanced Load

Consider a system shown in Fig. 8.47 with the following parameters:

$$V_{AN} = 120\angle 10° \ V$$

$$V_{BN} = 110\angle 150° \ V$$

```
% Wye connected four wire unbalanced load
% Phase voltages:
% V_AN=120 V angle 10 deg;V_BN=110 V angle 150 deg;V_CN=150 V angle -50 deg
% Impedances:
% Z1=10 Ohms angle 10 deg;Z2=15 Ohms angle -25 deg;Z3=20 Ohms angle -10 deg;
% Find: Line to line voltages V_AB, V_BC, V_CA
% Find: Phase currents I_AN, I_BN, I_CN
% Find: Line currents I_A, I_B, I_C

clc;clear;
% Line to line voltages
V_AN=120*cos(10)+i*120*sin(10);
V_BN=110*cos(150*(pi/180))+i*110*sin(150*(pi/180));
V_CN=150*cos(-50*(pi/180))+i*150*sin(-50*(pi/180));
% Impedances
Z1=10*cos(10*(pi/180))+i*10*sin(10*(pi/180));
Z2=15*cos(-25*(pi/180))+i*15*sin(-25*(pi/180));
Z3=20*cos(-10*(pi/180))+i*20*sin(-10*(pi/180));
% Line to line voltages
V_AB=V_AN-V_BN;
V_BC=V_BN-V_CN;
V_CA=V_CN-V_AN;
V_AB_mag=abs(V_AB);
V_AB_ang=angle(V_AB)*180/pi;
V_BC_mag=abs(V_BC);
V_BC_ang=angle(V_BC)*180/pi;
V_CA_mag=abs(V_CA);
V_CA_ang=angle(V_CA)*180/pi;
fprintf('Line to line voltages:\n');
fprintf('V_AB= %.3f A      Angle= %.3f degree\n',V_AB_mag,V_AB_ang);
fprintf('V_BC= %.3f A      Angle= %.3f degree\n',V_BC_mag,V_BC_ang);
fprintf('V_CA= %.3f A      Angle= %.3f degree\n\n',V_CA_mag,V_CA_ang);
% Phase currents
I_AN=V_AN/Z1;
I_BN=V_BN/Z2;
I_CN=V_CN/Z3;
I_AN_mag=abs(I_AN);
I_AN_ang=angle(I_AN)*180/pi;
I_BN_mag=abs(I_BN);
I_BN_ang=angle(I_BN)*180/pi;
I_CN_mag=abs(I_CN);
I_CN_ang=angle(I_CN)*180/pi;
fprintf('Phase currents:\n');
fprintf('I_AN= %.3f A      Angle= %.3f degree\n',I_AN_mag,I_AN_ang);
fprintf('I_BN= %.3f A      Angle= %.3f degree\n',I_BN_mag,I_BN_ang);
fprintf('I_CN= %.3f A      Angle= %.3f degree\n\n',I_CN_mag,I_CN_ang);
% Line currents
fprintf('Line currents:\n');
fprintf('I_A= %.3f A     Angle= %.3f degree\n',I_AN_mag,I_AN_ang);
fprintf('I_B= %.3f A     Angle= %.3f degree\n',I_BN_mag,I_BN_ang);
fprintf('I_C= %.3f A     Angle= %.3f degree\n',I_CN_mag,I_CN_ang);
```

Fig. 8.48 Code—A wye connected four wire unbalanced load

```
Command Window
  Line to line voltages:
   V_AB= 120.405 A          Angle= -92.583 degree
   V_BC= 256.144 A          Angle= 138.446 degree
   V_CA= 203.258 A          Angle= -14.131 degree

   Phase currents:
   I_AN= 12.000 A           Angle= -157.042 degree
   I_BN= 7.333 A            Angle= 175.000 degree
   I_CN= 7.500 A            Angle= -40.000 degree

   Line currents:
   I_A= 12.000 A            Angle= -157.042 degree
   I_B= 7.333 A             Angle= 175.000 degree
   I_C= 7.500 A             Angle= -40.000 degree
```

Fig. 8.49 Output – A wye connected four wire unbalanced load

$$V_{CN} = 150 \angle -50^\circ \; V$$

$$Z_1 = 10 \angle 10^\circ$$

$$Z_2 = 15 \angle -25^\circ$$

$$Z_3 = 20 \angle -10^\circ$$

Determine:

(a) Line-to-line voltages V_{AB}, V_{BC}, and V_{CA}
(b) Phase currents I_{AN}, I_{BN}, and I_{CN}
(c) Line currents I_A, I_B, and I_C

The MATLAB code for this example is given in Fig. 8.48 with its output in Fig. 8.49.

Output

8.3.4.7   Wye-Connected Four-Wire Balanced Load

A wye-connected four-wire balanced load is shown in Fig. 8.50, where the impedances are distributed evenly in each phase to make the system balanced. In addition, the common point is connected to a neutral, which explains the reason for naming it a four-wire system.

Fig. 8.50 A wye connected
four wire balanced system

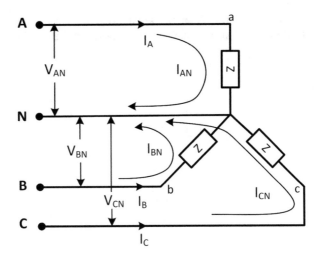

Table 8.7 Parameters in a wye-connected four-wire balanced load system

Line-to-line voltage	Line to neutral voltage/phase voltage	Line current	Phase current
V_{AB}	$V_a = V_{AN}$	I_A	I_{AN}
V_{BC}	$V_b = V_{BN}$	I_B	I_{BN}
V_{CA}	$V_c = V_{CN}$	I_C	I_{CN}

Table 8.8 Relationship between the parameters in a wye-connected four-wire balanced load system

Phase voltage	Line-to-line voltage	Phase current and line current
$V_{AN} = I_{AN} \times Z$ $V_{BN} = I_{BN} \times Z$ $V_{CN} = I_{CN} \times Z$ $\mid V_{AN} \mid = \mid V_{BN} \mid = 1$ $V_{CN} = V_P$	**Considering acb sequence,** $V_{AB} = V_{AN} - V_{BN}$ $= V_P \angle 0° - V_P \angle 120°$ $= V_P(1 \angle 0° - 1 \angle 120°$ $= \sqrt{3} V_P \angle - 30°$ $V_{BC} = V_{BN} - V_{CN}$ $= V_P \angle 120° - V_P \angle - 120°$ $= V_P(1 \angle 120° - 1 \angle - 120°$ $= \sqrt{3} V_P \angle 90°$ $V_{CA} = V_{CN} - V_{AN}$ $= V_P \angle - 120° - V_P \angle 0°$ $= V_P(1 \angle - 120° - 1 \angle 0°$ $= \sqrt{3} V_P \angle - 150°$ **Therefore,** $\mid V_{AB} \mid = \mid V_{BC} \mid = \mid V_{CA} \mid = \sqrt{3} V_P = V_L$	$I_{AN} = \frac{V_{AN}}{Z} = I_A$ $I_{BN} = \frac{V_{BN}}{Z} = I_B$ $I_{CN} = \frac{V_{CN}}{Z} = I_C$ $\mid I_{AN} \mid = \mid I_{BN} \mid = \mid I_{CN} \mid = \mid \frac{V_P}{Z} \mid$

The relevant parameters of a wye-connected four-wire balanced load system are enlisted in Table 8.7.

The relationship among the parameters for a wye-connected four-wire balanced load system can be summarized as shown in Table 8.8.

```
% Wye connected four wire balanced load
% Phase voltages:
% V_AN=110 V angle 0 deg;V_BN=110 V angle 120 deg;V_CN=110 V angle 240 deg
% Impedances:
% Z=10 Ohms angle 10 deg;
% Find: Line to line voltages V_AB, V_BC, V_CA
% Find: Phase currents I_AN, I_BN, I_CN
% Find: Line currents I_A, I_B, I_C

clc;clear;
% Line to line voltages
V_AN=110*cos(0)+i*110*sin(0);
V_BN=110*cos(120*(pi/180))+i*110*sin(120*(pi/180));
V_CN=110*cos(240*(pi/180))+i*110*sin(240*(pi/180));
% Impedances
Z=10*cos(10*(pi/180))+i*10*sin(10*(pi/180));
% Line to line voltages
V_AB=V_AN-V_BN;
V_BC=V_BN-V_CN;
V_CA=V_CN-V_AN;
V_L=abs(V_AB);
V_AB_ang=angle(V_AB)*180/pi;
V_BC_ang=angle(V_BC)*180/pi;
V_CA_ang=angle(V_CA)*180/pi;
fprintf('Line to line voltages:\n');
fprintf('V_AB= %.3f A      Angle= %.3f degree\n',V_L,V_AB_ang);
fprintf('V_BC= %.3f A      Angle= %.3f degree\n',V_L,V_BC_ang);
fprintf('V_CA= %.3f A      Angle= %.3f degree\n\n',V_L,V_CA_ang);
% Phase currents
I_AN=V_AN/Z;
I_BN=V_BN/Z;
I_CN=V_CN/Z;
I_AN_mag=abs(I_AN);
I_AN_ang=angle(I_AN)*180/pi;
I_BN_mag=abs(I_BN);
I_BN_ang=angle(I_BN)*180/pi;
I_CN_mag=abs(I_CN);
I_CN_ang=angle(I_CN)*180/pi;
fprintf('Phase currents:\n');
fprintf('I_AN= %.3f A      Angle= %.3f degree\n',I_AN_mag,I_AN_ang);
fprintf('I_BN= %.3f A      Angle= %.3f degree\n',I_BN_mag,I_BN_ang);
fprintf('I_CN= %.3f A      Angle= %.3f degree\n\n',I_CN_mag,I_CN_ang);
% Line currents
fprintf('Line currents:\n');
fprintf('I_A= %.3f A      Angle= %.3f degree\n',I_AN_mag,I_AN_ang);
fprintf('I_B= %.3f A      Angle= %.3f degree\n',I_BN_mag,I_BN_ang);
fprintf('I_C= %.3f A      Angle= %.3f degree\n',I_CN_mag,I_CN_ang);
```

Fig. 8.51 Code—A wye connected four wire balanced load

```
Command Window
  Line to line voltages:
  V_AB= 190.526 A          Angle= -30.000 degree
  V_BC= 190.526 A          Angle= 90.000 degree
  V_CA= 190.526 A          Angle= -150.000 degree

  Phase currents:
  I_AN= 11.000 A           Angle= -10.000 degree
  I_BN= 11.000 A           Angle= 110.000 degree
  I_CN= 11.000 A           Angle= -130.000 degree

  Line currents:
  I_A= 11.000 A            Angle= -10.000 degree
  I_B= 11.000 A            Angle= 110.000 degree
  I_C= 11.000 A            Angle= -130.000 degree
```

Fig. 8.52 Output—A wye connected four wire balanced load

8.3.4.8   MATLAB Example 8.17: Wye-Connected Four-Wire Balanced Load

Consider a system shown in Fig. 8.50 with the following parameters:

$$V_{AN} = 110\angle 10° \ V$$

$$V_{BN} = 110\angle 150° \ V$$

$$V_{CN} = 110\angle -50° \ V$$

$$Z = 10\angle 10°$$

Determine:

(a) Line-to-line voltages V_{AB}, V_{BC}, and V_{CA}
(b) Phase currents I_{AN}, I_{BN}, and I_{CN}
(c) Line currents I_A, I_B, and I_C.

The MATLAB code for this example is given in Fig. 8.51 with its output in Fig. 8.52.

Output

Here, from the output, the phase and line currents of a wye-connected four-wire balanced system are equal.

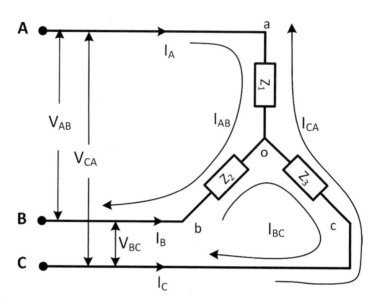

Fig. 8.53 A wye connected three wire unbalanced system

Table 8.9 Parameters in a wye-connected three-wire balanced load system

Line-to-line voltage	Phase voltage	Line current	Phase current
V_{AB}	V_a	I_A	I_{ao}
V_{BC}	V_b	I_B	I_{bo}
V_{CA}	V_c	I_C	I_{co}

Table 8.10 Relationship between the parameters in a wye-connected three-wire balanced load system

Line to common point voltage	Line-to-line voltage	Line current and phase current
$V_{ao} = I_A \times Z_1$	$V_{AB} = V_{ao} - V_{bo}$	$I_A = I_{ao}$
$V_{bo} = I_B \times Z_2$	$V_{BC} = V_{bo} - V_{co}$	$I_B = I_{bo}$
$V_{co} = I_C \times Z_3$	$V_{CA} = V_{co} - V_{ao}$	$I_C = I_{co}$

8.3.4.9   Wye-Connected Three-Wire Unbalanced Load

A wye-connected three-wire unbalanced load system is shown in Fig. 8.53, where the impedances are different in each phase to make the system unbalanced. In addition, the common point is not connected to a neutral; therefore, it will be regarded as a common point, not a neutral point.

The relevant parameters of a wye-connected three-wire balanced load system are enlisted in Table 8.9.

The relationship among the parameters for a wye-connected three-wire balanced load system can be summarized as shown in Table 8.10.

8.3.4.10 MATLAB Example 8.18: Wye-Connected Three-Wire Unbalanced Load

Consider a system shown in Fig. 8.53 with the following parameters:

```
% Wye connected three wire unbalanced load
% Phase voltages:
% V_ao=120 V angle 10 deg;V_bo=110 V angle 150 deg;V_co=150 V angle -50 deg
% Impedances:
% Z1=10 Ohms angle 10 deg;Z2=15 Ohms angle -25 deg;Z3=20 Ohms angle -10 deg;
% Find: Line to line voltages V_AB, V_BC, V_CA
% Find: Line currents I_A, I_B, I_C
clc;clear;
% Line to line voltages
V_ao=120*cos(10)+i*120*sin(10);
V_bo=110*cos(150*(pi/180))+i*110*sin(150*(pi/180));
V_co=150*cos(-50*(pi/180))+i*150*sin(-50*(pi/180));
% Impedances
Z1=10*cos(10*(pi/180))+i*10*sin(10*(pi/180));
Z2=15*cos(-25*(pi/180))+i*15*sin(-25*(pi/180));
Z3=20*cos(-10*(pi/180))+i*20*sin(-10*(pi/180));
% Line to line voltages
V_AB=V_ao-V_bo;
V_BC=V_bo-V_co;
V_CA=V_co-V_ao;
V_AB_mag=abs(V_AB);
V_AB_ang=angle(V_AB)*180/pi;
V_BC_mag=abs(V_BC);
V_BC_ang=angle(V_BC)*180/pi;
V_CA_mag=abs(V_CA);
V_CA_ang=angle(V_CA)*180/pi;
fprintf('Line to line voltages:\n');
fprintf('V_AB= %.3f A      Angle= %.3f degree\n',V_AB_mag,V_AB_ang);
fprintf('V_BC= %.3f A      Angle= %.3f degree\n',V_BC_mag,V_BC_ang);
fprintf('V_CA= %.3f A      Angle= %.3f degree\n\n',V_CA_mag,V_CA_ang);
% Line currents
I_A=V_ao/Z1;
I_B=V_bo/Z2;
I_C=V_co/Z3;
I_A_mag=abs(I_A);
I_A_ang=angle(I_A)*180/pi;
I_B_mag=abs(I_B);
I_B_ang=angle(I_B)*180/pi;
I_C_mag=abs(I_C);
I_C_ang=angle(I_C)*180/pi;
fprintf('Line currents:\n');
fprintf('I_A= %.3f A      Angle= %.3f degree\n',I_A_mag,I_A_ang);
fprintf('I_B= %.3f A      Angle= %.3f degree\n',I_B_mag,I_B_ang);
fprintf('I_C= %.3f A      Angle= %.3f degree\n\n',I_C_mag,I_C_ang);
```

Fig. 8.54 Code—A wye connected three wire unbalanced load

```
Command Window

  Line to line voltages:
  V_AB= 120.405 A        Angle= -92.583 degree
  V_BC= 256.144 A        Angle= 138.446 degree
  V_CA= 203.258 A        Angle= -14.131 degree

  Line currents:
  I_A= 12.000 A          Angle= -157.042 degree
  I_B= 7.333 A           Angle= 175.000 degree
  I_C= 7.500 A           Angle= -40.000 degree
```

Fig. 8.55 Output—A wye connected three wire unbalanced load

$$V_{ao} = 120\angle 10^\circ \ V$$

$$V_{bo} = 110\angle 150^\circ \ V$$

$$V_{co} = 150\angle -50^\circ \ V$$

$$Z_1 = 10\angle 10^\circ$$

$$Z_2 = 15\angle -25^\circ$$

$$Z_3 = 20\angle -10^\circ$$

Determine:

(a) Line-to-line voltages V_{AB}, V_{BC}, and V_{CA}
(b) Line currents I_A, I_B, and I_C

The MATLAB code for this example is given in Fig. 8.54 with its output in Fig. 8.55.

Output

8.3.4.11   Wye-Connected Three-Wire Balanced Load

A wye-connected three-wire balanced load system is shown in Fig. 8.56, where the impedances are equal in each phase to make the system balanced. In addition, the

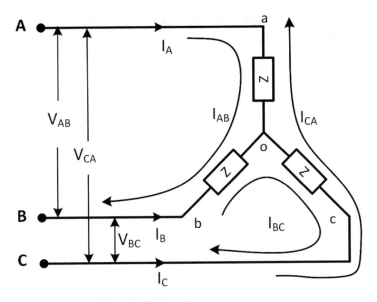

Fig. 8.56 A wye connected three wire balanced system

Table 8.11 Parameters in a wye-connected three-wire balanced load system

Line-to-line voltage	Phase voltage	Line current	Phase current
V_{AB}	V_a	I_A	I_{ao}
V_{BC}	V_b	I_B	I_{bo}
V_{CA}	V_c	I_C	I_{co}

Table 8.12 Relationship between the parameters in a wye-connected three-wire balanced load system

Line to common point voltage	Line-to-line voltage	Line current and phase current
$V_{ao} = I_A \times Z$ $V_{bo} = I_B \times Z$ $V_{co} = I_C \times Z$ $\|V_{ao}\| = \|V_{bo}\| = \|$ $V_{co}\| = V_P$	$V_{AB} = V_{ao} - V_{bo}$ $V_{BC} = V_{bo} - V_{co}$ $V_{CA} = V_{co} - V_{ao}$ $\|V_{AB}\| = \|V_{BC}\| = \|V_{CA}\| = V_L = \sqrt{3}V_P$	$I_A = I_{ao}$ $I_B = I_{bo}$ $I_C = I_{co}$

common point is not connected to a neutral; therefore, it will be regarded as a common point, not a neutral point.

All the relevant parameters in this system are summarized in Table 8.11 in accordance with Fig. 8.56.

The relationships to determine each parameter are enlisted in Table 8.12 to make the system more understandable.

8.3.4.12   MATLAB Example 8.19: Wye-Connected Three-Wire Balanced Load

Consider a system shown in Fig. 8.56 with the following parameters:

$$V_{ao} = 120\angle 0° \ V$$

```
% Wye connected three wire balanced load
% Phase voltages:
% V_ao=120 V angle 0 deg;V_bo=120 V angle 120 deg;V_co=120 V angle 240 deg
% Impedances:
% Z=10 Ohms angle 10 deg;
% Find: Line to line voltages V_AB, V_BC, V_CA
% Find: Line currents I_A, I_B, I_C
clc;clear;
% Line to line voltages
V_ao=120*cos(0)+i*120*sin(0);
V_bo=120*cos(120*(pi/180))+i*120*sin(120*(pi/180));
V_co=120*cos(240*(pi/180))+i*120*sin(240*(pi/180));
% Impedances
Z=10*cos(10*(pi/180))+i*10*sin(10*(pi/180));
% Line to line voltages
V_AB=V_ao-V_bo;
V_BC=V_bo-V_co;
V_CA=V_co-V_ao;
V_AB_mag=abs(V_AB);
V_AB_ang=angle(V_AB)*180/pi;
V_BC_mag=abs(V_BC);
V_BC_ang=angle(V_BC)*180/pi;
V_CA_mag=abs(V_CA);
V_CA_ang=angle(V_CA)*180/pi;
fprintf('Line to line voltages:\n');
fprintf('V_AB= %.3f A      Angle= %.3f degree\n',V_AB_mag,V_AB_ang);
fprintf('V_BC= %.3f A      Angle= %.3f degree\n',V_BC_mag,V_BC_ang);
fprintf('V_CA= %.3f A      Angle= %.3f degree\n\n',V_CA_mag,V_CA_ang);
% Line currents
I_A=V_ao/Z;
I_B=V_bo/Z;
I_C=V_co/Z;
I_A_mag=abs(I_A);
I_A_ang=angle(I_A)*180/pi;
I_B_mag=abs(I_B);
I_B_ang=angle(I_B)*180/pi;
I_C_mag=abs(I_C);
I_C_ang=angle(I_C)*180/pi;
fprintf('Line currents:\n');
fprintf('I_A= %.3f A      Angle= %.3f degree\n',I_A_mag,I_A_ang);
fprintf('I_B= %.3f A      Angle= %.3f degree\n',I_B_mag,I_B_ang);
fprintf('I_C= %.3f A      Angle= %.3f degree\n\n',I_C_mag,I_C_ang);
```

Fig. 8.57 Code—A wye connected three wire balanced load

```
Command Window
  Line to line voltages:
  V_AB= 207.846 A          Angle= -30.000 degree
  V_BC= 207.846 A          Angle= 90.000 degree
  V_CA= 207.846 A          Angle= -150.000 degree


  Line currents:
  I_A= 12.000 A            Angle= -10.000 degree
  I_B= 12.000 A            Angle= 110.000 degree
  I_C= 12.000 A            Angle= -130.000 degree
```

Fig. 8.58 Output—A wye connected three wire balanced load

$$V_{bo} = 110\angle 120^\circ \ V$$

$$V_{co} = 150\angle 240^\circ \ V$$

$$Z = 10\angle 10^\circ$$

Determine:

(a) Line-to-line voltages V_{AB}, V_{BC}, and V_{CA}
(b) Line currents I_A, I_B, and I_C.

The MATLAB code for this example is given in Fig. 8.57 with its output in Fig. 8.58.

Output

The magnitude of each phase voltage is 120 V. By multiplying this value with $\sqrt{3}$ provides the magnitude of each line-to-line voltages, which is 207.846 V. The magnitudes of the line currents are also similar in a wye-connected three-wire balanced system that can be realized from the above output as well.

8.4 Operational Amplifier

An operational amplifier is an active device that can amplify any input signals; can perform mathematical operations such as addition, multiplication, differentiation, and integration; and can be utilized for filtering purposes as well. The block diagram of an operational amplifier (Op-amp) is provided in Fig. 8.59. A standard Op-amp has five important ports, where port 1 and port 2 signify the inverting and the

Fig. 8.59 Pin diagram of an
operational amplifier

1. **Inverting Input 1**
2. **Non-inverting Input 2**
3. **Positive Voltage, V+**
4. **Negative Voltage, V-**
5. **Output**

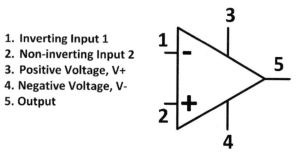

non-inverting input signals; and port 5 is for the output signal. In port 3 and port 4, the positive and negative voltage connection is provided.

8.4.1   Inverting Amplifier

In an inverting amplifier, the gain of the amplifier is negative. The gain of an amplifier is the ratio of output to input. A configuration of an inverting amplifier is given in Fig. 8.60, where a positive voltage source is connected with the negative input port of an Op-amp, and the positive input port is grounded.

$$V_{out} = -\frac{R_2}{R_1} \cdot V_{in} \qquad (8.30)$$

Fig. 8.60 Circuit diagram
of an inverting amplifier

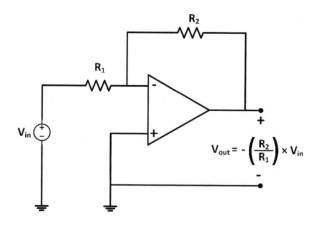

```
% Inverting amplifier
% Input voltage: V_in=40 V;
% Resistances: R1=4 Ohms; R2=2 Ohms;
% Find: Output voltage, V_out
% Find: Gain, G
clc; clear;
V_in=40; R1=4; R2=2;
V_out=-(R2/R1)*V_in;
G=V_out/V_in;
fprintf('Output voltage: %.2f V\n',V_out);
fprintf('Gain: %.2f\n',G);
```

Fig. 8.61 Code—Inverting amplifier

Fig. 8.62 Output—
Inverting amplifier

```
Command Window

    Output voltage: -20.00 V
    Gain: -0.50
```

8.4.1.1   MATLAB Example 8.20: Inverting Amplifier

Consider an inverting amplifier as shown in Fig. 8.60 with 40 V input and $R_1 = 4\Omega$, $R_2 = 2\Omega$. Determine the gain and the output voltage of the amplifier.

The MATLAB code for this example is given in Fig. 8.61 with its output in Fig. 8.62.

Output

8.4.2   Non-inverting Amplifier

In a non-inverting amplifier, the gain of the amplifier is positive. The gain of an amplifier is the ratio of output to input. A configuration of a non-inverting amplifier is given in Fig. 8.63, where a positive voltage source is connected with the positive input port of an Op-amp, and the negative input port is grounded.

$$V_{\text{out}} = \left(1 + \frac{R_2}{R_1}\right) \cdot V_{\text{in}} \tag{8.31}$$

Fig. 8.63 Circuit diagram of a non-inverting amplifier

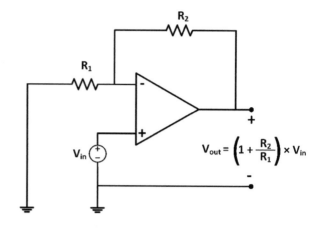

```
% Non-inverting amplifier
% Input voltage: V_in=40 V;
% Resistances: R1=4 Ohms; R2=2 Ohms;
% Find: Output voltage, V_out
% Find: Gain, G
clc; clear;
V_in=40; R1=4; R2=2;
V_out=(1+(R2/R1))*V_in;
G=V_out/V_in;
fprintf('Output voltage: %.2f V\n',V_out);
fprintf('Gain: %.2f\n',G);
```

Fig. 8.64 Code—Non-inverting amplifier

Fig. 8.65 Output—Non-inverting amplifier

Command Window

    Output voltage: 60.00 V
    Gain: 1.50

8.4.2.1   MATLAB Example 8.21: Non-inverting Amplifier

Consider a non-inverting amplifier as shown in Fig. 8.63 with 40 V input and $R_1 = 4\Omega$, $R_2 = 2\Omega$. Determine the gain and the output voltage of the amplifier.

The MATLAB code for this example is given in Fig. 8.64 with its output in Fig. 8.65.

Output

Fig. 8.66 Circuit diagram
of a follower circuit

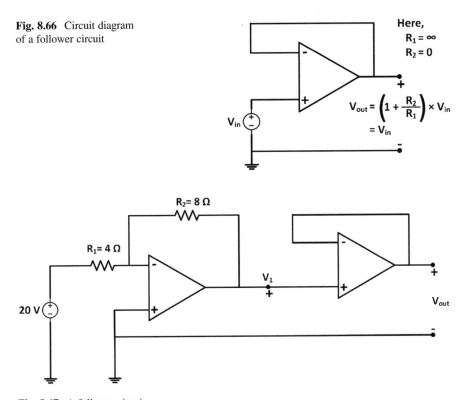

Here,

$R_1 = \infty$

$R_2 = 0$

$V_{out} = \left(1 + \dfrac{R_2}{R_1}\right) \times V_{in}$

$= V_{in}$

Fig. 8.67 A follower circuit

8.4.3 Follower Circuit

In a follower circuit, the gain is always unity. The feedback resistance is short-circuited, and the input impedance is open-circuited. Therefore, the input impedance of a follower circuit is infinity, or very high; and the output impedance is zero, or very small. A configuration of the follower circuit is given in Fig. 8.66.

8.4.3.1 MATLAB Example 8.22: Follower Circuit

Consider the circuit in Fig. 8.67, where the output of an inverting amplifier is connected as an input of a follower circuit. Determine the values of V_1 and V_{out} from the circuit.

The MATLAB code for this example is given in Fig. 8.68 with its output in Fig. 8.69.

```
% Follower circuit
% Input voltage: V_in=20 V;
% Resistances: R1=4 Ohms; R2=8 Ohms;
% Find: Output of the inverting amplifier, V1
% Find: Final output voltage, V_out
clc; clear;
V_in=20; R1=4; R2=8;
V1=-(R2/R1)*V_in;
fprintf('V1: %.2f V\n',V1);
V_out=V1;
fprintf('Final output voltage, V_out: %.2f V\n',V_out);
```

Fig. 8.68 Code—Follower circuit

Fig. 8.69 Output—
Follower circuit

Command Window
V1: -40.00 V
Final output voltage, V_out: -40.00 V

Output

8.4.4   Differentiator Circuit

A differentiator circuit is shown in Fig. 8.70. For a given signal input $V_{in}(t)$, the output of the Op-amp will be a differentiation of the input with the multiplication of resistance and capacitance value. As it is an inverting amplifier, the output will be inverted. The output of a differentiator circuit can be defined using the following formula:

$$V_{out}(t) = -RC\frac{dV_{in}}{dt} \tag{8.32}$$

The current through the resistance R and the capacitor C can be represented by the following formula:

$$I_R = \frac{V_{out}}{R} \tag{8.33}$$

$$I_C = C\frac{dV_{in}}{dt} \tag{8.34}$$

Fig. 8.70 Circuit diagram
of a differentiator circuit

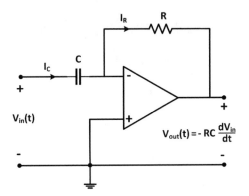

$$V_{out}(t) = -RC \frac{dV_{in}}{dt}$$

```
% Differentiator circuit
% R= 5 Ohms, C= 0.5 F
% Input signal, v(t)=2sin(t);
% Find: Output signal, v_out(t)
% Find: Output at t= 0.1 sec.
% FInd: I_R and I_C at t= 0.1 sec.
clc; clear;
R=5; C=0.5;
syms t
v= @(t) 2*sin(t);
v_out=@(t) -R*C*diff(v,t);
fprintf('The output signal:\n');
disp(v_out(t))
v_out= limit(v_out,t,0.1);
fprintf('The output voltage at t=0.1 sec: %.5f V\n',v_out);
I_R= -v_out/R;
I_C= limit(C*diff(v,t),t,0.1);
fprintf('\n');
fprintf('I_R at t=0.1 sec: %.5f A\n',I_R);
fprintf('\n');
fprintf('I_C at t=0.1 sec: %.5f A\n',I_C);
```

Fig. 8.71 Code—Differentiator circuit

8.4.4.1   MATLAB Example 8.23: Differentiator Circuit

Consider the circuit in Fig. 8.70, where $R = 5\Omega$ and $C = 0.5\ F$. If the input signal is $v(t) = 2 \sin (t)$, determine:

(a) The output signal, $v_{out}(t)$
(b) The output of the circuit at $t = 0.1$ s
(c) I_R and I_C at $t = 0.1$ s

The MATLAB code for this example is given in Fig. 8.71 with its output in Fig. 8.72.

```
Command Window
  The output signal:
  -5*cos(t)

  The output voltage at t=0.1 sec: -4.97502 V

  I_R at t=0.1 sec: 0.99500 A

  I_C at t=0.1 sec: 0.99500 A
```

Fig. 8.72 Output—Differentiator circuit

Output

8.4.5   Integrator Circuit

An integrator circuit is shown in Fig. 8.73. For a given signal input $V_{in}(t)$, the output of the Op-amp will be an integration of the input with the multiplication of some certain value. As it is an inverting amplifier, the output will be inverted. The output of a differentiator circuit can be defined using Eq. (8.35):

$$V_{out}(t) = -\frac{1}{RC} \int V_{in}(t)dt \qquad (8.35)$$

The current through the resistance R and the capacitor C can be represented by the following formula:

$$I_R = \frac{V_{out}}{R} \qquad (8.36)$$

$$I_C = C\frac{dV_{in}}{dt} \qquad (8.37)$$

8.4.5.1   MATLAB Example 8.24: Integrator Circuit

Consider the integrator circuit in Fig. 8.73, where $R = 5\Omega$ and $C = 0.5\,F$. If the input signal is $v(t) = -5\cos(t)$, determine:

Fig. 8.73 Circuit diagram
of an integrator circuit

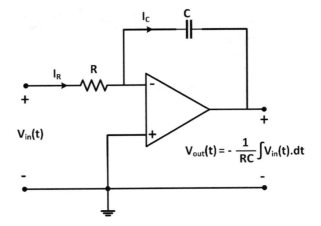

$$V_{out}(t) = -\frac{1}{RC}\int V_{in}(t).dt$$

```
% Integrator circuit
% R= 5 Ohms, C= 0.5 F
% Input signal, v(t)=-5cos(t);
% Find: Output signal, v_out(t)
% Find: Output at t= 0.1 sec.
% FInd: I_R and I_C at t= 0.1 sec.
clc; clear;
R=5; C=0.5;
syms t
v= @(t) -5*cos(t);
v_out=@(t) (-1/R*C)*int(v,t);
fprintf('The output signal:\n');
disp(v_out(t))
v_out= limit(v_out,t,0.1);
fprintf('The output voltage at t=0.1 sec: %.5f V\n',v_out);
I_R= -v_out/R;
I_C= limit(C*diff(v,t),t,0.1);
fprintf('\n');
fprintf('I_R at t=0.1 sec: %.5f A\n',I_R);
fprintf('\n');
fprintf('I_C at t=0.1 sec: %.5f A\n',I_C);
```

Fig. 8.74 Code—Integrator circuit

(a) The output signal, $v_{out}(t)$
(b) The output of the circuit at $t = 0.1$ s
(c) I_R and I_C at $t = 0.1$ s

The MATLAB code for this example is given in Fig. 8.74 with its output in
Fig. 8.75.

```
Command Window

  The output signal:
  sin(t)/2

  The output voltage at t=0.1 sec: 0.04992 V

  I_R at t=0.1 sec: -0.00998 A

  I_C at t=0.1 sec: 0.24958 A
```

Fig. 8.75 Output—Integrator circuit

Output

8.5   Transistor Circuit

A transistor is a semiconductor-based device that has two *pn* junctions. By sandwiching either a n-type with two p-types or a p-type with two n-types, a transistor is made. Therefore, a transistor has three terminals and three sections named emitter, base, and collector. The two types of transistors based on the *pn* junction, termed as $p - n - p$ transistor and $n - p - n$ transistors, are shown in Fig. 8.76.

In a transistor-based circuit, the connections can be made in three ways:

- Common emitter (CE) connection
- Common base (CB) connection
- Common collector (CC) connection

The circuit configurations of these three connection types are illustrated in Fig. 8.77.

In any transistor circuit, there is a relation of current among emitter, base, and collector currents that can be defined as follows:

$$I_E = I_B + I_C \tag{8.38}$$

Here, I_E, I_B, and I_C are the emitter current, base current, and collector current, respectively.

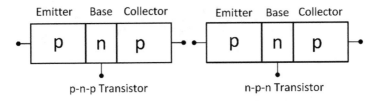

Fig. 8.76 p-n-p and n-p-n transistor

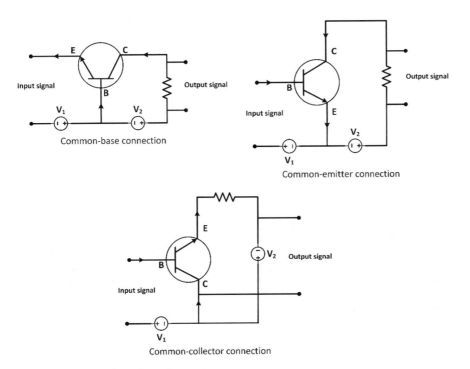

Fig. 8.77 Three configurations of a transistor

Two common terminologies used in transistor circuits are as follows:

Current amplification factor, α

The current amplification factor is the ratio of output current to input current, generally regarded as α. For a transistor with a common base connection, the current amplification factor can be written as:

$$\alpha = \frac{\text{Collector current}}{\text{Emitter current}} = \frac{I_C}{I_E} \tag{8.39}$$

Base current amplification, β

The base current amplification factor is the ratio of output current to base current, generally regarded as β. For a transistor with a common emitter connection, the current amplification factor can be written as:

$$\beta = \frac{\text{Collector current}}{\text{Base current}} = \frac{I_C}{I_B} \qquad (8.40)$$

8.5.1   MATLAB Example 8.25: Transistor Circuit

Consider a transistor with a common base connection. If the emitter current, $I_E = 10$ mA, and $\alpha = 0.8$, determine the collector current and the base current.

The MATLAB code for this example is given in Fig. 8.78 with its output in Fig. 8.79.

Output

8.6   Conclusion

In this chapter, the readers will be able to learn how to utilize MATLAB to solve different electrical circuit problems. The chapter describes some of the basic theorems of electrical circuits along with their implementations in MATLAB that are useful to solve different circuit problems. In DC circuit analysis, Ohm's law, Kirchhoff's theorem, Thevenin's theorem, and the maximum power transfer theorem are presented along with the formulas of voltage and current divider rules. The methods of determining equivalent resistance and delta-wye conversions are also

```
% Transistor problem
% Common base connection
% Emitter current, I_E= 10 mA; Alpha=0.8
% Determine: Collector current, I_C;
% Determine: Base current, I_B;
clc;clear;
I_E=10; Alpha=0.8;
I_C=Alpha*I_E;
I_B= I_E-I_C;
fprintf('Collector current: %.3f mA\n',I_C);
fprintf('Base current: %.3f mA\n',I_B);
```

Fig. 8.78 Code—Transistor

Fig. 8.79 Output—
Transistor

```
Command Window

  Collector current: 8.000 mA
  Base current: 2.000 mA
```

Fig. 8.80 A resistive
electrical circuit

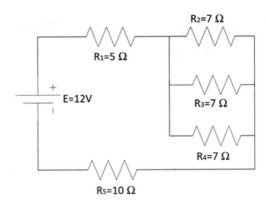

presented in this section. All of these topics are explained with necessary examples
that are shown by implementing in MATLAB individually. In AC circuit analysis,
definitions of some of the relevant terminologies, impedance, and power triangle are
covered. In addition, three-phase AC circuit analysis is also added in this section.
Due to the vast applications of Op-amp and transistor devices, both of their studies
are included in this chapter. In Op-amp, different categories and their applications
are demonstrated using MATLAB. However, in transistor circuits, this chapter
provides an overview of the different types of transistors and their structures with
limited details. For electrical engineers, the contents of this chapter will be very
essential to create fundamental knowledge on circuit analysis.

Exercise 8

1. Using MATLAB, determine the equivalent resistance of the following circuit in
   Fig. 8.80. Hence, verify Ohm's law if the voltage varies from 1 V to 12 V.
   The graph should look as shown in Fig. 8.81.

2. Consider the following circuit in Fig. 8.82, where $R_1 = 0.2$ ohm, $R_2 = 0.5$ ohm,
   $R_3 = 1$ ohm, $R_4 = 0.8$ ohm, and $R_5 = 1.44$ ohm.

   (a) Determine the equivalent resistance of the circuit using MATLAB.
   (b) Determine the current using MATLAB if the voltage is 6 V.

3. (a) Create a function *voltdiv()* which will calculate the divided voltages in the
   following circuit in Fig. 8.83 given the values of the resistances (R_1, R_2, and R_3)
   and voltage, V. Test the function with $R_1 = 2$ ohm, $R_2 = 4$ ohm, $R_3 = 8$ ohm, and
   $V = 24$ V.

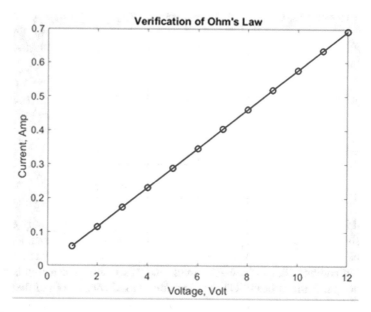

Fig. 8.81 Expected graphical output for Question 1

Fig. 8.82 A delta
connected resistive
electrical circuit

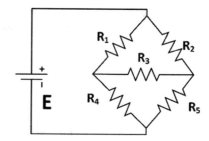

Fig. 8.83 An electrical
circuit with three resistors in
series

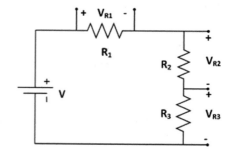

Fig. 8.84 An electrical
circuit with two resistors in
parallel

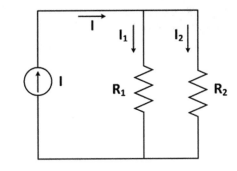

Fig. 8.85 An electrical
circuit with three resistors

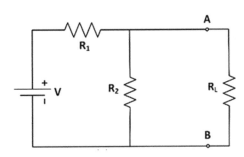

(b) Create a function *curdiv*() which will calculate the divided currents in the
following circuit in Fig. 8.84 given the values of the resistances (R_1 and R_2) and
current, I. Test the function with $R_1 = 2$ ohm, $R_2 = 4$ ohm, and $I = 8$A.

4. In the following circuit in Fig. 8.85, $R_1 = 4$ ohm, $R_2 = 9$ ohm, and load resistance,
   $R_L = 5$ ohm. The circuit is running at a voltage of 12 V. Using MATLAB,
   determine:

   (i) Thevenin's voltage, V_{th}
   (ii) Thevenin's equivalent resistance, R_{th}
   (iii) The current flowing through the load resistance, R_L
   (iv) From the calculated Thevenin's circuit, vary the load resistance starting
        from 1 to 20 ohms, and determine the output power for all scenarios to
        prove the maximum power transfer theorem; also determine the maximum
        output power.

5. Using MATLAB and Example 8.13 as a reference, determine the apparent power,
   S, and the power factor of a series RLC circuit with:

   (a) $P = 50$ W and $Q = 13$ Var
   (b) $P = 12$ W and $Q = 2.3$ Var

Fig. 8.86 An electrical
circuit with delta connection

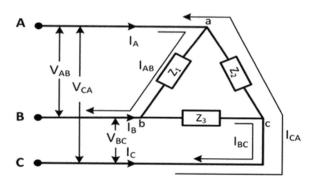

6. Consider the following delta-connected system as shown in Fig. 8.86.

(a) The following parameters for the system are given:

$$V_{AB} = 100\angle 0^\circ \ V$$

$$V_{BC} = 110\angle 120^\circ \ V$$

$$V_{CA} = 120\angle 240^\circ \ V$$

$$Z_1 = 8\angle 25^\circ$$

$$Z_2 = 14\angle 55^\circ$$

$$Z_3 = 18\angle -23^\circ$$

Determine if the system is balanced or unbalanced. Hence, calculate:
   (i) Phase currents I_{AB}, I_{BC}, and I_{CA}
   (ii) Line currents I_A, I_B, and I_C

(b) Now consider the following parameters for the same system:

$$V_{AB} = 100\angle 0^\circ \ V$$

$$V_{BC} = 110\angle 120^\circ \ V$$

$$V_{AB} = 120\angle 240^\circ \ V$$

$$Z_1 = Z_2 = Z_3 = Z = 5\angle 30^\circ$$

Determine if the system is balanced or unbalanced. Hence, calculate:

Fig. 8.87 An inverting amplifier

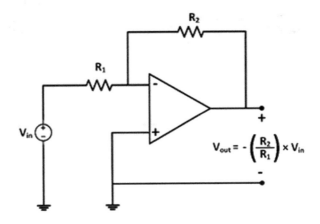

Fig. 8.88 A non-inverting amplifier

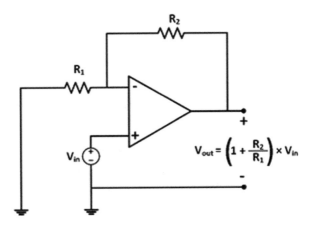

(i) Phase currents I_{AB}, I_{BC}, and I_{CA}.
(ii) Line currents I_A, I_B, and I_C

7. (a) Determine the gain and output voltage of the inverting amplifier as shown in Fig. 8.87, where $R_1 = 10$ ohm, $R_2 = 14$ ohm, and input voltage $V_{in} = 24$ V.

   (b) What is the gain and output voltage of the non-inverting amplifier if the diagram as shown in Fig. 8.88 has the same R_1, R_2, and input voltage of 5 ohm, 7 ohm, and 12 V, respectively?

8. (a) Design a differentiator circuit as shown in Fig. 8.89 using MATLAB, where you have an input of $v(t) = 6\cos^2(t)$, a resistor of 10 ohm, a capacitor of 0.5 Farads. What will be the output signal, $v_{out}(t)$ from your differentiator? What will be the output voltage, I_R, and I_C of the circuit at 0.1 s?

   (b) Design an integrator circuit as shown in Fig. 8.90 using MATLAB, where you have an input of $v(t) = \cos^2(t)/\sin(t)$, a resistor of 12 ohm, a capacitor of 0.2 Farads. What will be the output signal, $v_{out}(t)$ from your integrator? What will be the output voltage, I_R, and I_C of the circuit at 0.5 s?

Fig. 8.89 A differentiator circuit

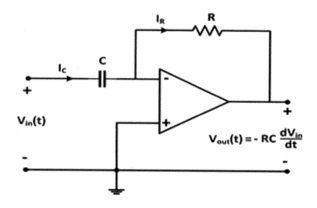

$$V_{out}(t) = - RC \frac{dV_{in}}{dt}$$

Fig. 8.90 An integrator circuit

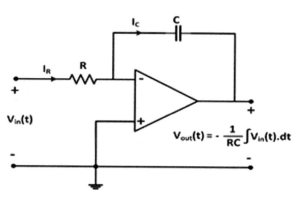

$$V_{out}(t) = - \frac{1}{RC} \int V_{in}(t).dt$$

Chapter 9
Control System and MATLAB

9.1 Introduction

In the engineering domain, the study of control systems is required to govern the behavior of any physical system. The behavior of a physical system can be regarded as the response or the output of that system. In control systems, one of the objectives is to regulate that response with respect to the input of that system. The response of a physical system can be represented mathematically either in the frequency domain or in the time domain. The concepts of both of these domains in control systems will be explained with practical illustration in this chapter using MATLAB. To implement the concepts of control systems, it is required to convert any physical system to a mathematical representation. The opposite can also be necessary for various aspects. In this chapter, the concept of state-space representation, which can be used to convert any physical system to a mathematical model, will be demonstrated with MATLAB implementation. In control system study, controllability, observability, and stability analysis hold special significance. Hence, all of these topics will be illustrated by means of MATLAB with proper theoretical guidance and practical examples.

9.2 Frequency Response Overview

Control systems are mostly associated with signals, because by understanding the signals and their response by the system, one can easily control different aspects of the complete system. These systems may be exposed to the signals of different frequencies, the response for which is important to comprehend and eventually control the system. The frequency response can therefore be stated as the output of a system to a waveform of a particular frequency. More specifically, the frequency response can provide more information on the amplitude response and phase response of a system.

A system in control system is commonly represented either in a time domain, an s-domain, or a frequency domain. In the s-domain, the system is represented by a new parameter called s, which is represented by a transfer function, and which can be derived from a time domain system through Laplace transform. A frequency domain, on the other hand, provides specific details on the amplitude and phase of the system. This can be determined from the s-domain by replacing s with $j\omega$, where ω is the frequency of the input signal. This section will discuss the formation and components of a transfer function, domain transformation using Laplace and inverse Laplace transformation, and analysis and operation on transfer function through partial fraction decomposition, giving insights to dive into more complex operations to understand the determination of frequency responses in control systems.

9.2.1   Linear Time-Invariant System

Linear time-invariant (LTI) systems refer to a certain group of systems with two distinctive characteristics—linearity and time invariance. Linearity is the characteristic that signifies that the output of the system is linearly related to the output. Consider Fig. 9.1 to understand the concept of linearity. The first output of the first is $y_1(t)$, and the input is $x_1(t)$, whereas the second output and the input of the system is $y_2(t)$ and $x_2(t)$ respectively. The linear characteristic of the system can be realized by observing the third inputs and outputs. Here, the given input is the combination of the previous two inputs—(a) and (b). Due to the linearity property, the output of the system changes linearly according to the changes made in the inputs, i.e., the output has also become the combination of the outputs of (a) and (b). Such linear characteristic is one of the characteristics of the LTI system.

Another characteristic of an LTI system is its time invariance, which signifies that if the input is applied at different timing, the output will not be dependent on that timing. Consider a system that produces $y(t)$ output for an input $x(t)$. If a time shift

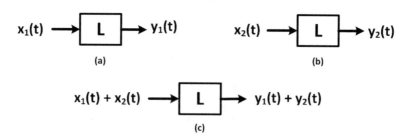

Fig. 9.1 Illustration of linearity

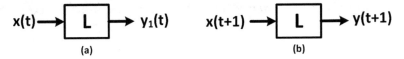

(a)   (b)

Fig. 9.2 Illustration of time invariance

Fig. 9.3 Illustration of
transfer function

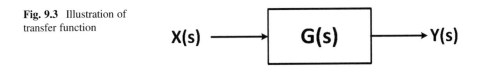

occurs in the input, such as $x(t + 1)$, the produced output will remain the same with the same time shift, i.e., $y(t + 1)$, as shown in Fig. 9.2.

The LTI system can also be defined in terms of a transfer function, which is an important characteristic of LTI systems, which will be discussed in later sections.

9.2.2   *Transfer Function*

Transfer function of a system can be defined as the ratio of the Laplace transform of output to the Laplace transform of input. Consider the following system in Fig. 9.3, where $Y(s)$ represents the Laplace transform of the output $y(t)$ and $X(s)$ is the Laplace transform of the input $x(t)$.

Hence, the transfer function of the system can be referred to as follows:

$$G(s) = \frac{Y(s)}{X(s)} \tag{9.1}$$

In MATLAB the transfer function can be created using $tf()$ function, where the input is a matrix representing the coefficients of the numerator and the denominator.

MATLAB command for transfer function:
$tf([Numerator], [Denominator])$

Here, in [*Numerator*] the coefficients of s need to be incorporated starting from the highest degree to the lowest degree as a row vector. For [*Denominator*], the same process will be repeated; however, the denominator represents the input and the numerator represents the output of the system both in the Laplace transformation forms.

An example is given below to make the above command more comprehensible.

9.2.2.1   MATLAB Example 9.1: Transfer Function

Create the following transfer function using MATLAB:

$$G(s) = \frac{s+50}{s^2+11s+12} \tag{9.2}$$

The MATLAB code for this example is given in Fig. 9.4 with its output in Fig. 9.5.

Output

The transfer function can also be defined as a function of s using MATLAB manually as below. The MATLAB code for this example is given in Fig. 9.6 with its output in Fig. 9.7.

Output

```
% Transfer function: (s+50)/(s^2+11s+12)
G=tf([1 50],[1 11 12]);
disp('Transfer function:')
G
```

Fig. 9.4 Code—Transfer function

Fig. 9.5 Output—Transfer function

```
Command Window

   Transfer function:

   G =

            s + 50
        ----------------
        s^2 + 11 s + 12

   Continuous-time transfer function.
```

```
% Transfer function
syms s
G=@(s)(s+50)/(s^2+11*s+12);
disp('Transfer function:')
disp(G(s))
```

Fig. 9.6 Code—Manually determining the transfer function

Fig. 9.7 Output—
Manually determining the
transfer function

```
Command Window

  Transfer function:
    (s + 50)/(s^2 + 11*s + 12)
```

9.2.3 Laplace Transform

Laplace transform is very essential in the study of the LTI system. The task of Laplace transformation is to convert any time domain input into a frequency domain, or s-domain output. One of the benefits of such conversion is that it can convert any differential equation into a simple algebraic equation in its frequency domain. Hence, the calculation becomes easier as the rules of algebra become applicable to such an equation. We already discussed that the transfer function of an LTI system is the ratio of the Laplace transform of both output and input. Therefore, the concept of Laplace transform needs further emphasis. In MATLAB, the Laplace transform of any time domain equation can be converted into an s-domain equation by using the following command:

> **MATLAB command for Laplace transform of g:**
> *laplace(g)*

9.2.3.1 MATLAB Example 9.2: Laplace Transform

Consider the following function for performing Laplace transform using MATLAB:

$$g(t) = e^{3t} \sin(6t) \tag{9.3}$$

The MATLAB code for this example is given in Fig. 9.8 with its output in Fig. 9.9.

```
% Laplace Transform
clc;clear;
syms t s
g=@(t) exp(3*t)*sin(6*t);
disp('Laplace transform:')
G(s)=laplace(g(t))
```

Fig. 9.8 Code—Laplace transform

Fig. 9.9 Output—Laplace transform

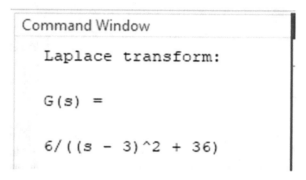

Command Window

Laplace transform:

G(s) =

6/((s - 3)^2 + 36)

Table 9.1 Differential terms and the corresponding Laplace transforms

Differential term	Laplace transform
y'	$L\{y'\} = sY(s) - y(0)$
y''	$L\{y''\} = s^2Y(s) - sy(0) - y'(0)$
y'''	$L\{y'''\} = s^3Y(s) - s^2y(0) - sy'(0) - y''(0)$

Output

Laplace transform can also be utilized for solving initial value problems involving differential equations. The Laplace transform of derivative terms can be determined using the following formula:

$$y^n(t) = s^n Y(s) - s^{n-1}y(0) - s^{n-2}y'(0) - s^{n-3}y''(0)\ldots\ldots - y^{n-1}(0) \qquad (9.4)$$

where $n = 1, 2, 3, \ldots\ldots$ Here, y^n represents the nth derivative of $y(t)$. Using the above formula, the Laplace transform of the first, second, and third derivative terms are listed in Table 9.1, as these are the most commonly used terms in the initial value problems.

9.2.3.2   MATLAB Example 9.3: Laplace Transform of Initial Value Problem with Differential Equation

Solve the following initial value problem using Laplace transform:

$$3y'''(t) + 2y''(t) + 3y(t) = 1; y(0) = y'(0) = 0; y''(0) = 1 \qquad (9.5)$$

The MATLAB code for this example is given in Fig. 9.10 with its output in Fig. 9.11.

Output

9.2.4   Inverse Laplace Transform

To convert the frequency domain output back into its original time domain input, inverse Laplace transform is required. The command for inverse Laplace transform in MATLAB is as follows:

```
%% Differential equation solve using Laplace transform
% 3*y'''(t) + 2*y''(t) + 3y(t) = 1
% Initial condition: y(0)=0;y'(0)=0;y''(0)=1;
clc;clear;
syms s Y
% Initial conditions
y0=0;dy0=0;dy20=1;
Y1=@(s) s*Y-y0;
Y2=@(s) s^2*Y-s*y0-dy0;
Y3=@(s) s^3*Y-s^2*y0-s*dy0-dy20;
% Differential equation
eqn=3*Y3(s)+2*Y2(s)+3*Y-laplace(1,s);
solve(eqn,Y)
```

Fig. 9.10   Code—Laplace transform of initial value problem with differential equation

Fig. 9.11   Output—Laplace transform of initial value problem with differential equation

```
Command Window

    ans =

    (1/s + 3)/(3*s^3 + 2*s^2 + 3)
```

MATLAB command for inverse Laplace transform of *G*:
ilaplace(G)

9.2.4.1   MATLAB Example 9.4: Inverse Laplace Transform

Consider the following function for performing inverse Laplace transform using MATLAB:

$$G(s) = \frac{6}{(s-3)^2 + 36} \tag{9.6}$$

The MATLAB code for this example is given in Fig. 9.12 with its output in Fig. 9.13.

Output

```
% Inverse laplace transform
clc;clear;
syms t s
G=@(s) 6/((s-3)^2+36);
disp('Inverse laplace transform:')
g(t) = ilaplace(G(s))
```

Fig. 9.12   Code—Inverse Laplace transform

Fig. 9.13   Output—Inverse Laplace transform

```
Command Window

   Inverse laplace transform:

   g(t) =

   sin(6*t)*exp(3*t)
```

9.2.5 *Partial Fraction*

Partial fraction is also an important concept that is required while performing Laplace, or inverse Laplace transform manually. It also appears in other relevant problems. Partial fraction decomposition is a method through which any rational fraction can be broken down in terms of simpler fractions to make the computations easier. This technique facilitates greatly while performing inverse Laplace of rational fractions. Before introducing the MATLAB command for performing partial fraction decomposition, let us have a look at the generalized format of a rational fraction and its partial fraction expansion as stated below:

$$F(s) = \frac{N(s)}{D(s)} = \frac{N_n.s^n + N_{n-1}s^{n-1} + \ldots + N_1.s + N_0}{D_m.s^m + D_{m-1}s^{m-1} + \ldots + D_1.s + D_0} \quad (9.7)$$

$$\text{Partial fraction expansion} = \frac{r_m}{s - p_m} + \frac{r_{m-1}}{s - p_{m-1}} + \ldots + \frac{r_1}{s - p_1} + k(s) \quad (9.8)$$

Here, $N(s)$ and $D(s)$ represent the numerator and denominator terms. The partial fraction expansion of the fraction $F(s)$ is also shown above.

In Eqs. (9.7)–(9.8), $N = [N_n\, N_{n-1}\ldots N_0]$ and $D = [D_n\, D_{n-1}\ldots D_0]$ indicate two row vectors containing the coefficients of s for numerator and denominator, respectively. These two vectors are the input of MATLAB for determining the residuals, poles, and coefficients of polynomials of the partial fraction expansion. Hence, MATLAB provides three row vectors, $r = [r_m\, r_{m-1}\ldots r_1]$; $p = [p_m\, p_{m-1}\ldots p_0]$; $k = [k_m\, k_{m-1}\ldots k_0]$, which represent the residuals, poles, and coefficients of polynomials, respectively. By incorporating the values of these three vectors, the partial fraction expansion can be derived as shown in the above equation.

The MATLAB command for partial fraction expansion is given below:

> **MATLAB command for partial fraction expansion:**
> $[r, p, k] = residue(N, D)$

Here, the input and output parameters have been kept the same as the discussion provided earlier for better comprehension.

A rational fraction can be proper or improper based on which the steps for determining partial fraction decompositions vary. The conditions of the proper and improper rational fraction are provided in Table 9.2.

For the third case, when the highest degree for both the numerator and the denominator becomes equal, the fraction can either be proper or improper based on certain conditions. To illustrate the conditions clearly, let's consider the following

Table 9.2 The conditions of the proper and improper rational fraction

Case	Decision
Number of the highest degree in the numerator < number of the highest degree in the denominator	Proper rational fraction
Number of the highest degree in the numerator > number of the highest degree in the denominator	Improper rational fraction
Number of the highest degree in the numerator = number of the highest degree in the denominator	Proper or improper

rational fraction $F(s)$, where the highest degree for both numerator and denominator is n:

$$F(s) = \frac{a_1 s^n + a_2 s^{n-1} + \ldots + a_{n-1} s + a_n s^0}{b_1 s^n + b_2 s^{n-1} + \ldots + b_{n-1} s + b_n s^0} \tag{9.9}$$

For the above-generalized fraction, the recognition of proper and improper fraction can be made by following the steps incorporated in Tables 9.3. Two examples are provided in Table 9.4.

9.2.5.1   MATLAB Example 9.5: Partial Fraction Expansion

Determine the partial fraction expansion of the following proper rational fraction, where the highest degree of the denominator is greater than the highest degree of numerator:

$$\frac{2s + 3}{s^2 + 2s} \tag{9.10}$$

The MATLAB code for this example is given in Fig. 9.14 with its output in Fig. 9.15.

Output

Table 9.3 Conditions for being proper and improper fraction when the highest degree for both in the numerator and the denominator is equal

Step 1:	Step 2:	$\cdots$	Step $(n-1)$:	Step n:
$If\,(a_1 > b_1)$	$If\,(a_2 > b_2)$	$\cdots$	$If\,(a_{n-1} > b_{n-1})$	$If\,(a_n > b_n)$
$Decision : Improper$	$Decision : Improper$	$\cdots$	$Decision : Improper$	$Decision : Improper$
$elseif\,(a_1 < b_1)$	$elseif\,(a_2 < b_2)$		$elseif\,(a_{n-1} < b_{n-1})$	$elseif\,(a_n < b_n)$
$Decision : Proper$	$Decision : Proper$		$Decision : Proper$	$Decision : Proper$
$else$	$else$		$else$	$else$
$Go \rightarrow Step\ 2$	$Go \rightarrow Step\ 3$		$Go \rightarrow Step\ n$	$Decision : Improper$
end	end		end	$(considering\ 1\ as$
				$Improper\ fraction)$
				end

Table 9.4 Examples of fractions

$\frac{2s^2+4s+2}{2s^2+3s+2}$	Step 1: 2 = 2
	Step 2: 4 > 3
	Decision: Improper fraction
$\frac{2s^4+3s^3}{2s^4+3s^3+4s^2+3s+1}$	Step 1: 2 = 2
	Step 2: 3 = 3
	Step 3: 0 < 4
	Decision: Proper fraction

```
%% Partial fraction
% Fraction: (2s+3)/(s^2+2s)
% Highest degree of Numerator < Highest degree of denominator
clc;clear;
syms s
N = [2 3];
D = [1 2 0];
disp('The residuals:')
[r,p,k] = residue(N,D)
Expan=@(s) r(1)/(s-p(1)) + r(2)/(s-p(2));
disp('The partial fraction expansion:')
disp(Expan(s))
```

Fig. 9.14 Code—Partial fraction

9.2.5.2   MATLAB Example 9.6: Partial Fraction Expansion

Determine the partial fraction expansion of the following improper rational fraction, where the highest degree of the denominator is equal to the highest degree of numerator:

$$\frac{2s^2 + 4s + 1}{s^2 + 2s} \tag{9.11}$$

The MATLAB code for this example is given in Fig. 9.16 with its output in Fig. 9.17.

Output

9.2.5.3   MATLAB Example 9.7: Partial Fraction Expansion

Determine the partial fraction expansion of the following proper rational fraction, where the highest degree of the denominator is equal to the highest degree of numerator:

```
Command Window

  The residuals:

  r =

          0.5000
          1.5000


  p =

          -2
           0


  k =

          []


  The partial fraction expansion:
  1/(2*(s + 2)) + 3/(2*s)
```

Fig. 9.15 Output—Partial fraction

```
%% Partial fraction-2
% Fraction:(2s^2+4s+1)/(s^2+2s)
% Highest degree of Numerator = Highest degree of denominator
clc;clear;
syms s
N = [2 4 1];
D = [1 2 0];
disp('The residuals:')
[r,p,k] = residue(N,D)
Expansion=@(s) r(1)/(s-p(1)) + r(2)/(s-p(2)) + k;
disp('The partial fraction expansion:')
disp(Expansion(s))
```

Fig. 9.16 Code—Partial fraction expansion 1

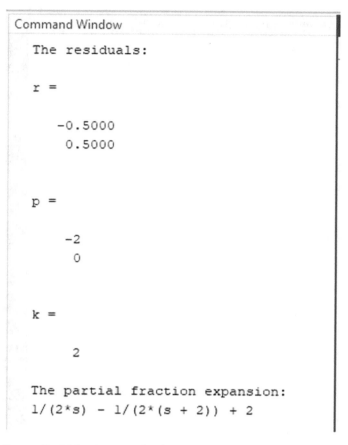

Fig. 9.17 Output—Partial fraction expansion 1

$$\frac{2s^2 + 2s + 1}{2s^2 + 4s + 3} \tag{9.12}$$

The MATLAB code for this example is given in Fig. 9.18 with its output in Fig. 9.19.

Output

```
%% Partial fraction-3
% Fraction:(2s^2+2s+1)/(2s^2+4s+3)
% Highest degree of Numerator = Highest degree of denominator
clc;clear;
syms s
N = [2 2 1];
D = [2 4 3];
disp('The residuals:')
[r,p,k] = residue(N,D)
Expansion=@(s) r(1)/(s-p(1)) + r(2)/(s-p(2)) + k;
disp('The partial fraction expansion:')
disp(Expansion(s))
```

Fig. 9.18 Code—Partial fraction expansion 2

```
Command Window
  The residuals:

  r =

      -0.5000
      -0.5000


  p =

    -1.0000 + 0.7071i
    -1.0000 - 0.7071i


  k =

      1


  The partial fraction expansion:
  1 - 1/(2*(s + (2^(1/2)*1i)/2 + 1)) - 1/(2*(s - (2^(1/2)*1i)/2 + 1))
```

Fig. 9.19 Output—Partial fraction expansion 2

9.2.5.4 MATLAB Example 9.8: Partial Fraction Expansion

Determine the partial fraction expansion of the following improper rational fraction, where the highest degree of denominator is less than the highest degree of numerator:

```
%% Partial fraction-4
% Fraction: (2s^3+4s^2+3s+2)/(s^2+2s+1)
% Highest degree of Numerator > Highest degree of denominator
clc;clear;
syms s
N = [2 4 3 3];
D = [1 2 1];
disp('The residuals:')
[r,p,k] = residue(N,D)
Expansion=@(s) r(1)/(s-p(1)) + r(2)/(s-p(2)) + (s*k(1)+k(2));
disp('The partial fraction expansion:')
disp(Expansion(s))
```

Fig. 9.20 Code—Partial fraction expansion 3

```
Command Window

  The residuals:

  r =

        1
        2


  p =

       -1
       -1


  k =

     2    0

  The partial fraction expansion:
  2*s + 3/(s + 1)
```

Fig. 9.21 Output—Partial fraction expansion 3

$$\frac{2s^3 + 4s^2 + 3s + 2}{s^2 + 2s + 1} \tag{9.13}$$

The MATLAB code for this example is given in Fig. 9.20 with its output in Fig. 9.21.

Output

9.2.6 DC Gain

In general, the value of a transfer function signifies gain, which is termed as AC gain due to the existence of the frequency term. When that frequency component becomes zero, the AC gain can be referred to as DC gain.

To be more precise, DC gain can be defined as the ratio of the steady-state step output or response to state input. It can also be regarded as the value of the transfer function solved at $s = 0$. Mathematically, it can be represented using the following formula:

$$\text{DC gain} = G(s). \tag{9.14}$$

Here, $G(s)$ represents the transfer function of a system. An example to determine the value of DC gain using MATLAB is shown below.

9.2.6.1 MATLAB Example 9.9: DC Gain

Determine the DC gain of the following transfer function:

```
% DC gain
% Transfer eqn: 20/(s^2+10*s+11)
clc;clear;
syms s
G=@(s) 20/(s^2+10*s+11);
DC_gain=limit(G(s),s,0);
fprintf('DC gain: %f\n',DC_gain)

DC gain: 1.818182
```

Fig. 9.22 DC gain example

$$G(s) = \frac{20}{s^2 + 10s + 11} \tag{9.15}$$

The MATLAB code for this example with its output is given in Fig. 9.22.

9.2.7 Initial Value and Final Value Theorem

The initial value theorem is used to determine the value of a time domain function, $g(t)$, at $t = 0$ given the Laplace transform of that function. On the other hand, the final value theorem helps to determine the final value of the function at $t = \infty$. Both of these theorems are regarded together as the limiting theorem.

The initial and the final value theorem can be written as follows:

$$\textbf{Initial value theorem} : g(t) = \lim_{s \to \infty} sG(s) \tag{9.16}$$

$$\textbf{Final value theorem} : g(t) = \lim_{s \to 0} sG(s) \tag{9.17}$$

Here, $G(s)$ is the Laplace transform of the time domain function $g(t)$.

9.2.7.1 MATLAB Example 9.10: Initial and Final Value Theorem

Consider the following transfer function:

$$G(s) = \frac{2 + 6s + 2s^2}{2s(s + 2)^2} \tag{9.18}$$

```
%% Intial value problem
% Transfer eqn: (2 + 6*s + 2*s^2)/(2*s*(s+2)^2)
clc;clear;
syms s
G=@(s) (2 + 6*s + 2*s^2)/(2*s*(s+2)^2);
Initial_val=limit(s*G(s),s,Inf);
fprintf('Inital value: %.3f\n',Initial_val);

Inital value: 1.000
```

Fig. 9.23 Initial value problem

```
%% Final value problem
% Transfer function: (2 + 6*s + 2*s^2)/(2*s*(s+2)^2)
clc;clear;
syms s
G=@(s) (2 + 6*s + 2*s^2)/(2*s*(s+2)^2);
Final_val=limit(s*G(s),s,0);
fprintf('Final value: %.3f\n',Final_val);

Final value: 0.250
```

Fig. 9.24 Final value problem

Determine the initial and final value of $g(t)$ by using the initial and final value theorem in MATLAB. Here, $G(s)$ is the Laplace transform $g(t)$.

The MATLAB code and output for this example are given in Fig. 9.23 and Fig. 9.24.

9.2.8 Poles/Zeros

The poles are the roots of the denominator of the transfer function of a system. On the other hand, zeros are the roots of the numerator of the transfer function of a system. In MATLAB, the commands for determining poles and zeros of a system from its transfer function are listed below:

MATLAB command for determining poles from transfer function, G:
  $pole(G)$
MATLAB command for determining zeros from transfer function, G:
  $zero(G)$
MATLAB command for pole-zero mapping from transfer function, G:
  $pzmap(G)$

9.2.8.1 MATLAB Example 9.11: Poles and Zeros

Consider the following transfer function to determine the poles and zeros:

```
%% Poles/zeros
% Transfer function: (s+50)/(s^2+11s+12)
clc;clear;
G=tf([1 50],[1 11 12]);
disp('Transfer function:')
G
poles=pole(G)
zeros=zero(G)
% Pole-zero map
pzmap(G)
grid on
```

Fig. 9.25 Code—Poles and zeros

```
Command Window

    Transfer function:

    G =

            s + 50
        ---------------
        s^2 + 11 s + 12

    Continuous-time transfer function.


    poles =

        -9.7720
        -1.2280


    zeros =

        -50
```

Fig. 9.26 Output—Poles and zeros

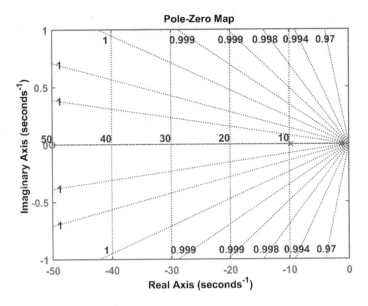

Fig. 9.27 Graphical output—Pole zero map

$$G(s) = \frac{s + 50}{s^2 + 11s + 12} \qquad (9.19)$$

The MATLAB code for this example is given in Fig. 9.25 with its output in Figs. 9.26 and 9.27.

Output

9.2.9   Laplace Transform in Electrical Circuit

The concept of Laplace transform can be utilized for electrical circuit analysis. We can convert the ratio of output to input into Laplace transformation to determine its transfer function. Later, from the transfer function, we can perform frequency domain analysis. Consider the RLC circuit drawn in Fig. 9.28 to determine the following aspects using MATLAB:

(a) Transfer function
(b) Poles and zeros
(c) DC gain
(d) Initial and final value

Fig. 9.28 An RLC circuit

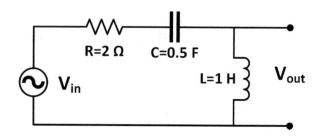

(a) When converting from the time domain to the frequency domain, a capacitive element is represented as $\frac{1}{sC}$ and an inductive element is represented as sL, while the resistive element stays the same.

Transfer function of the circuit:

$$G(s) = \frac{V_{\text{out}}}{V_{\text{in}}} = \frac{sL}{R + sL + \frac{1}{sC}} = \frac{s}{2 + s + \frac{1}{0.5s}} = \frac{s^2}{s^2 + 2s + 2}$$

(b) Poles and zeros:

Poles = The roots of the denominator $s^2 + 2s + 2$

$$\text{Poles} = \frac{-2 \pm \sqrt{2^2 - 4 \times 1 \times 2}}{2 \times 1} = -1 \pm i$$

Zeros = The roots of the numerator s^2

$$\text{Zeros} = 0, 0$$

$$\text{DC gain : DC gain} = G(s) = \left(\frac{s^2}{s^2 + 2s + 2}\right) = 0$$

(c) Initial and final value:

$$\text{Initial value theorem : } g(t) = sG(s) = \left(s \cdot \frac{s^2}{s^2 + 2s + 2}\right) = \infty$$

$$\text{Final value theorem : } g(t) = sG(s) = \left(s \cdot \frac{s^2}{s^2 + 2s + 2}\right) = 0$$

MATLAB Implementation

The MATLAB code for this example is given in Fig. 9.29 with its output in Fig. 9.30.

Output

9.3 Time Response Overview

Time response signifies the output of a time domain function that characterizes any dynamic system. To determine the time response of a system, the nature of the input and the mathematical model of the system need to be known.

The time response of a system can be categorized into two components—transient response and steady-state response.

Transient response is the early part of the time response which exists only for a short period of time, and approaches toward zero as time continues to proceed. Conversely, a steady-state response is the stable response of the system that happens right after the transient response dries out.

```
%% Transfer function: s^2/(s^2+2*s+2)
clc;clear;
disp('Transfer function:')
G=tf([1 0 0],[1 2 2])
% Poles/zeros
poles=pole(G)
zeros=zero(G)
% Pole-zero map
pzmap(G)
grid on
% DC gain
syms s
G=@(s) s^2/(s^2+2*s+2);
DC_gain=limit(G(s),s,0);
fprintf('DC gain: %f\n',DC_gain)
% Intial value
Initial_val=limit(s*G(s),s,Inf);
fprintf('Inital value: %.3f\n',Initial_val);
% Final value
Final_val=limit(s*G(s),s,0);
fprintf('Final value: %.3f\n',Final_val);
```

Fig. 9.29 Code—Laplace transform in electrical circuit

```
Command Window

  Transfer function:

  G =

          s^2
     ---------------
     s^2 + 2 s + 2

  Continuous-time transfer function.


  poles =

      -1.0000 + 1.0000i
      -1.0000 - 1.0000i


  zeros =

          0
          0


  DC gain: 0.000000
  Inital value: Inf
  Final value: 0.000
```

Fig. 9.30 Output—Laplace transform in electrical circuit

Table 9.5 Some input signals and their functions in time and frequency domain

Input signal	Time domain function	Frequency domain function
Unit step function	$R(t) = 1; t \geq 0$	$L\{R(t)\} = \frac{1}{s}$
Unit impulse function	$\delta_n(t) = \frac{1}{n}; 0 \leq t \leq n$	$L\{\delta(t)\} = 1$
Unit ramp function	$R(t) = t; t \geq 0$	$L\{R(t)\} = \frac{1}{s^2}$
Sinusoidal function	$\sin \omega t = Img[e^{j\omega t}]$	$L\{\sin \omega t\} = \frac{\omega}{s^2 + \omega^2}$
	$\cos \omega t = Real[e^{j\omega t}]$	$L\{\cos \omega t\} = \frac{s}{s^2 + \omega^2}$

Before discussing more about time response, it is required to first illustrate some of the basic input signals that are used in different control systems. In Table 9.5, some of the input signal functions with their time and frequency domain characteristics are listed.

9.3.1 First-Order System

The general representation of the transfer function of a first-order system can be generalized using the following formula:

$$G(s) = \frac{K}{1 + \tau s} \tag{9.20}$$

Here, $G(s)$ represents the transfer function of the first-order system, K signifies the DC gain, and τ depicts the time constant.

Consider the following transfer function of a first-order system:

$$G(s) = \frac{8}{2 + 5s} \tag{9.21}$$

Determine the following parameters:

(a) DC gain
(b) Time constant

Solution

$$G(s) = \frac{K}{1 + \tau s} = \frac{8}{2 + 5s} = \frac{4}{1 + \frac{5}{2}s}$$

Hence, DC gain, $K = 4$; time constant, $\tau = \frac{5}{2}$.

9.3.1.1 Specific Characteristics of First-Order Systems

Rise Time The time required by a signal to reach 90% of its final value starting from 10% is regarded as the rise time. Consider the following step response of a first-order system:

$$y(t) = K\left(1 - e^{\frac{-t}{\tau}}\right), \tag{9.22}$$

where K is the DC gain and τ is the time constant.

The time required to reach 10% of the final value can be determined by the following formula:

$$t_{10\%} = -\tau \ln(0.1) \tag{9.23}$$

The time required to reach 90% of the final value can be determined by the following formula:

$$t_{90\%} = -\tau \ \ln(0.9) \tag{9.24}$$

Therefore, the rise time of a first-order system can be determined by using the following formula:

$$\text{Rise time}, T_R = t_{90\%} - t_{10\%} = \tau \ \ln(9) \approx 2.2 \ \tau \tag{9.25}$$

Settling Time

The time required for a signal to reach and remain steady within 2%–5% of its final value is regarded as the settling time.

The formula for determining settling time for 2% criteria can be written as:

$$\text{Maximum settling time}, T_{S\_max} = -\tau \ln\ln(0.02) \approx 4 \ \tau \tag{9.26}$$

The formula for determining settling time for 5% criteria can be written as:

$$\text{Minimum settling time}, T_{S\_min} = -\tau \ln\ln(0.05) \approx 3 \ \tau \tag{9.27}$$

Delay Time   Delay time can be defined as the required time for a response to reach 50% of its final value during the first half cycle of the waveform.

The formula for determining delay time can be represented as follows:

$$\text{Delay time}, T_D = -\tau \ln\ln(0.5) \approx 0.7 \ \tau \tag{9.28}$$

9.3.2 Second-Order System

The general format of the transfer function of a second-order system can be represented using Eq. (9.29):

$$G(s) = \frac{K\omega_n^2}{s^2 + 2\zeta\omega_n s + \omega_n^2} \tag{9.29}$$

Here, K is the DC gain; ω_n is the natural frequency; and ζ represents the damping ratio of the second-order system.

The roots of the denominator of the transfer function indicate the poles of the system, based on which the stability of a system can be determined. Hence, the poles of a second-order system can be determined using the formula in Eq. (9.30):

$$\text{poles} = \frac{-2\zeta\omega_n \pm \sqrt{4\zeta^2\omega_n{}^2 - 4 \times 1 \times \omega_n{}^2}}{2 \times 1} = -\zeta\omega_n \pm \omega_n\sqrt{\zeta^2 - 1} \qquad (9.30)$$

The damping ratio of a second-order system, ζ, creates an opportunity to classify different systems based on damping, which represents the oscillation characteristics of a system.

9.3.2.1 Specific Characteristics of Second-Order Systems

Delay Time The definition of delay time is the same as mentioned earlier for a first-order system. However, the delay time of a second-order system is quite different from a first-order system, and is represented by a different formula in Eq. (9.31):

$$\text{Delay time, } T_D = \frac{0.7\zeta + 1}{\omega_n\sqrt{1 - \zeta^2}} \qquad (9.31)$$

Rise Time Rise time of the response of a second-order system can be defined as the required time of the response to reach from its 10% final value to 90% final value during the first cycle of the response. It is to be noted that the definition works when the damping ratio is greater than 1. The formula of the rise time calculation for a second-order system can be written as in Eq. (9.32):

$$\text{Rise time, } T_R = \frac{\pi - \frac{\sqrt{1-\zeta^2}}{\zeta}}{\omega_n\sqrt{1 - \zeta^2}} \qquad (9.32)$$

Peak Time The time required for the response of a second-order system to reach its peak or maximum value during the first cycle is called the peak time. The peak time of a second-order system can be determined by using Eq. (9.33):

$$\text{Peak time, } T_P = \frac{\pi}{\omega_n\sqrt{1 - \zeta^2}} \qquad (9.33)$$

Settling Time The definition of settling time is the same as before mentioned in the first-order system. The formula for determining settling time for the response of a second-order system can be represented by Eqs. (9.34)–(9.35):

$$\text{Maximum settling time, } T_{S\_max} = \frac{4}{\zeta\omega_n} \qquad (9.34)$$

$$\text{Minimum settling time, } T_{S\_min} = \frac{3}{\zeta\omega_n} \qquad (9.35)$$

Percent of Overshoot Percent of overshoot of a second-order system can be determined mathematically by using Eq. (9.36):

$$\%\text{Overshoot} = 100e^{\frac{\zeta\pi}{\sqrt{1-\zeta^2}}} \tag{9.36}$$

9.3.3 Impact of Damping Ratio

The value of damping ratio categorizes second-order systems in four different cases, which are named as below:

(i) Overdamped system
(ii) Critically damped system
(iii) Underdamped system
(iv) Negative damped system

9.3.3.1 Overdamped System

When the damping ratio of a system is greater than 1, i.e., $\zeta > 1$, the system is regarded as an overdamped system. It occurs when the poles of the system are real, unequal, and negative.

9.3.3.2 MATLAB Example 9.12: Overdamped System

Consider the following second-order system:

$$G(s) = \frac{K\omega_n^2}{s^2 + 2\zeta\omega_n s + \omega_n^2} = \frac{50}{s^2 + 15s + 25} \tag{9.37}$$

Here, $K = 2$, $\zeta = 1.5$, and $\omega_n = 5$. The response of the above-mentioned overdamped system is produced using MATLAB in the example below. The MATLAB code for this example is given in Fig. 9.31 with its output in Figs. 9.32, 9.33, and 9.34.

Output

```
%% Overdamped system : zeta=1.5
clc;clear;
K=2;
omega_n=5;
zeta=1.5;
s=tf('s');
disp('Transfer function:')
G=(K*omega_n^2)/(s^2+2*zeta*omega_n*s+omega_n^2)
step(G);
grid on;
ylim([0 2.5]);
disp('Parameters:')
disp(stepinfo(G))
% Pole-zero map
figure(2);
pzmap(G)
grid on
```

Fig. 9.31   Code—Overdamped system

```
Command Window
    Transfer function:

    G =

              50
        ---------------
        s^2 + 15 s + 25

    Continuous-time transfer function.

    Parameters:
              RiseTime: 1.1717
           SettlingTime: 2.1309
            SettlingMin: 1.8024
            SettlingMax: 1.9999
              Overshoot: 0
             Undershoot: 0
                   Peak: 1.9999
               PeakTime: 5.1997
```

Fig. 9.32   Output—Overdamped system

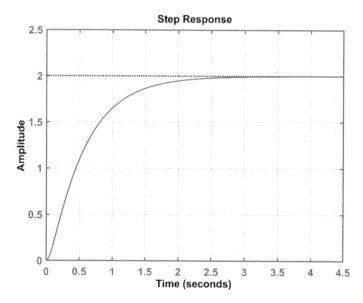

Fig. 9.33 Graphical output—Step response of overdamped system

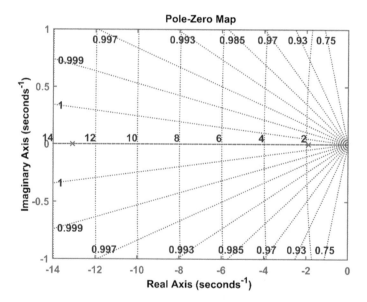

Fig. 9.34 Graphical output—Pole zero map of overdamped system

9.3.3.3   Critically Damped System

Critically damped system represents those systems that have a damping ratio, $\zeta = 1$. For such systems, the poles are real, equal, and negative.

9.3.3.4   MATLAB Example 9.13: Critically Damped System

Consider the following second-order system:

$$G(s) = \frac{K\omega_n^2}{s^2 + 2\zeta\omega_n s + \omega_n^2} = \frac{50}{s^2 + 10s + 25} \tag{9.38}$$

Here, $K = 2$, $\zeta = 1$, and $\omega_n = 5$. The response of the above-mentioned critically damped system is produced using MATLAB in the example below. The MATLAB code for this example is given in Fig. 9.35 with its output in Figs. 9.36, 9.37, and 9.38.

Output

9.3.3.5   Underdamped System

Underdamped system represents those systems that have a damping ratio, $0 < \zeta < 1$. For such systems, the poles are complex numbers with negative real parts.

9.3.3.6   MATLAB Example 9.14: Underdamped System

Consider the following second-order system:

```
%% Critically damped system : zeta=1
clc;clear;
K=2;
omega_n=5;
zeta=1;
s=tf('s');
disp('Transfer function:')
G=(K*omega_n^2)/(s^2+2*zeta*omega_n*s+omega_n^2)
step(G);
grid on;
ylim([0 2.5]);
disp('Parameters:')
disp(stepinfo(G))
% Pole-zero map
figure(2);
pzmap(G)
grid on
```

Fig. 9.35 Code—Critically damped system

```
Command Window
  Transfer function:

  G =

              50
       ----------------
       s^2 + 10 s + 25

  Continuous-time transfer function.

  Parameters:
            RiseTime: 0.6717
        SettlingTime: 1.1668
         SettlingMin: 1.8016
         SettlingMax: 1.9998
           Overshoot: 0
          Undershoot: 0
                Peak: 1.9998
            PeakTime: 2.3900
```

Fig. 9.36 Output—Critically damped system

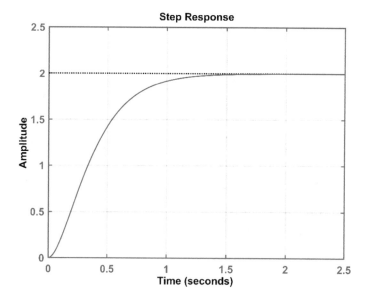

Fig. 9.37 Graphical output—Step response of critically damped system

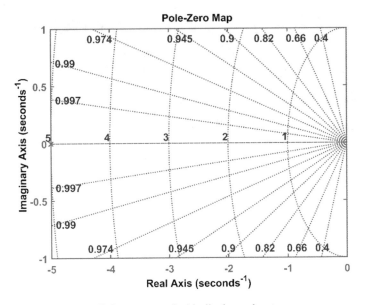

Fig. 9.38 Graphical output—Pole zero map of critically damped system

$$G(s) = \frac{K\omega_n^2}{s^2 + 2\zeta\omega_n s + \omega_n^2} = \frac{50}{s^2 + 5s + 25} \tag{9.39}$$

Here, $K = 2$, $\zeta = 0.5$, and $\omega_n = 5$. The response of the above-mentioned underdamped system is produced using MATLAB in the example below. The MATLAB code for this example is given in Fig. 9.39 with its output in Figs. 9.40, 9.41, and 9.42.

Output

9.3.3.7   Undamped System

Undamped system has a damping ratio, $\zeta = 0$. In an undamped system, the poles are complex; however, the real parts become zero in such cases.

```
%% Underdamped system : zeta=0.5
clc;clear;
K=2;
omega_n=5;
zeta=0.5;
s=tf('s');
disp('Transfer function:')
G=(K*omega_n^2)/(s^2+2*zeta*omega_n*s+omega_n^2)
step(G);
grid on;
%xlim([0 5]);
ylim([0 2.5]);
disp('Parameters:')
disp(stepinfo(G))
% Pole-zero map
figure(2);
pzmap(G)
grid on
```

Fig. 9.39 Code—Underdamped system

```
Command Window

   Transfer function:


   G =


              50
        --------------
        s^2 + 5 s + 25


   Continuous-time transfer function.

   Parameters:
              RiseTime: 0.3278
          SettlingTime: 1.6152
           SettlingMin: 1.8630
           SettlingMax: 2.3259
             Overshoot: 16.2929
            Undershoot: 0
                  Peak: 2.3259
              PeakTime: 0.7184
```

Fig. 9.40 Output—Underdamped system

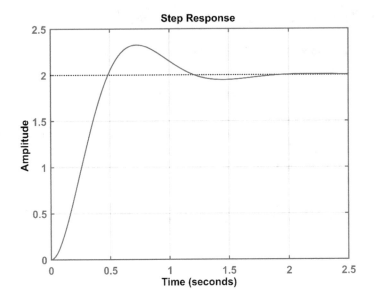

Fig. 9.41 Graphical output—Step response of underdamped system

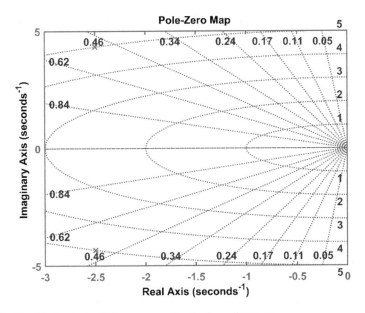

Fig. 9.42 Graphical output—Pole zero map of underdamped system

9.3.3.8   MATLAB Example 9.15: Undamped System

Consider the following second-order system:

$$G(s) = \frac{K\omega_n^2}{s^2 + 2\zeta\omega_n s + \omega_n^2} = \frac{50}{s^2 + 25} \tag{9.40}$$

Here, $K = 2$, $\zeta = 0$, and $\omega_n = 5$. The response of the above-mentioned underdamped system is produced using MATLAB in the example below. The MATLAB code for this example is given in Fig. 9.43 with its output in Figs. 9.44, 9.45, and 9.46.

N.B. In the output shown in Fig. 9.44, the term Inf appears when the result is infinity; and the term NaN stands for "not a number."

Output

9.3.3.9   Negative Damped System

In a negative damped system, the damping ratio is negative, i.e., $\zeta < 0$. The poles of a negative damped system are positive real numbers, i.e., they are positioned in the right half-plane of the coordinate system. Therefore, such systems are always regarded as unstable systems.

```
%% Undamped system : zeta=0
clc;clear;
K=2;
omega_n=5;
zeta=0;
s=tf('s');
disp('Transfer function:')
G=(K*omega_n^2)/(s^2+2*zeta*omega_n*s+omega_n^2)
step(G);
grid on;
xlim([0 5]);
ylim([-0.5 5]);
disp('Parameters:')
disp(stepinfo(G))
% Pole-zero map
figure(2);
pzmap(G)
grid on
```

Fig. 9.43 Code—Undamped system

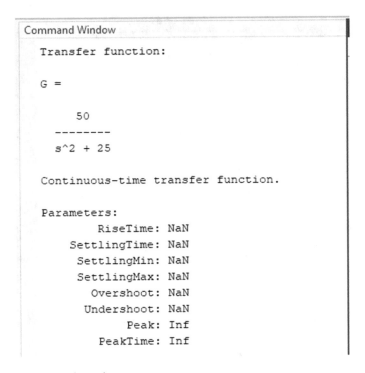

Fig. 9.44 Output—Undamped system

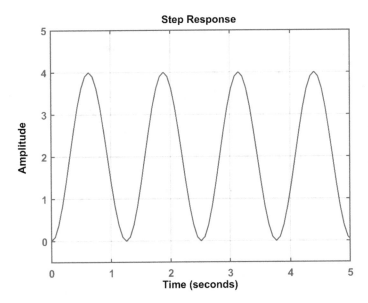

Fig. 9.45 Graphical output—Step response of undamped system

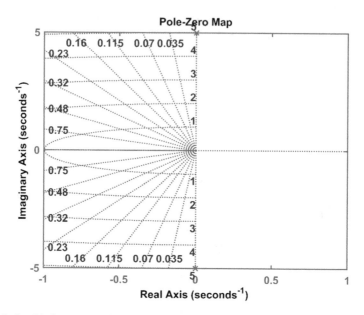

Fig. 9.46 Graphical output—Pole zero map of undamped system

9.3.3.10 MATLAB Example 9.16: Negative Damped System

Consider the following second-order system:

$$G(s) = \frac{K\omega_n^2}{s^2 + 2\zeta\omega_n s + \omega_n^2} = \frac{50}{s^2 - 20s + 25} \tag{9.41}$$

Here, $K = 2, \zeta = -2$, and $\omega_n = 5$. The response of the above-mentioned negative damped system is produced using MATLAB in the example below. The MATLAB code for this example is given in Fig. 9.47 with its output in Figs. 9.48, 9.49, and 9.50.

Output

```
%% Negative damped system : Zeta=-2
clc;clear;
K=2;
omega_n=5;
zeta=-2;
s=tf('s');
disp('Transfer function:')
G=(K*omega_n^2)/(s^2+2*zeta*omega_n*s+omega_n^2)
step(G);
grid on;
xlim([0 0.25]);
ylim([-0.5 2.5]);
disp('Parameters:')
disp(stepinfo(G))
% Pole-zero map
figure(2);
pzmap(G)
grid on
```

Fig. 9.47 Code—Negative damped system

```
Command Window
   Transfer function:

   G =

            50
   ---------------
   s^2 - 20 s + 25

   Continuous-time transfer function.

   Parameters:
            RiseTime: NaN
         SettlingTime: NaN
          SettlingMin: NaN
          SettlingMax: NaN
            Overshoot: NaN
           Undershoot: NaN
                 Peak: Inf
             PeakTime: Inf
```

Fig. 9.48 Output—Negative damped system

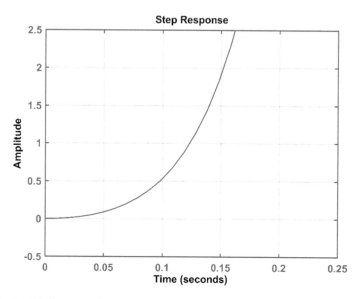

Fig. 9.49 Graphical output—Step response of negative damped system

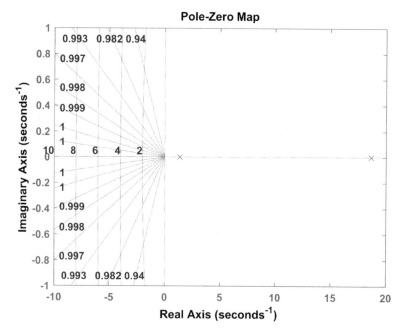

Fig. 9.50 Graphical output—Pole zero map of negative damped system

9.3.4 Steady-State Error

The steady-state error of a system can be defined by utilizing the following formula for a system illustrated in Fig. 9.51:

$$E_{ss} = \frac{sR(s)}{1 + H(s)G(s)} \tag{9.42}$$

Position Error Coefficient Position error coefficient can be defined by the following formula mathematically for the same system shown in Fig. 9.51:

$$K_p = \lim_{s \to 0} H(s) \cdot G(s) \tag{9.43}$$

Velocity Error Coefficient
Velocity error coefficient can be defined by the following formula mathematically for the same system shown in Fig. 9.51:

$$K_v = \lim_{s \to 0} s \cdot H(s) \cdot G(s) \tag{9.44}$$

Acceleration Error Coefficient
Acceleration error coefficient can be defined by the following formula mathematically for the same system shown in Fig. 9.51:

$$K_a = \lim_{s \to 0} s^2 \cdot H(s) \cdot G(s) \tag{9.45}$$

9.3.4.1 MATLAB Example 9.17: Steady-State Error

Consider the system shown in Fig. 9.51. Here, $G(s) = \frac{20(s+2)}{(s+3)(s+2)s}$ and $H(s) = 1$.

The input of the system in the time domain is $r(t) = 1 + 5t$.
Find the following using MATLAB:

(a) Position error coefficient
(b) Velocity error coefficient

Fig. 9.51 A closed-loop
feedback system

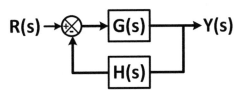

```
% Transfer eqn: 20(s+2)/(s^3+10*s+11)
% Position error coefficient: K_a
syms s t
G=@(s) 20*(s+2)/((s+3)*(s+2)*s);
H=1;
K_p=limit(H*G(s),s,0,'right');
fprintf('Position error coefficient: %f\n',K_p)

% Velocity error coefficient: K_v
K_v=limit(s*H*G(s),s,0);
fprintf('Velocity error coefficient: %f\n',K_v)

% Acceleration error coefficient: K_a
K_a=limit(s^2*H*G(s),s,0);
fprintf('Velocity error coefficient: %f\n',K_a)

% Steady-state error
r=@(t) 1+5*t;
R(s)=laplace(r(t));
E=(s*R(s))/(1+H*G(s));
Ess=limit(E,s,0,'right');
fprintf('Steady-state error: %f\n',Ess)
```

Fig. 9.52   Code—Steady-state error

```
Command Window

    Position error coefficient: Inf
    Velocity error coefficient: 6.666667
    Velocity error coefficient: 0.000000
    Steady-state error: 0.750000
```

Fig. 9.53   Output—Steady-state error

(c) Acceleration error coefficient
(d) Steady-state error

The MATLAB code for this example is given in Fig. 9.52 with its output in Fig. 9.53.

Output

9.4 A State-Space Representation for RLC Circuit

State-space representation signifies a mathematical model to describe any physical system in terms of input, output, variables, and the first-order derivative. In state-space representation, the variables that are used to define a physical system are called state variables. One of the advantages of state-space representation is that state-space modeling reduces any nth order mathematical model of dynamic systems into a simple first-order mathematical representation; hence, the computation becomes easier.

A simple illustration of any state-space representation of any dynamic system can be shown as in Fig. 9.54.

From Fig. 9.54, the state-space representation consists of three important parameters—inputs, state variables, and outputs. A generalized mathematical representation of a state-space model can be defined by the following sets of equations, where the first one is regarded as a state equation and the second one can be called as output equation:

$$x' = Ax + Bu \tag{9.46}$$

$$y = Cx + Du \tag{9.47}$$

Here, x is the vector representing the state variables; u and y are the input and output vector, respectively; A is the system matrix; B is regarded as the control input matrix; C is called the output matrix; and D is the direct matrix of the state-space representation.

State-space representation can be applied to electrical systems as well. Any RLC circuit can be defined in terms of state space. If we can define the state variables, inputs, and outputs, an RLC circuit can be represented by adopting state-space modeling to make the calculation more simplified.

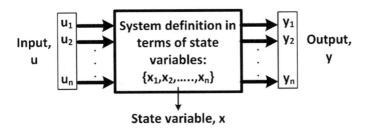

Fig. 9.54 Output—Steady-state error

9.4.1   State-Space Model and Response

Consider the following RLC series circuit, where the output is the voltage across the inductor, v_L, and the input is the voltage source, $v(t)$. Here, the state variable is the voltage across the capacitor, v_C, and the current of the circuit, i. The system is shown in Fig. 9.55 for more clarification.

In the RLC circuit, the storage elements are inductance L and capacitance C. The following two relationships are available for any electrical circuit:

$$C\frac{dv_C}{dt} = Cv_C' = i \tag{9.48}$$

$$L\frac{di}{dt} = Li' = v_L \tag{9.49}$$

$$v(t) = iR + v_C + v_L \tag{9.50}$$

Using the above equations, the state equations can be defined as follows:

$$v_C' = \left(\frac{1}{C}\right)i \tag{9.51}$$

$$i' = \frac{1}{L}(v_L) = \frac{1}{L}(v(t) - iR - v_C) = \left(-\frac{1}{L}\right)v_C + \left(-\frac{1}{R}\right)i + \left(\frac{1}{L}\right)v(t) \tag{9.52}$$

The output equation can be represented as follows by using Eq. (9.50):

$$v_L = (-1)v_C + (-R)i + v(t) \tag{9.53}$$

Finally utilizing state Eqs. (9.51) and (9.52), and output Eq. (9.53), the state-space representation of the RLC circuit can be written as follows:

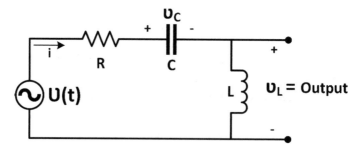

Fig. 9.55 An RLC series circuit

$$\begin{bmatrix} v'_C & i' \end{bmatrix} = \begin{bmatrix} 0 & \dfrac{1}{C} - \dfrac{1}{L} - \dfrac{1}{R} \end{bmatrix} \begin{bmatrix} v_C & i \end{bmatrix} + \begin{bmatrix} 0 & 1 \end{bmatrix} v(t) \tag{9.54}$$

$$v_L = \begin{bmatrix} -1 - R \end{bmatrix} \begin{bmatrix} v_C & i \end{bmatrix} + \begin{bmatrix} 1 \end{bmatrix} v(t) \tag{9.55}$$

9.4.2 State-Space Model to Transfer Function

From the state-space representation, it is possible to determine the transfer function of the system. Consider the state-space representation in Eqs. (9.46) and (9.47). The mathematical formula to determine the transfer function from the above-mentioned state-space representation can be written by the following equation:

$$\text{Transfer function, } G(s) = C * (sI - A)^{-1} * B + D \tag{9.56}$$

Here, I is the identity matrix of the same dimension as A matrix.

In MATLAB, by defining the matrices A, B, C, and D, the state-space representation of a system can be determined using $ss()$ as mentioned below:

MATLAB command for state-space representation:
$ss(A, B, C, D)$

To determine the transfer function from its state-space model, MATLAB provides an in-built function called $ss2tf()$.

MATLAB command for determining transfer function from state-space model:
$ss2tf(A, B, C, D)$

9.4.2.1 MATLAB Example 9.18: State-Space Model and Conversion into Transfer Function

Consider the circuit in Fig. 9.56, where the output and input are v_L and $v(t)$, respectively. The state variables are v_C and i.

(a) Determine the state-space representation of the RLC circuit.
(b) From the state-space model, determine the transfer function using $ss2tf()$.
(c) Verify the result of (b) by using the state-space to transfer function conversion formula manually.

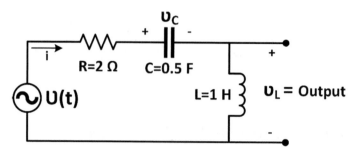

Fig. 9.56 An RLC series circuit

```
% State-space model to transfer function
% Resistance: R
% Inductance: L
% Capacitance: C
% Numerator: N
% Denominator: D
clc;clear;
R= 2; L=1; C=0.5;

% State-space metrices

A = [0 1/C; -1/L -R/L];
B = [0; 1/L];
C = [-1 -R];
D = [1];

% State-space model.
sys = ss(A, B, C, D);

% Transfer function
[Num Den] = ss2tf(A,B,C,D);
disp('Transfer function:')
TF=tf([Num],[Den])

% Verification
syms s
I=eye(2);
G1= C*inv(s*I-A)*B+D;
disp('Transfer function using formula:')
disp(simplify(G1))
```

Fig. 9.57 Code—State-space model to transfer function

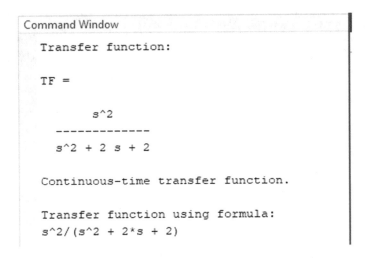

Fig. 9.58 Output—State-space model to transfer function

The MATLAB code for this example is given in Fig. 9.57 with its output in Fig. 9.58.

Output

9.4.3   Transfer Function to State-Space Model

From a given transfer function of a system, it is also possible to determine the state-space model of that system. The MATLAB command for such conversion is provided below:

MATLAB command for determining state-space model from transfer function from:
tf2ss([Numerator], [Denominator])

Here, [*Numerator*] represents a vector containing the coefficients of the numerator of the transfer function; [*Denominator*] is a vector incorporating the coefficient of the denominator of the transfer function.

```
% Transfer function to state-space model
% Numerator: N
% Denominator: D
clc;clear;
% Transfer function
N=[2];
D=[1 20 2];
disp('Transfer function:')
G=tf([N],[D])

% State-space metrices
[A,B,C,D]=tf2ss([N],[D]);
disp('State-space model:')
state_space = ss(A, B, C, D)
```

Fig. 9.59 Code—Transfer function to state-space model

9.4.3.1 MATLAB Example 9.19: Conversion into the State-Space Model from the Transfer Function

Consider the following transfer function to determine its state-space model using MATLAB:

$$G(s) = \frac{2}{s^2 + 20s + 2} \tag{9.57}$$

The MATLAB code for this example is given in Fig. 9.59 with its output in Fig. 9.60.

Output

9.5   Controllability and Observability of State-Space Model

9.5.1   Controllability

The controllability of a system requires to be identified, as it can help to determine whether the state output of the system can be controlled with a control input. A system is called controllable if the initial state of a system can be forwarded to the desired state by controlling the input of a system within a finite time.

Consider the state-space representation in Eqs. (9.46) and (9.47). If a control input $u(t)$ exists that can transfer any initial state of the system $x(t_o)$ to our desired state $x(t)$ in a finite time, the system can be regarded as controllable.

```
Command Window

 Transfer function:

 G =

            2
    ---------------
    s^2 + 20 s + 2

 Continuous-time transfer function.

 State-space model:

 state_space =

    A =
           x1     x2
      x1   -20    -2
      x2    1      0

    B =
           u1
      x1    1
      x2    0

    C =
           x1   x2
      y1    0    2

    D =
           u1
      y1    0

 Continuous-time state-space model.
```

Fig. 9.60 Output—Transfer function to state-space model

9.5.2   Testing for Controllability

To test the controllability, the rank of the controllability matrix needs to be equal to the rank of the system matrix. Therefore, the condition of controllability can be listed as follows:

If Rank (Controllability Matrix) = = Rank(System Matrix, A)
System : Controllable
else
System : Not controllable
end

9.5.3   Observability

A system is called observable if the initial state of a system can be determined by observing the output and the control input within a finite time. Consider the state-space representation in Eqs. (9.46) and (9.47). If a finite time T exists within which the initial state of the system $x(t_o)$ can be determined by observing the output $y(t)$ for a given input $u(t)$, the system can be regarded as observable.

9.5.4   Testing for Observability

To test observability, the rank of the observability matrix needs to be equal to the rank of the system matrix. Therefore, the condition of controllability can be listed as follows:

If Rank (Observability Matrix) = = Rank(System Matrix, A)
System : Observable
else
System : Not Observable
end

9.5.4.1 MATLAB Example 9.20: Controllability and Observability

Consider the following system:

$$x' = [2\ 0 - 1 - 3\ 4\ 0\ 10\ 6 - 8\ ]x + [-1\ 0\ 1\ ]u \tag{9.58}$$

$$y = [3\ 2\ 4\ ]x + [1]u \tag{9.59}$$

Determine the controllability and the observability of the system.

The MATLAB code for this example is given in Fig. 9.61 with its output in Fig. 9.62.

```
% Controllability & Observability
% State-spece representation:
% x' = Ax + Bu
% y  = Cx + Du
clc;clear;
A = [2 0 -1;-3 4 0;10 6 -8];
B = [-1;0;1];
C = [3 2 4];
D = [1];

% Controllability test
disp('Controllability matrix:');
Control_M = ctrb(A,B)
R_Con_M = rank(Control_M);
R_A = rank(A);

if (R_Con_M == R_A)
    disp('Comment: The system is controllable');
else
    disp('Comment: The system is not controllable');
end
fprintf('-------------------------------\n');
% Observability test
disp('Observability matrix:');
Observe_M = obsv(A,C)
R_Obs_M = rank(Observe_M);
R_A = rank(A);

if (R_Obs_M == R_A)
    disp('Comment: The system is observable');
else
    disp('Comment: The system is not observable');
end
```

Fig. 9.61 Code—Controllability and observability of state-space model

```
Command Window

  Controllability matrix:

  Control_M =

        -1     -3      12
         0      3      21
         1    -18     132

  Comment: The system is controllable
  ------------------------------------
  Observability matrix:

  Observe_M =

         3      2       4
        40     32     -35
      -366    -82     240

  Comment: The system is observable
```

Fig. 9.62 Code—Controllability and observability of state-space model

Output

9.6 Stability Analysis

The stability of a system can be defined by observing the output of a system that corresponds to a specific input signal. A system can be regarded as stable when the output is bounded for a specific bounded input. In control systems, it is crucial to determine the stability of a system. Hence, stability analysis is a very essential part of the control study.

In general, the stability analysis is followed by the rules mentioned below:

(a) *Stable system*: If all the poles of a system reside in the left-half plane of the coordinate system, the system is regarded as a stable system.
(b) *Unstable system*: If at least one of the poles of a system is positioned in the right-half plane, it will be referred to as an unstable system. Again, if the number of

zeros is greater than the number of poles of a system, the system becomes unstable.

(c) *Marginally stable system*: Marginally stable is another type of stability to define a system which occurs when at least one of the poles lies on the imaginary axis, and the others on the left-half plane of the coordinate system.

The above rules are applicable for any system having a closed-loop transfer function. There are many methods to conclude the stability of a system. In this section, the following methods will be covered to determine the stability of a system:

1. Routh criteria
2. Root locus
3. Bode plot
4. Nyquist plot

9.6.1 Routh Criteria

This is a method that yields stability information without the need to solve for the closed-loop system poles. Using this method, we can tell how many closed-loop system poles are in the left half-plane, in the right half-plane, and on the $j\omega$-axis. (Notice that we say how many, not where.) We can find the number of poles in each section of the s-plane, but we cannot find their coordinates. The method is called the Routh-Hurwitz criterion for stability.

The method requires two steps:

(i) Generate a data table called a Routh table.
(ii) Interpret the Routh table to tell how many closed-loop system poles are in the left half-plane, in the right half-plane, and on the $j\omega$ -axis.

Consider the following transfer function of a linear closed-loop system:

$$G(s) = \frac{Y(s)}{X(s)} = \frac{b_1 s^n + b_2 s^{n-1} + b_3 s^{n-2} + \ldots + b_N s^0}{a_1 s^m + a_2 s^{m-1} + a_3 s^{m-2} + \ldots + a_M s^0} \qquad (9.60)$$

Here, n and m represent the highest degree of the numerator and the denominator, respectively. $\{b_1, b_2, \ldots, b_N\}$ are the coefficients of the numerator, and $\{a_1, a_2, \ldots, a_M\}$ represent the coefficients of the denominator. The denominator of the transfer function is regarded as the characteristic equation, the solutions of which provide the poles of a system. For the Routh's criterion, the method is applied on this characteristic equation, which is rewritten as follows:

$$C(s) = a_1 s^m + a_2 s^{m-1} + a_3 s^{m-2} + \ldots + a_M s^0 \qquad (9.61)$$

Table 9.6 The Routh table from the characteristic equation in Eq. (9.61)

s^m	a_1	a_3	$\ldots$	$\ldots$
s^{m-1}	a_2	a_4	$\ldots$	a_M
s^{m-2}	$A_{m-2}^1 = \frac{1}{a_2}\left\lvert a_1\ a_3\ a_2\ a_4 \right\rvert$	$A_{m-2}^2 = \frac{1}{a_4}\left\lvert a_3\ a_5\ a_4\ a_6 \right\rvert$	$\ldots$	$A_{m-2}^M = \frac{1}{a_M}\left\lvert a_{M-1}\ a_{M+1}\ a_M\ a_{M+2} \right\rvert$
s^{m-3}	$A_{m-3}^1 = \frac{1}{A_{m-2}^1}\left\lvert a_2\ a_4\ A_{m-2}^1\ A_{m-2}^2 \right\rvert$	$A_{m-3}^2 = \frac{1}{A_{m-2}^2}\left\lvert a_4\ a_6\ A_{m-2}^2\ A_{m-2}^3 \right\rvert$	$\ldots$	$A_{m-3}^M = \frac{1}{A_{m-2}^M}\left\lvert a_M\ a_{M+2}\ A_{m-2}^M\ A_{m-2}^{M+1} \right\rvert$
$\cdots$	$\cdot\ \cdot\ \cdot$	$\cdot\ \cdot\ \cdot$	$\cdots$	$\cdot\ \cdot\ \cdot$
s^0	A_0^1			

The first step in Routh criteria is to generate the Routh table or array from the characteristic equation as shown in Table 9.6.

From the Routh table, the number of rows indicates the highest degree of the characteristic equation. In the first row, the coefficients on the odd position of the $C(s)$ are enlisted, whereas in the second row, the coefficients of the even positions take the place of the columns. It is to be noted that $C(s)$ needs to be ordered from the highest degree to the lowest. Starting from the third row, the first column value, A^1_{m-2}, is defined as shown in Eq. (9.62):

$$A^1_{m-2} = \frac{1}{a_2} |a_1 \, a_3 \, a_2 \, a_4| \tag{9.62}$$

Here, $|a_1 \, a_3 \, a_2 \, a_4|$ indicates the determinant of the matrix. It can be observed that, in the determinant matrix, the values are from the immediate previous two row values of two columns; and the division value is the immediate previous row value. The same pattern can be seen for all the other values as well.

After the Routh array or table is made, the values of the first column (marked green) are evaluated to determine the stability of the system.

According to Routh's criterion, the number of roots of the characteristic equation, i.e., the number of poles having positive real part, i.e., lie on the right-half plane, is equivalent to the number of sign changes that occur in the first green marked column of the Routh's array, or table.

Therefore, if the number of sign changes in the first green marked column is equal to zero, the system can be regarded as stable; otherwise, it is unstable. It is also to be noted that Routh's criterion is capable of only making a decision, whether the system is stable or unstable; it cannot provide any additional remarks, such as how to make an unstable system stable.

The MATLAB code for Routh's criterion is shown in the later examples. In Routh's criterion, there are two special cases as mentioned below:

Special Case 1: When the first element of any row is zero

Special Case 2: When the entire row is zero

If zero appears in any of the rows, the consecutive row value becomes undefined, or infinite. Therefore, during such special cases, the implementation of Routh's table is changed to accommodate such variation. These two special cases are beyond the scope of this book. Therefore, the deviation of the implementation of Routh's table during such special cases will not be included in this chapter. However, in such cases, the former value is replaced with 0.001 instead of 0 to proceed further with calculation for avoiding the complex methods of Routh's table implementations during special cases.

9.6.1.1 MATLAB Example 9.21: Routh's Criteria

Determine the stability of a system having the following characteristic equation:

```
% Characteristic polynomial
% C(s)= s^3+2s^2+10s^2+15
% Input: coeff = Vector of coefficients of the C(s); e.g., [1 2 10 15]

clc;clear;
coeff=input('Enter the coefficients:')
L=length(coeff);
if (rem(L,2)==0)
    Routh_array=zeros(L,L/2);
    for i=1:L/2
        Routh_array(1,i)=coeff(1,2*i-1);
        Routh_array(2,i)=coeff(1,2*i);
    end
else
    Routh_array=zeros(L,(L+1)/2);
    for i=1:(L+1)/2
        Routh_array(1,i)=coeff(1,2*i-1);
        if i==(L+1)/2
            break;
        end
        Routh_array(2,i)=coeff(1,2*i);
    end
end

for i=3:size(Routh_array,1)
    if Routh_array(i-1,1)==0
        Routh_array(i-1,1)=0.001;
    end
    for j=1:size(Routh_array,2)-1
        Routh_array(i,j)=(-1/Routh_array(i-1,1))*det([Routh_array(i-2,1) ...
            Routh_array(i-2,j+1);Routh_array(i-1,1) Routh_array(i-1,j+1)]);
    end
end
Routh_array
S=sign(Routh_array);
count=0;
for i=1:L
    if S(i,1)==1
        count=count+1;
    end
end
if count==L
    disp('The system is stable')
else
    disp('THe system is unstable')
end
% Verify
fprintf('\n');
disp('Verification:')
Roots=roots(coeff);
disp('Poles:')
disp(Roots)
```

Fig. 9.63 Code—Routh's criterion

```
Command Window

  Enter the coefficients: [1 2 10 15]

  coeff =

       1     2     10     15


  Routh_array =

         1.0000     10.0000
         2.0000     15.0000
         2.5000          0
        15.0000          0

  The system is stable

  Verification:
  Poles:
     -0.1989 + 3.0534i
     -0.1989 - 3.0534i
     -1.6021 + 0.0000i
```

Fig. 9.64 Output—Routh's criterion

$$C(s) = s^3 + 2s^2 + 10s + 15 \qquad\qquad (9.63)$$

In the following MATLAB code, an *input()* function is used, through which we can provide the coefficients of any characteristic equation manually to make the code applicable for any such problems. For the above problem, the input is a vector of the coefficients [1 2 10 15], which has been provided to test the stability.

The MATLAB code for the example is given in Fig. 9.63 with its output in Fig. 9.64. The code is created based on Table 9.6 described earlier.

Output

In the above example, the Routh array is produced for better understanding. From the Routh array, it can be observed that no sign changes have occurred. Therefore, the system does not have any poles which have positive real parts, i.e., all the poles reside in the left-half plane. Based on this information, it can be concluded that the system is stable, as shown in MATLAB as well. For verification, the poles of the system have been determined using *root*(), which determined the roots of the characteristic equation. From the output, it can be observed that all the poles have negative real parts and match with the previous hypothesis as well as the outcome.

9.6.2 Root Locus

Root locus represents a graphical method that illustrates the locus of the roots of a characteristic equation in an s-plane followed by the changes of system parameters. By observing the root locus of a system, the stability of a system can be decided. In the next example, the root loci of a stable system, an unstable system, and a marginally stable system are shown using the following MATLAB command.

MATLAB command for producing root locus from a transfer function:
rlocus(*sys*)
Here, *sys* **is the transfer function.**

9.6.2.1 MATLAB Example 9.22: Root Locus

Produce the root locus of the following systems and comment on the stability:

(a) $G(s) = \frac{50}{s^2+12s+11}$

(b) $G(s) = \frac{s+1}{s^3+9s^2}$

(c) $G(s) = \frac{s+1}{s^3-20s^2-10s+1}$

The MATLAB code for the example is given in Fig. 9.65.

```
%% Root locus:
% Example 1: Stable system
sys1 = tf([50],[1 12 11]);
rlocus(sys1)
% Example 2: Marginally Stable
sys2 = tf([1 1],[1 9 0 0]);
rlocus(sys2)
% Example 3: Unstable
sys3 = tf([1 1],[1 -20 -10 1])
rlocus(sys3)
```

Fig. 9.65 Code—Root locus

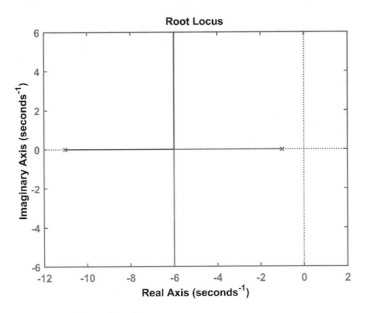

Fig. 9.66 Output—Root locus for stable system

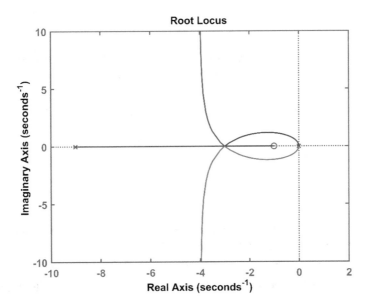

Fig. 9.67 Output—Root locus for marginally stable system

The root loci of the systems (a), (b), and (c) are provided in Figs. 9.66, 9.67, and 9.68, respectively.

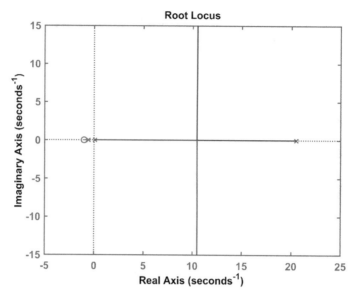

Fig. 9.68 Output—Root locus for unstable system

Comment for Fig. 9.66 The system is stable, as the poles lie on the right-half plane.

Comment for Fig. 9.67 The system is marginally stable, as one of the poles lies on the imaginary axis, and the rest reside in the left-half plane.

Comment for Fig. 9.68 The system is unstable, as one of the poles lies in the right-half plane.

9.6.3   Bode Plot

The Bode plot is a graphical representation of the frequency response of a system, which consists of two parts—Bode magnitude plot and Bode phase plot. To understand the Bode plot, some terminologies need to be defined first, such as:

Phase Crossover Frequency The phase crossover frequency, ω_{pc}, is the frequency where the phase shift is equal to -180°.

Gain Crossover Frequency The gain crossover frequency, ω_{gc}, is the frequency where the amplitude ratio is 1, or when log modulus is equal to 0.

Gain Margin The gain margin of a system can be defined as in Eq. (9.64):

$$\text{Gain Margin} = \frac{1}{\left|G\left(i\omega_{pc}\right)\right|} \tag{9.64}$$

Phase Margin The phase margin of a system can be defined as in Eq. (9.65):

$$\text{Phase Margin} = 180^{\circ} + arg\left(G\left(i\omega_{gc}\right)\right). \tag{9.65}$$

The stability conditions according to the Bode plot are given below:

- *Stable System*: Both the margins should be positive or the phase margin should be greater than the gain margin.
- *Marginal Stable System*: Both the margins should be zero or the phase margin should be equal to the gain margin.
- *Unstable System*: If any of them is negative or phase margin should be less than the gain margin.

The MATLAB command for producing Bode plot of a system is as follows:

MATLAB command for producing bode plot from a transfer function:
margin(sys)
Here, *sys* is the transfer function.

9.6.3.1 MATLAB Example 9.23: Bode Plot

Produce the Bode plot of the following systems and comment on the stability:

```
%% Bode plot
% Example 1: Stable system
G1 = tf([50],[1 12 11]);
figure(1)
margin(G11);
grid on;
% Example 2: Unstable
G2 = tf([1 1],[1 -20 -10 1]);
figure(3)
margin(G2);
grid on;
```

Fig. 9.69 Code—Bode plot

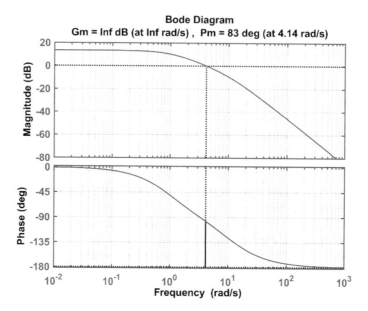

Fig. 9.70 Bode diagram for stable system

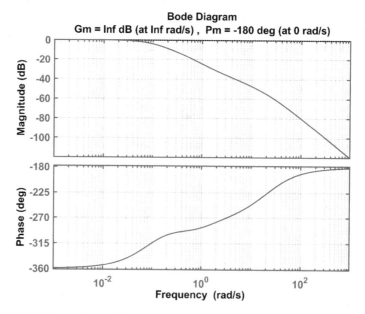

Fig. 9.71 Bode diagram for unstable system

(a) $G(s) = \frac{50}{s^2+12s+11}$

(b) $G(s) = \frac{s+1}{s^3-20s^2-10s+1}$

The MATLAB code for the example is given in Fig. 9.69.

The Bode plots of the systems (a) and (b) are provided in Figs. 9.70 and 9.71, respectively.

Comment for Fig. 9.70 The system is stable, as both the margins are positive.

Comment for Fig. 9.71 The system is unstable, as the phase margin is negative.

9.6.4 Nyquist Plot

Nyquist plot is a graphical representation of the frequency response of a system drawn in polar coordinates. In the polar coordinate, the gain of the transfer function of a system indicates the radial points, whereas the phase represents the angular coordinates in a Nyquist plot.

The stability criteria of the Nyquist plot can be defined by the following rules:

- **If the contour does not enclose $(-1, 0)$ point, and the number of poles in the right-half plane is zero, the system is stable.**
- **If the number of clockwise encirclements by the contour of $(-1,0)$ point is equivalent to the number of poles in the right-half plane, the system is stable.**

  In summary, if the following formula is satisfied, the system is stable; otherwise, it is unstable:

$$Z = N + P$$

Here,
Z = number of zeroes in the right-half plane
P = number of poles in the right-half plane
N = number of encirclements of $(-1,0)$ point by the Nyquist contour

The MATLAB command for producing the Nyquist plot of a system is as follows:

MATLAB command for producing Nyquist plot from a transfer function:
nyquist(sys)
Here, *sys* is the transfer function.

9.6.4.1   MATLAB Example 9.24: Nyquist Plot

Produce the Nyquist plot of the following systems and comment on the stability:

(a) $G(s) = \frac{50}{s^2 + 12s + 11}$

(b) $G(s) = \frac{400}{s^3 - 4s^2 - 50s + 45}$

The MATLAB code for the example is given in Fig. 9.72.

The Nyquist plots of the systems (a) and (b) are provided in Figs. 9.73 and 9.74, respectively.

Comment for Fig. 9.73   The contour does not encircle $(-1, 0)$; hence, $N = 0$. In addition, $Z = 0$ and $P = 0$. The system satisfies the Nyquist stability criteria: $Z = N + P$; therefore, the system is stable.

Comment for Fig. 9.74   The contour encircles $(-1, 0)$ two times; hence, $N = 2$. In addition, $Z = 0$ and $P = 0$. The system does not satisfy the Nyquist stability criteria: $Z = N + P$. Therefore, the system is unstable.

9.7   Conclusion

In this chapter, various aspects of control systems are presented with the combination of both theoretical knowledge and practical implementations via MATLAB. The chapter initiates by introducing both frequency and time domain responses of physical systems followed by the concept of state-space representations. In addition, controllability and observability are also covered here. The chapter is concluded by presenting an overview of the stability analysis of any physical system. While introducing all of these topics of control systems, the chapter not only takes the

```
%% Nyquist plot
% Example 1: Stable system
G1= tf([50],[1 12 11]);
figure(1)
nyquist(G1)
grid on;
% Example 2: Unstable
G2 = tf([400],[1 4 50 45]);
figure(2)
nyquist(G2)
```

Fig. 9.72   Code—Nyquist plot

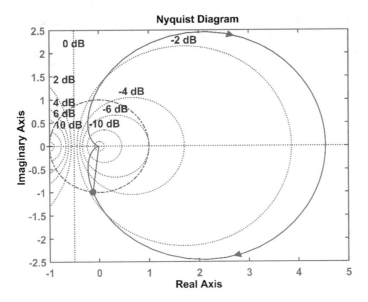

Fig. 9.73 Nyquist diagram for stable system

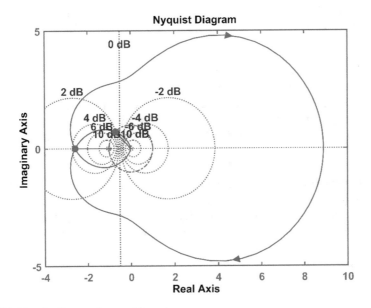

Fig. 9.74 Nyquist diagram for unstable system

theoretical approach but also provides MATLAB illustrations to facilitate the reader to understand. The objective of this chapter is to help the reader to visualize the important concepts of control systems and realize the applications in the engineering domain with the aid of MATLAB.

Exercise 9

1. Define transfer function. How does a transfer function mathematically represent a control system?
2. What are the engineering applications of Laplace transform and inverse Laplace transform?
3. Explain the following functions with examples:

    (a) laplace()
    (b) residue()
    (c) limit()
    (d) ss()
    (e) pzmap()

4. Consider the given transfer functions:

    (a) $G(s) = \frac{s-12}{s^2-4s+1}$
    (b) $G(s) = \frac{s^2-6s+52}{2s^2-15s-25}$
    (c) $G(s) = \frac{s^3+5s+12}{6s^2-4s+1}$

    Perform the following using MATLAB:

    (i) Determine the pole and zero of G(s).
    (ii) Demonstrate the pole-zero map.
    (iii) Determine the inverse Laplace of G(s). Perform Laplace on the result to verify if the output returns G(s). (The resulting fractions may be hard to comprehend; use *pretty*() function to display the results in a more convenient manner.)
    (iv) What is the DC gain for the transfer function?
    (v) Decompose the transfer function into partial fractions (sometimes, the result may return large fraction values. To convert the large fractions with symbols into decimals of *x* significant values, use the function *vpa(function, x)*. To convert large numeric fractions into decimals, use the function *double*().

Fig. 9.75 An RLC series
circuit

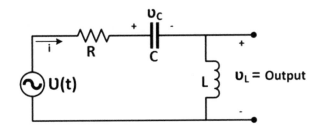

5. The following transfer functions have $K = 2$. From the functions, classify which
   are overdamped, critically damped, underdamped, or underdamped. Determine
   the step response, pole-zero diagram, rise time, settling time, overshoot, and peak
   time using MATLAB.

   (a) $G(s) = \frac{18}{s^2+12s+9}$

   (b) $G(s) = \frac{196}{2s^2+14s+98}$

   (c) $G(s) = \frac{8}{s^2+4s+4}$

   (d) $G(s) = \frac{6}{3s^2+3}$

6. In the following circuit in Fig. 9.75, given that $R = 2$ ohm, $C = 0.6\ F$,
   and $L = 1.5\ H$. The output and input are v_L and $v(t)$, respectively. The state
   variables are v_C and i.

   (a) Determine the state-space representation of the circuit.
   (b) Determine the transfer function of the system from the state-space model.
   (c) Verify the result of (b) by using the state-space to transfer function conversion
       formula manually.

7. Using the MATLAB code used to determine the Routh's criteria in Example 9.21,
   show whether the following characteristic polynomials are stable. Perform verification of the result by determining the root of the characteristic equation:

   (a) $s^5 + 3s^4 + 27s^3 + 45s^2 - 60$
   (b) $s^4 + 21s^3 + 36s^2 + 5s + 1$.

8. Produce the (i) root locus, (ii) Bode plot, and (iii) Nyquist plot of the following
   systems and comment on the stability based on their respective graphical output:

   (a) $G(s) = \frac{36}{2s^2+14s+61}$

   (b) $G(s) = \frac{4s+1}{s^3+2s+6}$

   (c) $G(s) = \frac{s^2-2s+1}{7s^3+11s^2-5s+1}$

Chapter 10
Optimization Problem

10.1 Introduction

Optimization is a pivotal aspect in the field of engineering. To determine the best possible outcome subjected to different conditions or constraints, optimization is one of the best processes. When a certain problem does not have a perfect solution, optimization has the capability to come with different good answers to follow. In the optimization process, the very first step is to define the problem mathematically, which leads to two basic parts of any optimization problem—objective function and condition. The objective function refers to our objective in terms of the mathematical equation with single or multiple decision variables that affect the outcome of the objective function. The goal of any optimization algorithm is to either minimize or maximize the outcome of the objective function by optimizing the values of decision variables. However, to maximize or minimize a function, there is another part that needs to be incorporated in an optimization problem, which is the condition. For a certain boundary or range of the variables, the optimization problem can be solved. An optimization problem can also be subjected to different constraints. Based on the number of variables of an objective function, it can be classified into two classes in general—one-dimensional optimization and multidimensional optimization. The mathematical nature of objective function can also be used to categorize optimization problems into three classes—linear programming, quadratic programming, and nonlinear programming. To implement optimization problems, MATLAB can be utilized efficiently, as it provides in-built functions that can be used to solve different optimization problems.

© The Author(s), under exclusive license to Springer Nature Switzerland AG 2022   283
E. Hossain, *MATLAB and Simulink Crash Course for Engineers*,
https://doi.org/10.1007/978-3-030-89762-8_10

10.2   One-Dimensional Optimization

In one-dimensional optimization, the objective function has a single decision variable. The goal is to find the value of the decision variable for which the objective function will be maximized or minimized. Mathematically, such an optimization problem can be defined as follows:

$$Obj = f(x);$$

$$\text{subject to } x_{upper} < x < x_{lower} \tag{10.1}$$

Here, $f(x)$ is the objective function with decision variable x. And the inequality term is the condition that needs to be satisfied for the optimization solution.

To solve the one-dimensional optimization problem, different algorithms are available, such as golden section search, Fibonacci search, parabolic interpolation search, etc. In MATLAB, one-dimensional optimization can be performed by utilizing the built-in function *fminbd*(). The parameters of this function are listed below:

MATLAB command for one-dimensional optimization:

$$[x, \text{value}] = fminbd\left(obj, x_{low}, x_{up}\right)$$

Inputs: obj refers to the objective function; x_{low} and x_{up} are the lower and upper limits of the decision variable x.
Outputs: x is the value of the decision variable for which obj is minimized. value indicates the minimized value of obj.

It is to be noted that *fminbd*() uses a combination of both golden section search and parabolic interpolation search algorithms for solving one-dimensional optimization problems.

10.2.1   MATLAB Example 10.1: One-Dimensional Optimization

Consider the following one-dimensional optimization problem for finding the value of the optimized decision variable and the minimized value of the objective function.

$$Obj = 2x + e^x;$$

```
% One dimensional optimization
% Objective function: minimize obj(x)= 2*x + exp(x)
% Condition: -5<x<10
% Lower limit, x_low = -5;
% Upper limit, x_up = 10;
clc;clear;
syms x;
obj= @(x) 2*x + exp(x);
x_low=-5;
x_up=10;
[x,value]=fminbnd(obj,x_low,x_up);
fprintf('Optimized value of the decision variable: %.5f\n',x);
fprintf('Minimized value of the objective function: %.5f\n',value);
```

Fig. 10.1 Code—One-dimensional optimization

```
Command Window

  Optimized value of the decision variable: -4.99994
  Minimized value of the objective function: -9.99315
```

Fig. 10.2 Output—One-dimensional optimization

$$\text{subject to} - 5 < x < 10 \tag{10.2}$$

The MATLAB code and its output for the above optimization problem are provided in Figs. 10.1 and 10.2, respectively.

Output

It is to be noted that the lower limit and the upper limit of the decision boundary do not include the boundary values in the condition. Therefore, while we are defining the boundary of x from –5 to 10, the *fminbd()* considers the values of x as any values within the range, but does not include the boundary values itself, i.e., –5 and 10.

10.3 Multidimensional Optimization

In a multidimensional optimization problem, the objective function is a function of multiple decision variables. The objective function can be subjected to different linear or nonlinear constraints, with boundary conditions. The generalized form of a multidimensional optimization problem can be defined as follows, with different types of constraints and boundary conditions:

$$\text{obj} = f(x(1), x(2), \ldots \ldots x(n))$$

subject to :

Boundary condition : $x_{low} \leq x \leq x_{up}$

Linear inequality constraint : $A \cdot x \leq B$

Linear equality constraint : $A_{EQ} \cdot x = B_{EQ}$

Nonlinear inequality constraint : $C \cdot x \leq 0$

Nonlinear equality constraint : $C_{EQ} \cdot x = 0$ (10.3)

To solve such an optimization problem, MATLAB has a built-in function called *fmincon()*. For the above-mentioned generalized format of a multidimensional optimization problem, the MATLAB command can be written as follows:

MATLAB command for multidimensional optimization:

$$[x, \text{value}] = fmincon\left(\text{obj}, x_o, A, B, A_{EQ}, B_{EQ}, x_{low}, x_{up}, \text{nonLinearConstraint}\right)$$

Inputs:
obj refers to the objective function with multiple variables, which is represented by a vector or matrix, x.

x_o is the initial values of the decision variables, which is also represented by a vector or matrix.

A is a matrix containing the coefficients of x of a linear inequality constraint.

B is a vector containing the constant of the linear inequality constraint.

A_{EQ} is a matrix incorporating the coefficients of x of a linear equality constraint.

B_{EQ} is a vector containing the constant of the linear equality constraint.

x_{low} and x_{up} are lower and upper limits of the decision variables x, which is given input as two matrices.

nonLinearConstraint indicates the nonlinear equality constraint, C_{EQ}, and the nonlinear inequality constraints, C, of x.

Outputs: x is the vector or matrix of the decision variables for which obj is minimized.

value indicates the minimized value of obj that satisfies all the linear and nonlinear constraints with decision boundary conditions.

```
function [C,C_EQ]=nonlinear_constraint(x)
% Non-linear constraints
% x(1)^2 + x(2)^2 + x(3)^2 = 12;
% x(1)*x(2) + x(2)*x(3) <= 30;
C=x(1)*x(2) + x(2)*x(3)-30;
C_EQ=x(1)^2 + x(2)^2 + x(3)^2 - 12;
end
```

Fig. 10.3 Code—Creating function for multidimensional optimization

```
% Multidimensional optimization
% Objective function:
% min obj(x)=x(1)^2 + 2*x(1)*x(2) + x(3)^2 + exp(x(2));
% Limits: -1<=x(1)<= 5; 0<=x(2)<= 5; 0<=x(3)<= 7;
% Linear inequality constraint: x(1)+ x(2) + x(3) < 10;
% Linear equality constraint: x(1)+ 2*x(3)= 4;
% Non-linear equality constraint: x(1)^2 + x(2)^2 + x(3)^2 = 12;
% Non-linear inequality constraint: x(1)*x(2) + x(2)*x(3) <= 30;
% Initial values: xo= [-1,0,0];
clc;clear;
obj=@(x) x(1)^2 + 2*x(1)*x(2) + x(3)^2 + exp(x(2));
x_low=[-1,0,0];
x_up=[5,5,7];
xo=[-1,0,0];
A=[1,1,1];
B=[10];
A_EQ=[1,0,2];
B_EQ=[4];
nonLinearConstraint= @nonlinear_constraint;
[x,value] = fmincon(obj,xo,A,B,A_EQ,B_EQ,x_low,x_up,nonLinearConstraint);
fprintf('Optimized value of the decision variable:\n');
fprintf('x1: %.5f\n',x(1));
fprintf('x2: %.5f\n',x(2));
fprintf('x3: %.5f\n\n',x(3));
fprintf('Minimized value of the objective function: %.5f\n',value);
```

Fig. 10.4 Code—Multidimensional optimization

10.3.1   MATLAB Example 10.2: Multidimensional Optimization

Solve the following multidimensional optimization problem using MATLAB:

$$\text{obj}(x) = x_1^2 + 2x_1x_2 + x_3^2 + e^{x_2}$$

$$\text{subject to :}$$

$$-1 \leq x_1 \leq 5; 0 \leq x_2 \leq 5; 0 \leq x_3 \leq 7;$$

```
Command Window

  Local minimum found that satisfies the constraints.

  Optimization completed because the objective function is non-decreasing in
  feasible directions, to within the value of the optimality tolerance,
  and constraints are satisfied to within the value of the constraint tolerance.

  <stopping criteria details>
  Optimized value of the decision variable:
  x1: -1.00000
  x2: 2.17945
  x3: 2.50000

  Minimized value of the objective function: 11.73254
```

Fig. 10.5 Output—Multidimensional optimization

$$x_1 + x_2 + x_3 < 10;$$
$$x_1 + 2x_3 = 4;$$
$$x_1^2 + x_2^2 + x_3^2 = 12;$$
$$x_1 x_2 + x_2 x_3 \le 30; \tag{10.4}$$

To solve the above-mentioned problem, a function has been created to address the nonlinear constraints, which must be satisfied while optimizing. C and C_EQ represent a value to check whether the optimized values of the decision variables are satisfying the conditions or not. The function has been named "nonlinear_constraint. m," which will be used as an input in the later *fmincon()* function.

The MATLAB code for the optimization problem is given in Figs. 10.3 and 10.4, and the output is given in Fig. 10.5.

Output

10.4 Linear Programming Optimization

The mathematical nature of the objective function can be utilized to classify optimization problems into three categories—linear programming, quadratic programming, and nonlinear programming optimization.

Linear programming optimization deals with linear objective function with linear constraints for optimization. The generalized format of linear programming optimization can be defined as follows:

$$obj = obj(x)$$

$$\text{subject to :}$$

$$\text{Boundary condition} : x_{\text{low}} \le x \le x_{\text{up}}$$

$$\text{Linear inequality constraint} : A \cdot x \le B$$

$$\text{Linear equality constraint} : A_{EQ} \cdot x = B_{EQ} \qquad (10.5)$$

In MATLAB, there is a built-in function for solving linear programming optimization, which is shown below:

MATLAB command for linear programming optimization:

$$[x, \text{value}] = linprog\left(\text{obj}, A, B, A_{EQ}, B_{EQ}, x_{\text{low}}, x_{\text{up}}\right)$$

Inputs:
obj refers to the objective function with multiple variables, which is represented by a vector or matrix, x.

A is a matrix containing the coefficients of x of a linear inequality constraint.

B is a vector containing the constant of the linear inequality constraint.

A_{EQ} is a matrix incorporating the coefficients of x of a linear equality constraint.

B_{EQ} is a vector containing the constant of the linear equality constraint.

x_{low} and x_{up} are lower and upper limits of the decision variables x, which is given input as two matrices.

Outputs: x is the vector or matrix of the decision variables for which obj is minimized.

value indicates the minimized value of obj that satisfies all the linear and nonlinear constraints with decision boundary conditions.

10.4.1   MATLAB Example 10.3: Linear Programming Optimization

Solve the following multidimensional optimization problem using MATLAB.

$$\text{obj}(x) = 2x_1 + 3x_2 + x_3$$

$$\text{subject to :}$$

$$-1 \le x_1 \le 5; 0 \le x_2 \le 10; 0 \le x_3 \le 15;$$

```
% Linear programming optimization
% Objective function:
% min obj(x)=2*x(1) + 3*x(2) + x(3);
% Limits: -1<=x(1)<= 5; 0<=x(2)<= 10; 0<=x(3)<= 15;
% Linear inequality constraint: x(1)+ x(2) + x(3) <= 15;
% Linear inequality constraint: x(1)- 4*x(2) + x(3) <= 8;
% Linear equality constraint: x(1)+ 2*x(3)= 4;
clc;clear;
obj=[2 3 1];
x_low=[-1,0,0];
x_up=[5,10,15];
A=[1 1 1;1 -4 1];
B=[15 8];
A_EQ=[1 0 2];
B_EQ=[4];
[x,value] = linprog(obj,A,B,A_EQ,B_EQ,x_low,x_up);
fprintf('Optimized value of the decision variable:\n');
fprintf('x1: %.5f\n',x(1));
fprintf('x2: %.5f\n',x(2));
fprintf('x3: %.5f\n\n',x(3));
fprintf('Minimized value of the objective function: %.5f\n',value);
```

Fig. 10.6 Code—Linear programming optimization

```
Command Window

  Optimal solution found.

  Optimized value of the decision variable:
  x1: -1.00000
  x2: 0.00000
  x3: 2.50000

  Minimized value of the objective function: 0.50000
```

Fig. 10.7 Output—Linear programming optimization

$$x_1 + x_2 + x_3 \leq 15;$$

$$x_1 - 4x_2 + x_3 \leq 8;$$

$$x_1 + 2x_3 = 4; \tag{10.6}$$

The MATLAB code and the output for the optimization problem are given in Figs. 10.6 and 10.7, respectively.

Output

10.5   Quadratic Programming Optimization

In quadratic optimization, the objective function is quadratic in nature; however, the constraints are linear. The general format of a quadratic programming optimization can be represented as follows:

$$\mathrm{obj}(x) = 0.5x^T H x + F^T x$$

subject to :

Boundary condition : $x_{\mathrm{low}} \le x \le x_{\mathrm{up}}$

Linear inequality constraint : $A \cdot x \le B$

Linear equality constraint : $A_{EQ} \cdot x = B_{EQ}$ $\qquad$ (10.7)

In MATLAB, there is a built-in function for solving quadratic programming optimization, which is shown below:

MATLAB command for quadratic programming optimization:

$$[x, \mathrm{value}] = quadprog\left(H, F, A, B, A_{\mathrm{EQ}}, B_{\mathrm{EQ}}, x_{\mathrm{low}}, x_{\mathrm{up}}\right)$$

Inputs:
H refers to the Hessian matrix, which is determined by using the following formula:

$$H_{i,j}^F = \frac{\partial^2 F}{\partial x_i \partial x_j}.$$

F is a vector containing the coefficients of the linear variables.
A is a matrix containing the coefficients of x of a linear inequality constraint.
B is a vector containing the constant of the linear inequality constraint,
A_{EQ} is a matrix incorporating the coefficients of x of a linear equality constraint.
B_{EQ} is a vector containing the constant of the linear equality constraint.
x_{low} and x_{up} are lower and upper limits of the decision variables x, which is given input as two matrices.
Outputs: x is the vector or matrix of the decision variables for which obj is minimized.
value indicates the minimized value of obj that satisfies all the linear and nonlinear constraints with decision boundary conditions.

10.5.1   MATLAB Example 10.4: Quadratic Programming Optimization

Consider the following objective function in Eq. (10.8) and optimize it using MATLAB:

$$\text{obj}(x) = 2x_1^2 + 3x_2^2 + 0.5x_1x_2 - 4x_1 + x_2$$

subject to:

$$-1 \leq x_1 \leq 5; 0 \leq x_2 \leq 10;$$

$$x_1 + x_2 \leq 15;$$

$$x_1 - 4x_2 \leq 8;$$

$$x_1 + 2x_2 = 4. \tag{10.8}$$

The MATLAB code and the output for the optimization problem are given in Figs. 10.8 and 10.9, respectively.

Output

```
% Quadratic programming optimization
% Objective function:
% min obj(x)=2*x(1)^2 + 3*x(2)^2 + 0.5*x(1)*x(2)- 4*x(1) + x(2);
% Limits: -1<=x(1)<= 5; 0<=x(2)<= 10;
% Linear inequality constraint: x(1)+ x(2) <= 15;
% Linear inequality constraint: x(1)- 4*x(2) <= 8;
% Linear equality constraint: x(1)+ 2*x(2)= 4;
clc;clear;
H=[4 0.5;0.5 6];
F=[-4;1];
x_low=[-1,0];
x_up=[5,10];
A=[1 1;1 -4];
B=[15 8];
A_EQ=[1 2];
B_EQ=[4];
[x,value] = quadprog(H,F,A,B,A_EQ,B_EQ,x_low,x_up);
fprintf('Optimized value of the decision variable:\n');
fprintf('x1: %.5f\n',x(1));
fprintf('x2: %.5f\n',x(2));
fprintf('Minimized value of the objective function: %.5f\n',value);
```

Fig. 10.8   Code—Quadratic programming optimization

```
Command Window

 Minimum found that satisfies the constraints.

 Optimization completed because the objective function is non-decreasing in
 feasible directions, to within the value of the optimality tolerance,
 and constraints are satisfied to within the value of the constraint tolerance.

 <stopping criteria details>
 Optimized value of the decision variable:
 x1: 1.90000
 x2: 1.05000
 Minimized value of the objective function: 4.97500
```

Fig. 10.9 Code—Quadratic programming optimization

10.6   Nonlinear Programming Optimization

For nonlinear programming optimization, the objective function needs to be a nonlinear function subjected to either linear constraints or nonlinear constraints. In this type of optimization problem, both linear and nonlinear constraints may also be included. A nonlinear programming optimization problem can be solved by using *fmincon*() function, as explained earlier in Sect. 10.3. The difference between multidimensional and nonlinear programming optimization is that in multidimensional optimization, it is not required for the objective function to be nonlinear; conversely, in nonlinear programming, the objective function must have to be nonlinear.

10.7   Li-ion Battery Optimization Problem and Solutions

Consider a Li-ion battery system, whose sizing needs to be minimized in terms of minimized modules' number. Here, the minimized modules' number indicates the most optimized combination of battery modules connected in row and column to form a pack, which is capable of providing a predetermined discharge rate by maintaining an acceptable nominal voltage range. For this particular problem, the battery pack needs to be designed in such a way, so that the discharge rate always remains 100 kWh with an overall output voltage of 150–400 V. A small variation of $\pm 1\%$ of the discharge rate may be allowed for this problem. The aim is to solve an optimization problem for the battery sizing, where the number of rows and columns for the modules to form a battery pack will be minimized by satisfying the required conditions. a simple illustration of a battery pack is given in Fig. 10.10.

The following parameters are going to be considered for the problem:

a. Number of cells per module, $\text{cell}_{\text{mod}} = 4$
b. The number of rows in a battery pack, row
c. The number of columns in a battery pack, col

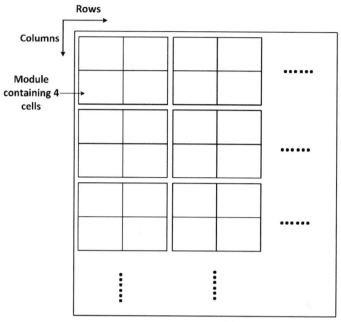

Fig. 10.10 A Li-ion battery pack

```
function [C,C_EQ]=nonLin_Constraint(x)
% Non-linear constraints
% Non-linear ineqaulity condition:
% 99*1000 Wh <=dischage_cell*x(1)*x(2)*cell_mod<= 101*1000 Wh
% Here, discharge_mod = 90 W; cell_mod=4
dischage_cell=90; cell_mod=4;
C=[dischage_cell*x(1)*x(2)*cell_mod-101*1000; ...
    -dischage_cell*x(1)*x(2)*cell_mod+99*1000];
C_EQ=[];
end
```

Fig. 10.11 Code—Creating function for the Li-ion battery optimization problem

d. The nominal voltage per module, $vol_{mod} = 12$ V
e. The maximum discharge rate per module, $discharge_{cell} = 90$ W

The optimization problem can be defined mathematically, by utilizing the above-mentioned parameters, as follows:

```
% Li-ion battery sizing
% Decision variables:
% Number of rows, x(1); Number of columns, x(2);
% Objective function:
% Minimize obj = x(1)*x(2)
% Subject to:
% 99*1000 Wh <=discharge_cell*x(1)*x(2)*cell_mod<= 101*1000 Wh;
% 150 V <=vol_mod*x(2)<= 400 V
% Here, discharge_cell = 90 W; cell_mod = 4;vol_mod = 12 V;
clc;clear;
obj=@(x) x(1)*x(2);
discharge_cell = 90;
cell_mod=4;
vol_mod = 12;
x_low = [4,150/(vol_mod)];
x_up = [14,400/(vol_mod)];
A=[];
B=[];
A_EQ=[];
B_EQ=[];
xo=[4,14];
nonLinear_Constraint=@nonLin_Constraint;
[x,value]=fmincon(obj,xo,A,B,A_EQ,B_EQ,x_low,x_up,nonLinear_Constraint);
fprintf('Battery size:\n');
fprintf('Row = %d   Column = %d\n',round(x(1)),round(x(2)));
fprintf('Size = %d x %d = %d\n', round(x(1)),round(x(2)),round(value));
```

Fig. 10.12 Code—Li-ion battery optimization problem

$$obj(row, col) = row \times col$$

$$Subject\ to :$$

$$150\ V \leq vol_{mod} \times col \leq 400\ V$$

Command Window

```
Local minimum found that satisfies the constraints.

Optimization completed because the objective function is non-decreasing in
feasible directions, to within the value of the optimality tolerance,
and constraints are satisfied to within the value of the constraint tolerance.

<stopping criteria details>
Battery size:
Row = 14   Column = 20
Size = 14 x 20 = 275
```

Fig. 10.13 Output—Li-ion battery optimization problem

$$99.5 \text{ kWh} \leq \text{discharge}_{\text{cell}} \times \text{row} \times \text{col} \times \text{cell}_{\text{mod}} \leq 100.5 \text{ kWh} \qquad (10.9)$$

The MATLAB code for solving the battery optimization problem is given in Figs. 10.11 and 10.12, and the output of the code is given in Fig. 10.13.

MATLAB Code

Output

10.8   Conclusion

From this chapter, the readers can gain knowledge on optimization problems, and their solving procedures via MATLAB. The categorizations of different optimization problems are enlisted here based on the variables and the nature of objective functions. For each of the optimization problems, MATLAB has built-in functions, which have been explained with necessary illustrations in this chapter. A sample optimization problem directly related to the engineering field is incorporated at the end of the chapter to walk through how a real optimization problem can be solved utilizing MATLAB. This chapter will greatly help the readers to attain a strong fundamental knowledge of various optimization problems.

Exercise 10

1. What are the differences between one-dimensional optimization and multidimensional optimization?
2. (a) Consider a function $f(x) = 6x^4 - 11x + 10$. Determine the optimized value of the decision variable and minimized value of the objective function at a range of $-12 < x < 12$.
   (b) Consider a function $f(x) = -x^5 + 4x^3 + 7x^2 - 15x$. Determine the optimized value of the decision variable and minimized value of the objective function at a range of $0 < x < 1$.
3. Using the same function *nonlinear _ constraint*() and the same linear and nonlinearity inequality and equality constrains, and initial values, replicate Example 10.2 for the following limits:

   a. $-4 < x_1 < 7; -2 < x_2 < 9; -1 < x_3 < 10$
   b. $1 < x_1 < 4; -3 < x_2 < 3; -7 < x_3 < -3$

Mention which of the programs will successfully run, and which one will show errors. For the successful programs, show the optimized values of the decision variables, and the minimized values of the objective function. For the programs that stopped prematurely or converged to an infeasible point, explain why these errors occurred.

4. Solve the following multidimensional optimization problem using *linprog()*. Comment on the feasibility of the solution.

a. $obj(x) = 4x_1 + 6x_2 + 2x_3$

subject to :

$$0 \le x_1 \le 10; \ -3 \le x_2 \le 9; 0 \le x_3 \le 12;$$
$$4x_1 + 5x_2 + 8x_3 \le 30;$$
$$7x_1 + 12x_2 + 3x_3 \le 65;$$
$$2x_1 + 3x_2 + 5x_3 = 11;$$

b. $obj(x) = 5x_1 + 7x_2 - 2x_3$

subject to :

$$-3 \le x_1 \le 4; \ -2 \le x_2 \le 7; 2 \le x_3 \le 11;$$
$$2x_1 + x_2 + 3x_3 \le 20;$$
$$-4x_1 + 2x_2 \le 10;$$
$$3x_1 + x_2 - 2x_3 = 16;$$

c. $obj(x) = 4x_1 + 9x_2 + x_3$

subject to:

$$2 \le x_1 \le 6; \ -10 \le x_2 \le 10; 0 \le x_3 \le 22;$$
$$x_1 + x_2 + x_3 \le 26;$$
$$8x_2 - 3x_3 \le 15;$$
$$x_1 + 9x_2 + 4x_3 = 18;$$

5. (a) Given the Hessian matrix of the quadratic objective function [10 3; 3 7] with a vector [−5;3] which contains the coefficients of the linear variables, determine if the function has the minimum that satisfies the constrains. If so, find the optimized value of the decision variables x_1 and x_2 for the function, and determine the minimized value of the objective function. The function maintains the following constraints:

subject to:

$$0 \le x_1 \le 8; \; -5 \le x_2 \le 5;$$

$$4x_1 + 5x_2 \le 21;$$
$$3x_1 - 9x_2 \le 15;$$
$$5x_1 + 3x_2 = 12;$$

(b) Given the Hessian matrix of the quadratic objective function [1 0.24; 0.24 5] with a vector [1;6] which contains the coefficients of the linear variables, determine if the function has the minimum that satisfies the constrains. If so, find the optimized value of the decision variables x_1 and x_2 for the function, and determine the minimized value of the objective function. The function maintains the following constraints:

subject to:

$$-4 \le x_1 \le 4; 0 \le x_2 \le 10;$$
$$6x_1 - 7x_2 \le 17;$$
$$x_1 + 12x_2 \le 10;$$
$$-3x_1 + 13x_2 = 21;$$

Chapter 11
App Designer and Graphical User Interface in MATLAB

11.1 Introduction

App Designer is a product of MATLAB that provides an environment to create professional apps. It is an interactive environment with a sophisticated component library that can be used to build any desired app. The programming platform enables the users to define the behavior of the app. It is a new modified and improved version of the previous Graphical User Interface Design Environment (GUIDE). In future releases of MATLAB, GUIDE will no longer be a part of MATLAB and will be completely replaced by App Designer (https://www.mathworks.com/products/matlab/app-designer/comparing-guide-and-app-designer.html). However, the GUIDE apps will run in MATLAB without any editing feature.

11.2 App Designer

To initiate App Designer in MATLAB, the first task is to open the App Designer window by typing *appdesigner* command in the command window (Fig. 11.1). After pressing "Enter," the "App Designer Start Page" (Fig. 11.2) will appear.

From the start page, any example can be loaded for the first-time user. To start a new blank app, click on the "Blank App" option from the start page. By clicking "Open" from the left panel of the start page, any existing app can be loaded using the App Designer. The file extension of an App Designer is ".mlapp."

Fig. 11.1 appdesigner command

Fig. 11.2 App Designer Start Page

11.2.1 Basic Layout of App Designer

After creating a blank app, the graphical user interface of the App Designer will appear. In the basic layout of the App Designer, two different views are available—design view and code view. In the design view, the structure of an app can be designed by utilizing the components. In the code view, the individual building blocks, or components, can be customized and programmed by writing MATLAB code. The basic layout in these two view modes is illustrated in Figs. 11.3 and 11.4.

In the design view layout of App Designer, there is a Component Library tab at the left panel, where all the available components for designing an app can be found. Using this library, an app can be designed in the "Design window." The components dragged on the Design window will appear in the component browser. By selecting each component from the component browser, the individual parameters can be defined from the "Inspect" tab.

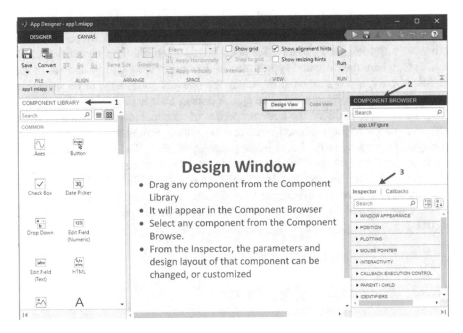

Fig. 11.3 App Designer Design Window

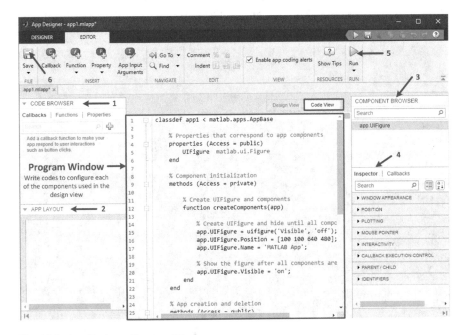

Fig. 11.4 App Designer Program Window

After completing the design of an app in the design view mode, the next step is to shift the layout to the code view mode. In the code view layout, there is a code browser in the left panel. Below the code browser tab, the app layout feature is available, where the designed app created in the design view will be shown. In the middle, there is a program window, where the relevant codes need to be written to configure each component used in an app. A default code is already available in the program window to define the basic properties of an app. However, when a particular app needs to be designed, each of the added components requires to be configured by updating the codes in the program window. The steps of creating an entire app will be discussed in more detail in the later sections.

At the end of the app design, the "Run" button is available to test the entire app. The app can be saved by utilizing the "Save" button. These two buttons are available on the upper strip of the app design available both in the design view and code view modes.

11.2.2   Components of App Designer

In the Component Library of the App Designer in design view mode, all the components are listed and categorized among Common, Containers, Figure tools, Instrumentation, and Aerospace. In each of these sections, multiple components are available for designing an app. In Fig. 11.5, all the available components are listed.

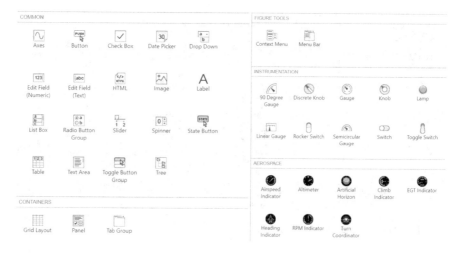

Fig. 11.5 Components of App Designer

Fig. 11.6 Error Tab

11.2.3   Detecting and Correcting Errors

After pressing the "Run" button, the designed app will be demonstrated in the App Designer, where the user can test the app. If errors occurred in the coding part, MATLAB will automatically detect the errors and will explain the reason for such error occurrence. An example is given in Fig. 11.6, where an error tab appears on the exact place where the error occurs, and it also explains the reason for this error.

If the error can be detected automatically, and the reason is explained, it is easier to find a solution to correct the errors. Based on the reasoning, the corrections are made which will differ in each case.

11.2.4   Designing and Programming a GUI with App Designer

In the previous section, the layout and the components are explained. In this section, some apps will be designed to explain the entire process of app designing step-by-step.

Example 11.1 Designing a calculator *Design View*

Step 1: From the component library, drag the following components into the design window:

a. Edit field (numeric)
b. Button
c. Panel

Step 2: Select each component from the component browser, and edit the parameters from the Inspection tab. For example, the labels of the three Edit Field (Numeric) components have been renamed as "Enter Number," "Enter Number," and "Output." Four push buttons have been renamed as "ADD," "SUB," "MUL," and "DIV." In the component browser, the push button will appear as "app. ADDButton," "app.SUBButton," "app.MULButton," and "app.DIVButton." The three Edit Field (Numeric) components have also been renamed in the component browser as "app.Num1," "app.Num2," and "app.out" for convenience.

Step 3: Select each of the components from the component browser to configure them by coding. For example, consider configuring the "ADD" push button. As the name of the push button was renamed as "ADD" earlier, it will appear as "app.

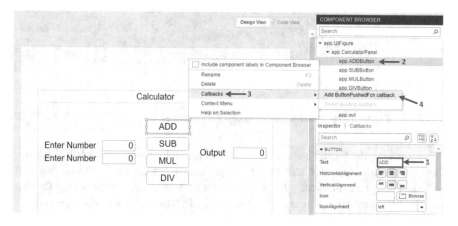

Fig. 11.7 Component Browser

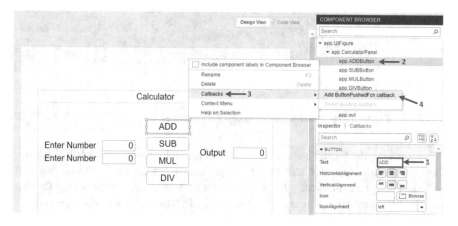

Fig. 11.8 App layout of calculator app

ADDButtton" in the component browser. The next goal is to write a callback function, whose objective is to perform a certain task whenever the button is pushed. To configure such a program, right-click on the "app.ADDButton" and select the option "Callbacks" followed by "AddButtonPushedFcn callback," as shown in Fig. 11.7.

Step 4: The previous step will shift the layout to code view mode, where a function will be created automatically for the "ADDButton." To configure this pushed button, the user can write their codes within the function. In the design view, the overall design of the app can be made. For configuring each component, the user has to shift to the coding view mode. The app layout of the calculator app is given in Fig. 11.8.

In the above app, the user can provide two numbers as input. Four push buttons are available for performing addition, subtraction, multiplication, and division. The output will provide the result that corresponds to the pushed button. The entire layout has been added in a panel, which has been dragged from the Component Library, and renamed as "Calculator." In the later part, this configuration will be ensured via MATLAB coding.

Code View

There are four push buttons, which need to be configured in such a way, so that they can perform one of the four basic operations—addition, subtraction, multiplication, and division—individually. In the previous section, it has been shown how to create a callback function for a specific component. To recall, the step is to select a specific component, such as "app.ADDButton" from the component browser, right-clicking on it, and navigate to Callbacks→AddButtonPushedFcn callback. This step will automatically create an "ADDButtonPushed" function as shown below in code view mode (Fig. 11.9), where the white space indicates the place for writing our code to assign a specific task that will be performed whenever the ADDButton will be pushed.

In the above code, app.Num1.Value and app.Num2.Value indicate the two values that will be provided as input by the user. The objective of the ADDButton is to perform the addition operation between these two input numerical values, and assign the output to the app.out.Value.

In the same manner, the other three push buttons can be configured by defining the functions for each component as shown in Fig. 11.10.

```
% Button pushed function: ADDButton
function ADDButtonPushed(app, event)
    app.out.Value=app.Num1.Value + app.Num2.Value
end
```

Fig. 11.9 Code—Addition Button

```
% Button pushed function: SUBButton
function SUBButtonPushed(app, event)
    app.out.Value=app.Num1.Value - app.Num2.Value
end
```

```
% Button pushed function: MULButton
function MULButtonPushed(app, event)
    app.out.Value=app.Num1.Value * app.Num2.Value
end
```

```
% Button pushed function: DIVButton
function DIVButtonPushed(app, event)
    app.out.Value=app.Num1.Value / app.Num2.Value
end
```

Fig. 11.10 Code—Subtraction, Multiplication, and Division Button

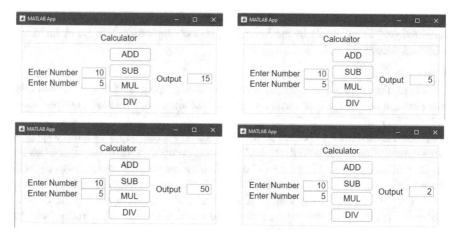

Fig. 11.11 Output—Calculator app

Output

The last step is to press the "Run" button to create the final app, and to test the different outputs. To test whether the app is working in our designed way, two numeric inputs have been provided, and four individual push buttons are pressed one by one to check the different outputs. The results are shown in Fig. 11.11, which justifies that the calculator app is working perfectly.

Example 11.2 Risk warning app In a risk warning app, the objective is to show warning through different colors of lamps indicating the level of risks. The risks are assessed via a knob which indicates the level of risk starting from "Low" to "High" including the "Off" condition as well. For each state of the knob, lamps of different colors are utilized to demonstrate the level of warning. When the knob indicates the "High" risk, a red alert lamp will get lit. For the "Medium" and "Low" levels in the knob, the lamp colors will be yellow and green, respectively. During the "Off" level, all the lights will be in dark mode.

Design View

In the design view, the components that have been chosen for the design of this app are one knob, three lamps, and one panel. The design of the app is shown in Fig. 11.12.

Coding View

After the design, the knob is configured by choosing the callback option, which creates a function named knobValueChanged () in the code view. To address the objective of the app, the following code (Fig. 11.13) is written to define the task of the knobValueChanged () function:

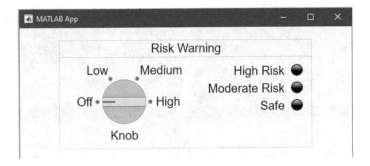

Fig. 11.12 App layout of risk warning app

```
% Value changed function: Knob
function KnobValueChanged(app, event)
    value = app.Knob.Value;
    if isequal(value,'Off')
        app.red.Color=[0 0 0];
        app.yellow.Color=[0 0 0];
        app.green.Color=[0 0 0];
    elseif isequal(value,'Low')
        app.red.Color=[0 0 0];
        app.yellow.Color=[0 0 0];
        app.green.Color=[0 1 0];
    elseif isequal(value,'Medium')
        app.red.Color=[0 0 0];
        app.yellow.Color=[1 1 0];
        app.green.Color=[0 0 0];
    elseif isequal(value,'High')
        app.red.Color=[1 0 0];
        app.yellow.Color=[0 0 0];
        app.green.Color=[0 0 0];
    else
        app.red.Color=[0 0 0];
        app.yellow.Color=[0 0 0];
        app.green.Color=[0 0 0];
    end
end
```

Fig. 11.13 Code—Risk levels in the risk warning app

Output

After running the app, it is tested for different scenarios. In the first scenario, the knob is in the "Off" state, and all the lamps are in black, indicating no warning. In the second scenario, the knob is in the "Low" level, which initiates a green lamp indicating a safe zone. The yellow lamp and the red lamp can be seen, when the knob touches the "Medium" and "High" level, respectively. The yellow demonstrates moderate risk, and the red refers to high risk. The output is shown in Fig. 11.14.

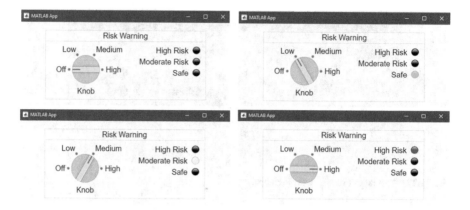

Fig. 11.14 Output—Risk warning app

Example 11.3 App for step response of a second-order system The generalized formula of the transfer function of a second-order system can be represented as follows:

$$G(s) = \frac{K\omega_n}{s^2 + 2\zeta\omega_n s + \omega_n^2} \tag{11.1}$$

In Eq. (11.1), K is the gain; ω_n is the natural frequency; and ζ refers to the damping ratio. Based on the different values of ζ, the second-order systems can be categorized among different types, such as overdamped, underdamped, critically damped, and negative damped systems. The goal of this app is to plot the step response of a second-order system with the given inputs of K, ω_n, and ζ, and also showing the types of the system based on damping ratio.

Design View

The components used for the design of the app layout are as follows:

a. Axes: for plotting the step response
b. Two edit field (numeric): for taking input K, ω_n
c. Slider: to vary the value of ζ
d. Edit field (numeric): to display the value of ζ that will be selected by the slider
e. Edit field (string): To display the type of the system based on the damping ratio
f. Panel: To incorporate the entire design in a single panel

The app layout design is shown in Fig. 11.15.

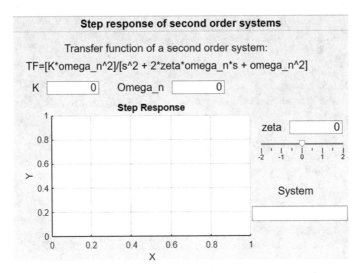

Fig. 11.15 App layout of step response of second-order systems app

Coding View

The value of K and ω_n (omega_n) will be considered as the values of the two "Edit Field (Numeric)," which are defined in the following code as "app.K.Value" and "app.omega_n.Value." In the AmplitudeSliderValueChanged (), the value of the slider is assigned as the value of zeta. All these values are utilized to calculate the transfer function by using the generalized transfer function formula of the second-order system. Later, the plot command is used to generate the step response and show it in the axes component. Another "Edit Field (String)" is defined to show the type of the system, which can be determined based on the selected value of zeta. According to the code, five different types of system are defined based on the values of zeta, which are:

1. Overdamped system: $\zeta > 1$
2. Underdamped system: $0 < \zeta < 1$
3. Critically damped system: $\zeta = 1$
4. Undamped system: $\zeta = 0$
5. Negative damped system: $\zeta < 0$

The added code to configure the app is provided in Figs. 11.16 and 11.17.

Output

After pressing the "Run" button, the step response of the second-order systems app will appear. To test the app, the values of K and omega_n are chosen as 2 and 5, respectively. After providing the input values of K and omega_n, the slider position is changed to select different values of zeta. Based on the change made by the slider, the specific value of the zeta that is selected will appear on the "Edit

```
% Value changed function: zetavalue
function zetavalueValueChanged(app, event)
    value=app.zetavalue.Value;
end

% Value changed function: K
function KValueChanged(app, event)
    value = app.K.Value;
end

% Value changed function: Omega_n
function Omega_nValueChanged(app, event)
    value = app.Omega_n.Value;
end

% Value changed function: System
function SystemValueChanged(app, event)
    value = app.System.Value;
end
```

Fig. 11.16 Code—Step response of second-order systems app

```
% Value changed function: AmplitudeSlider
function AmplitudeSliderValueChanged(app, event)
    app.zetavalue.Value = app.AmplitudeSlider.Value;
    K=app.K.Value;omega_n=app.Omega_n.Value;
    zeta=app.AmplitudeSlider.Value;
    s=tf('s');
    G=(K*omega_n^2)/(s^2+2*zeta*omega_n*s+omega_n^2);
    plot(app.UIAxes, step(G));
    if zeta>1
        app.System.Value='Overdamped';
    elseif zeta==1
        app.System.Value='Critically damped';
    elseif zeta>0 && zeta<1
        app.System.Value='Underdamped';
    elseif zeta==0
        app.System.Value='Undamped';
    else
        app.System.Value='Negative damped'
    end
end
```

Fig. 11.17 Code—Step response of second-order systems app

Field (Numeric)" named zeta. When the value of zeta will be selected, the step response of the second-order system will be plotted on the axes component. The type of the system will also appear on "Edit Field (String)" named System. The values of zeta are selected in different ranges that define different types of systems that have been described earlier. The step responses of four such types are produced one by one to verify the performance of the app. The produced outputs are shown in Fig. 11.18.

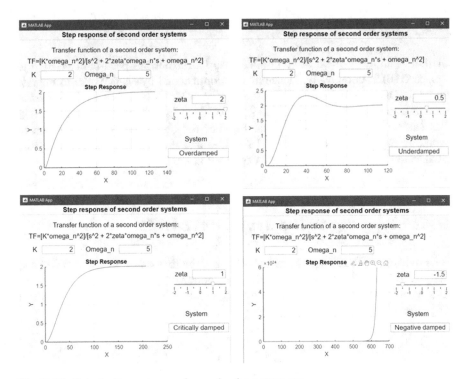

Fig. 11.18 Output—Step response of second-order systems app

11.3 App Designer vs GUIDE

App Designer is the new improved version of MATLAB that was launched in 2016. GUIDE is an old version of app building technology that was built based on third-party technologies. However, in the App Designer, MATLAB has improved the canvas and the environment with increased library components. The robustness of the programming in the App Designer is also significantly improved. Another feature of the App Designer is its web-based sharing options. Any improvements or updates of GUIDE have been stopped since 2016, and it will be completely removed from the future version of MATLAB, according to an announcement of MathWorks made in 2019 [1]. Therefore, it is imperative to migrate from GUIDE to App Designer.

11.4   GUIDE

GUIDE is an old version of app builder. App Designer has been built in 2016 to replace GUIDE completely. Hence, it is recommended by MathWorks to migrate from GUIDE to App Designer. For the current guide users, it is very important to learn how to migrate from GUIDE to App Designer. In addition, the current GUIDE files can be exported as M-files and Fig-files for future use. The procedures of exporting any GUIDE app as MATLAB files and the migration procedures from GUIDE to the App Designer will be explained in the later part in detail.

11.4.1   Exporting GUIDE App as MATLAB file

For exporting a GUIDE app as MATLAB files, the first step is to select "File" and navigate to the "Export to MATLAB-file" option, as shown in Fig. 11.19. By clicking this option, the app will be exported as two separate files—".m" file and ".fig" file.

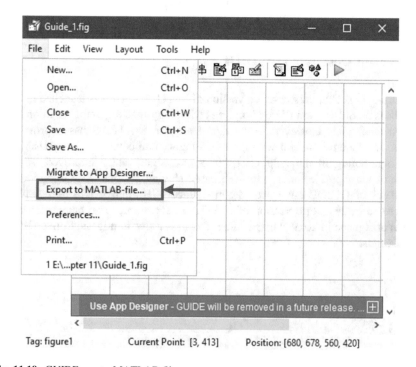

Fig. 11.19 GUIDE app to MATLAB file

11.4.2   Migrate GUIDE App to App Designer

As in the future version of MATLAB "GUIDE" will be completely removed, it is necessary to learn how to migrate to the App Designer from GUIDE. To migrate the GUIDE app to App Designer, the first step is to open GUIDE and create a new graphical user interface. By writing a command "guide" in the command window, the start page of GUIDE can be loaded. By clicking "Blank GUI (Default)," and browsing the directory, a new GUIDE file can be created, which is shown in Fig. 11.20.

After opening a GUIDE file, the migration can be started by using any of the two options shown in Fig. 11.21.

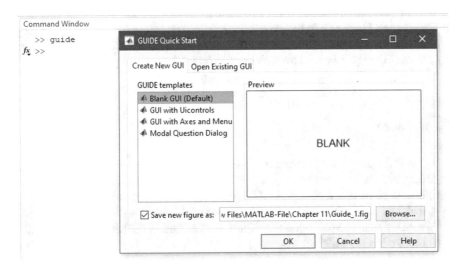

Fig. 11.20   GUIDE Quick Start

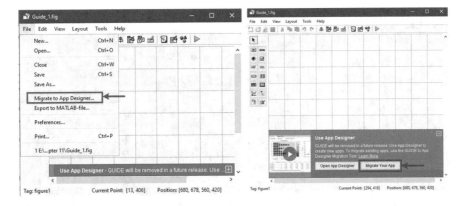

Fig. 11.21   GUIDE to App Designer

The above-mentioned step will automatically load the "Guide to App Designer Migration Tool," and the migration will occur automatically.

11.5   Conclusion

This chapter mainly concentrates on illustrating how to use the App Designer for app building. The chapter initiates by providing a brief overview of the basic layout and components of the App Designer for the sake of new users. Later, step-by-step guidance to build a calculator app is presented to help the readers understand how to design and program to create a complete app. To make the readers more comfortable in making more sophisticated apps, two more examples of app building are included in the chapter. For the GUIDE users, the migration procedure for shifting to the App Designer is incorporated in the last part of the chapter, as the GUIDE will be replaced by the App Designer in a future release of MATLAB.

Exercise 11

1. What is the difference between the design view and code view of MATLAB App Designer?
2. Mention some usage of MATLAB App Designer in engineering.
3. Create an app with a button and an edit field (text) to show the sentence "AppDesigner is Fun" once the "Click me" button is clicked. The app should look like in Fig. 11.22.

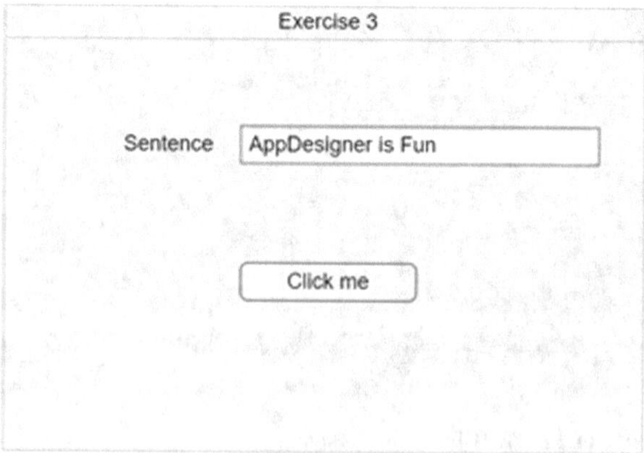

Fig. 11.22 App Designer is Fun

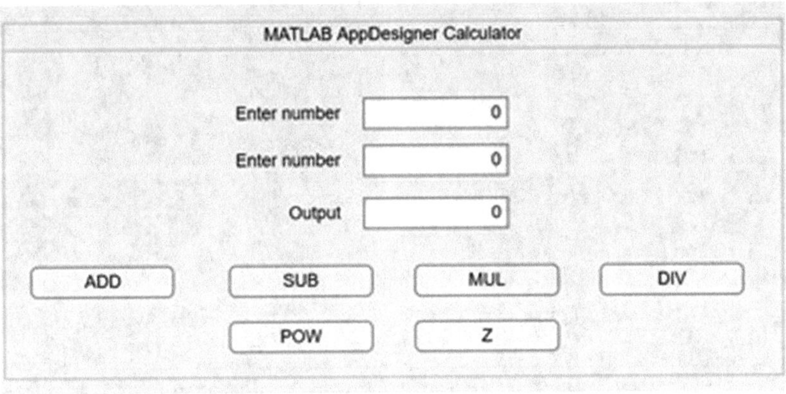

Fig. 11.23 MATLAB App Designer Calculator

Table 11.1 The status and lamp color of each knob level of the stereo system

Knob level	Status	Lamp color
0	None	No color
1–20	Low	Green
21–40	Mid-low	Green and cyan
41–60	Mid	Green, cyan, and yellow
61–80	Mid-high	Green, cyan, yellow, and magenta
81–100	High	Green, cyan, yellow, magenta, and red

4. Replicate the calculator application that was built in Example 11.1.

   a. Add another button with the title "POW" to calculate the power of the first number to the second one (Firstnumber$^{Second\ number}$).
   b. Add another button with the title "Z," which will perform the average of both the numbers and display the result.

The application should look like in Fig. 11.23.

5. In this exercise, you have to create a stereo system volume knob with five lamps, which will change the color as shown in Table 11.1.

The app should look like in Fig. 11.24 while changing the knob:

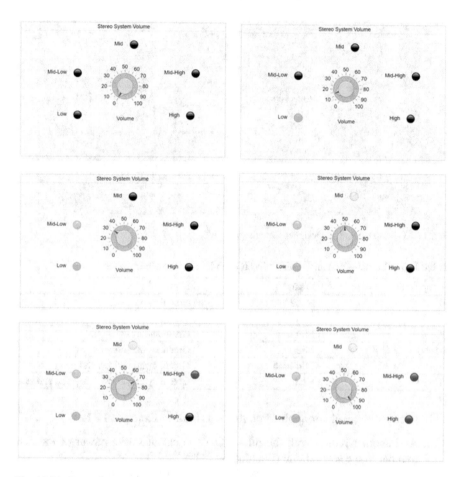

Fig. 11.24   Stereo System App Layout

Chapter 12
Introduction to Simulink

12.1   What Is Simulink?

Simulink is a platform that allows the modeling, design, and simulation of any dynamic physical system or embedded system. It provides graphically programmable blocks to create such designs. All the blocks can be customized according to users' preferences. Simulink and MATLAB can interact with each other, and thus the behavior of a model can be changed and simulated simultaneously. In Simulink Library Browser, several blocks are available based on numerous applications. The user can also create additional customizable blocks defined via MATLAB programming. In addition, data flow can also happen from Simulink to MATLAB and vice versa. A tested Simulink model can be deployed in any embedded system for practical use. Thus, Simulink works as a sophisticated platform to simulate any dynamic system in real time via graphical tools; and with the participation of MATLAB environment, such model can be analyzed and tested in-depth to improve the model for practical implementation.

12.2   Starting Simulink

A Simulink can be started from the header toolstrip of MATLAB. In the toolbar, click on the "Simulink" button as shown in Fig. 12.1.

Pressing on the "Simulink" button will create the appearance of the "Simulink Start Page" shown in Fig. 12.2, from where a blank model can be produced by selecting the "Blank Model→Create Model" option.

After choosing the "Create Model" option, the Simulink window will appear, as shown in Fig. 12.3. In the Simulink window, there is a "Model Design Window" as shown in the figure. This window is the core part, where any block diagram model can be designed by utilizing components that will be available in the "Library

© The Author(s), under exclusive license to Springer Nature Switzerland AG 2022   317
E. Hossain, *MATLAB and Simulink Crash Course for Engineers*,
https://doi.org/10.1007/978-3-030-89762-8_12

Fig. 12.1 Starting Simulink

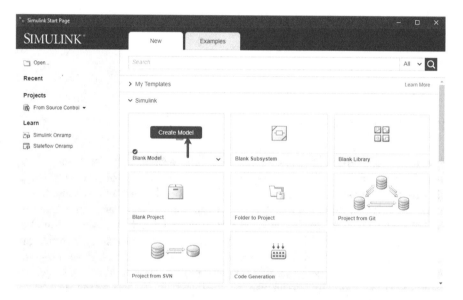

Fig. 12.2 Simulink Start Page

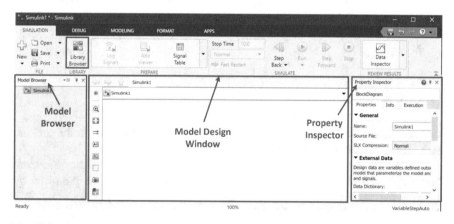

Fig. 12.3 Simulink Window

Browser" as marked in the header toolstrip of Simulink in Fig. 12.3. There are two other features in the Simulink starting window—Model Browser and Property Inspector, which can be minimized from the starting page as users' choice. Model Browser incorporates the name of the Simulink model available in the browser. Property Inspector can be used to customize the properties of the current Simulink model, such as changing the name of the Simulink file.

"Simulink Starter Page" can be accessed from the command window as well by typing the *simulink* command in there and pressing the *enter* key on the keyboard. The rest of the procedures to create a blank Simulink model are the same as mentioned earlier.

12.3   Basic Elements

In Simulink, two basic types of elements are available, with the use of which a Simulink model can be designed. These two elements are blocks and lines.

12.3.1   Blocks

In Simulink, a dynamic model can be designed and simulated, by utilizing different blocks available in Simulink Library Browser. In the Library Browser, multiple blocks are listed based on different applications, which can be dragged into the Model Design Window. Each of these blocks may have single or multiple ports based on its definition. They have certain properties which can be customized according to the users' requirements. Multiple blocks can be used and customized to create an entire model. In Fig. 12.4, a block named "Add" is dragged into the

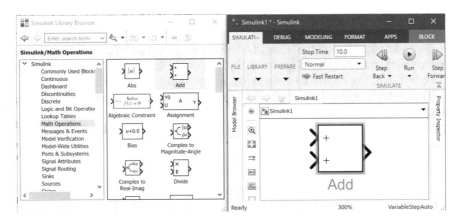

Fig. 12.4 Simulink Blocks

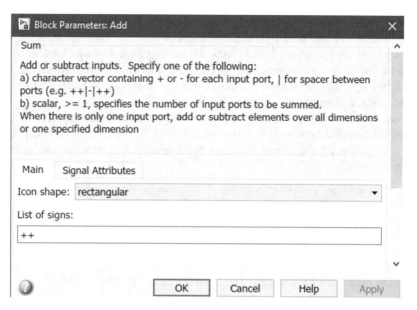

Fig. 12.5 Simulink Block Parameters

Simulink   Model   Design   window   from   "Simulink   Library
Browser→Simulink→Math Operations→Add." In the "Add" block, there are
three ports, two of which are the input ports (left side of the block) and one is the
output port (right side of the block).

By double-clicking on the "Add" block, the following window will appear, from
where a description of the block can be found. In addition, this window can be used
to change any parameters according to the requirement.

From Fig. 12.5, it can be observed that the "Add" block can be used to perform
both addition and subtraction. For doing so, two more input blocks and one output
block to simulate the result are required. Therefore, two "Constant" blocks and one
"Display" block are dragged into the Simulink Model Design window to complete
the design as shown in Fig. 12.6.

By double-clicking on the "Constant" block, the value can be changed manually.
In the above example, the values for the two input blocks are chosen as 5 and 10. A
"Display" block is chosen to observe the output result after the simulation. Thus,
blocks can be integrated into a Simulink Model Design window to create any desired
model.

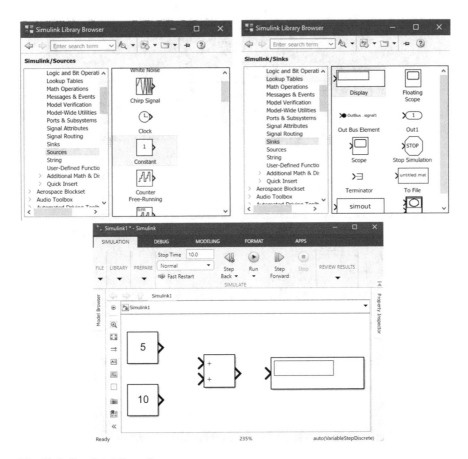

Fig. 12.6 Simulink Library Browser

12.3.2   Lines

To complete the design of a model, lines are very essential to make the relevant connections among different blocks. For example, in the previous model shown in Fig. 12.6, the goal is to perform the addition operation. Although all of the required blocks are gathered in the window, the model is not ready yet to run. All of the blocks are in float mode that needed to be connected via lines. A line connection can only be made between an input and an output port. Therefore, the output ports of the "Constant" blocks need to be connected with the input ports of the "Add" block. Similarly, the output port of the "Add" block is required to be connected with the input port of the "Display" block. The line connection can be made by dragging the mouse close to the ports. In Fig. 12.7, all the necessary line connections are made to complete the model design. The last part is to run the model for simulation which can be performed by clicking on the "Run" button from the Simulink header toolstrip.

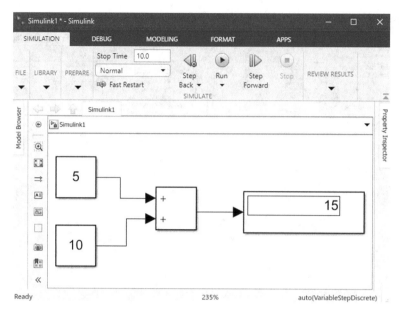

Fig. 12.7 Complete Simulink design for addition operation

After the simulation, the output will be appeared on the display, as shown in Fig. 12.7. Thus, blocks and lines constitute any basic model design in Simulink.

12.3.3   Other Features

Apart from blocks and lines, there are some other features which are very useful while creating any Simulink design. Some of the features, such as Annotation, Show Block Name, Fit to View, Area, Comment Out, and Uncomment, are covered in this section.

12.3.3.1   Annotation

Annotation is used to create captions of the parts of a design or an entire design. To write an annotation in the Simulink design window, the first step is to click on the "Annotation" option from the left strip of the design window as shown in Fig. 12.8. The second step is to click on the Simulink design window, which will create an appearance of a rectangular text box to write text, with a customization toolstrip above. Type the annotation and click outside the text box to complete the annotation. In Fig. 12.8, the final appearance of an annotation is shown in the right figure. The annotation can be dragged any place on the Simulink design window.

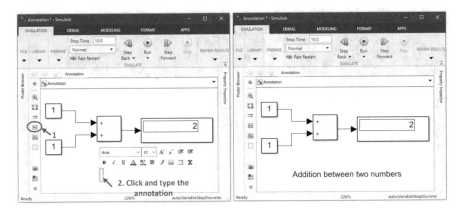

Fig. 12.8 Simulink Annotation

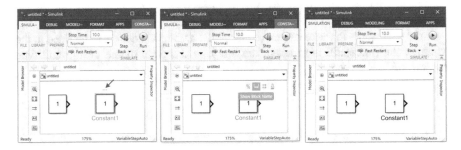

Fig. 12.9 Simulink Block Name

12.3.3.2   Show Block Name

In Simulink, after dragging a block in the Simulink design window, the name of the block disappears. To show the name of the block, select the block and move the mouse over the three blue dots as marked by a red arrow in the first figure of Fig. 12.9. When the mouse is moved over the three blue dots, a small tools strip will appear as shown in the second figure of Fig. 12.9. On that tool strip, move the mouse over each symbol, which will create the appearance of the name of each symbol. From there, select the option named "Show Block Name," which will make the appearance of the block name right below the blocks as shown in the third figure of Fig. 12.9.

12.3.3.3   Fit to View

A useful feature of the Simulink is the "Fit to View" option. Consider the first figure of Fig. 12.10, where the design does not fit to the entire window. By clicking on the "Fit to View" icon from the left tools strip, the entire design automatically fitted to

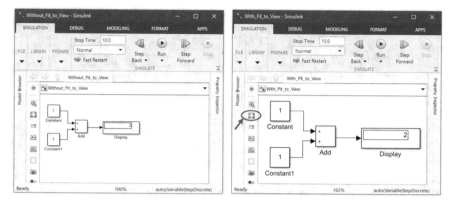

Fig. 12.10 Fit-to-view in Simulink

the design window. The alternative way to make this happen is to click the "Space button" after clicking on the Simulink design window.

If the user wants to make any particular block or part of a design to be fitted in the window, the procedure is to select the block or the parts of the design and type "Space + F."

12.3.3.4   Area

In Simulink, the Area feature is used to enclose a particular part of the design or the entire design. The enclosed part under an Area can be moved by only moving the Area window. In Fig. 12.11a, the procedure to enter an Area window into the Simulink design window is shown, where the "Area" icon is selected from the left tools strip. Later, the shape of the Area window can be adjusted according to the preference of the user. When a particular design is enclosed by an "Area" window, as shown in Fig. 12.11b, the entire design can be dragged by only dragging the Area window (Fig. 12.11c). The Area window is also used to highlight an important part of a design.

12.3.3.5   "Comment Out" and "Uncomment"

When designing a complete model in the Simulink, sometimes some blocks are required to make inactive as testing procedures. It happens especially when designing any complex model in the Simulink. As some blocks are required to make inactive for a short while, it is not an efficient way to delete the blocks for those

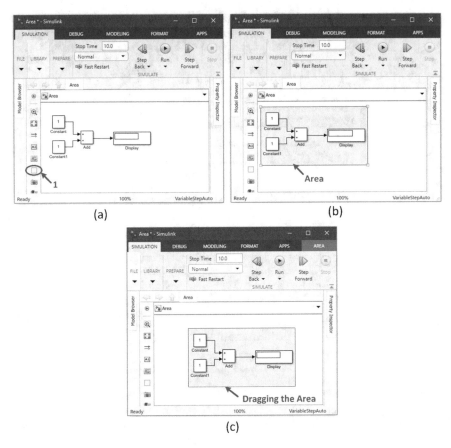

Fig. 12.11 Simulink Area

moments and reentered them again in the design. Another option is to leave the blocks unconnected; however, sometimes it creates some errors while simulating. Therefore, the best option in this scenario is to use the "Comment Out" and "Uncomment" features of the Simulink. By default, all the blocks remain in "Uncomment" mode while dragging to the Simulink design window. If the users "Comment Out" any block, it will become inactive even when simulating the model. When the users want to make the block active again, it can be done by changing the mode into "Uncomment." This is the most feasible option and becomes handy while designing complex model. To make particular blocks inactive, the first step is to select these blocks, which will create the appearance of three blue dots. Move the mouse over these blue dots, and a small ribbon of tools strip will appear with multiple tools symbols. Move the mouse over these tools one by one and find out the option named "Comment Out." Click on this option, and the selected blocks will become fade. An example is given in Fig. 12.12, where the first one is in the "Uncomment" mode, while the second one is in the "Comment Out" mode. Even

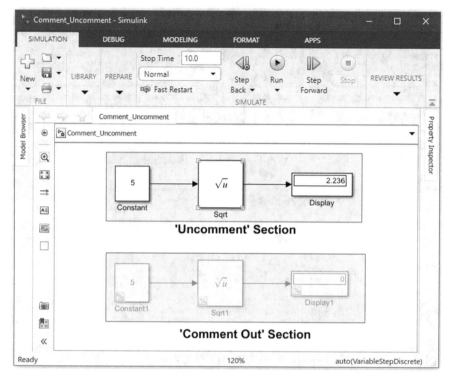

Fig. 12.12 Commenting and Uncommenting in Simulink

after running the simulation, the second one remains inactive in this mode as shown in Fig. 12.12.

12.4   Simulink Library Browser

A Simulink Library Browser contains all the blocks that are available in the Simulink arranged according to applications and operations. By clicking on the "Library Browser" option from the Simulink header toolstrip, it can be accessed. The appearance of the Simulink Library Browser window is provided in Fig. 12.13.

In the Simulink Browser window, different fields appeared on the left side, such as Simulink, Aerospace Blockset, Audio Toolbox, etc. In each of these fields, there are many subfields that incorporate several distinctive blocks. All of these blocks can be utilized to create a new model. Any of the blocks from this browser can be dragged to the Simulink Model Design window. There is a search option at the top of the browser, through which any particular block can be searched by the name as well.

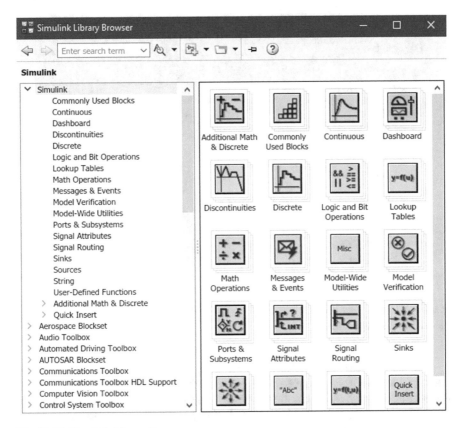

Fig. 12.13 Simulink Library Browser

12.5 Physical System Modeling

Like MATLAB which consists of several toolboxes, Simulink includes several modeling segments, which are specially created for specific applications. Event-based modeling, physical system modeling, systems engineering, code generation, application deployment, and reporting are some types of specialized products Simulink has to offer, which consist of different sets of libraries and syntaxes. Since this book will deal mostly with power and energy systems, Simulink's Simscape environment for modeling multi-domain physical systems will be utilized.

Simscape contains a new set of libraries and specialized simulation features to work with electrical systems, power electronics, mechanical devices, and control systems, which can be interconnected with any other Simulink blocks. The approach that Simscape takes to design physical system models is known as the Physical Network approach. Physical systems allow users with a unique experience to work with nondirectional devices and connect them like real physical components without specifying flow directions or information flow.

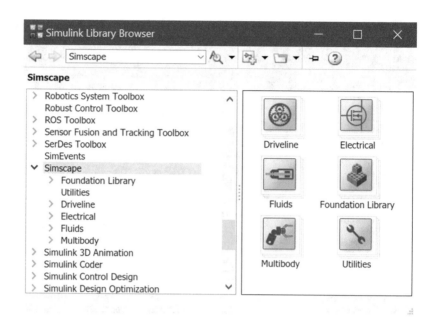

Fig. 12.14 Simscape Libraries

Simscape libraries are available in Simulink Library Browser. The names of the main libraries in Simscape are Driveline, Electrical, Fluids, Multibody, Utilities, and Foundation Library, as shown in Fig. 12.14. Most of the components in the library are in blue or green color to show distinction with other Simulink libraries.

The physical systems cannot be directly connected to the Simulink blocks because the physical systems are usually nondirectional like real physical components, whereas Simulink blocks require information about directionality. Converter blocks are needed to be used to convert the signal of a physical system into a Simulink output signal and vice versa. There are some Simscape blocks that serve the same purpose as other Simulink blocks. For example, the "voltage sensor" and "current sensor" blocks in Simscape measure voltage and current similar to a "voltage measurement" block and "current measurement" block. However, the output of the voltage sensor cannot be displayed in a display block in Simulink. In order to do so, a "PS-Simulink converter" is to be used. A "PS-Simulink converter" converts a physical signal into a Simulink output signal, which can be used to interconnect with any Simulink block available. A "Simulink-PS converter," on the other hand, converts a Simulink input signal into a physical system to be used by any Simscape blocks.

Figure 12.15 shows an example where the voltage and current in a DC circuit (with a voltage of 12 V and a resistor of 1000 Ω) are measured both in the physical environment (in the magenta area) and in the Simulink environment (in the cyan area). It is visible that each environment has its own counterpart for doing the same task. For example, DC voltage source and resistors exist in both environments but cannot be used interchangeably because of their difference in character. A "PS-Simulink converter" is used from the voltage and current sensors to convert the physical signal into the Simulink signal to be demonstrated using the "Display" Simulink block.

Physical models use two additional blocks to perform simulation, i.e., "powergui" and "solver configuration."

A powergui is an environment block that is used for designing electrical models. The block helps to use continuous and discretized methods to simulate a model. Since the model could be solved either by using a variable-step solver, or at a fixed time step, or through phasor solutions, the powergui block helps to select the appropriate method for the system. The block also aids the user in performing steady-state analysis and parameterizes the circuit for advanced designs. The block does not need to be connected to any component; rather, it is to be placed in the model file separate from other components, as shown in Fig. 12.15. The powergui block can be selected by going to Simscape→Electrical→Specialized Power Systems→Fundamental blocks. The types of simulation performed with this block are demonstrated in Fig. 12.16. Powergui block also consists of engineering tools that can facilitate advanced functionalities to a model such as steady-state and initial state calculation, line parameter and impedance calculation, fast Fourier transform analysis, line linear system analysis, hysteresis design, and load flow analysis. The tool can also generate reports and create custom blocks related to power systems. The window for the advanced powergui tools is shown in Fig. 12.17.

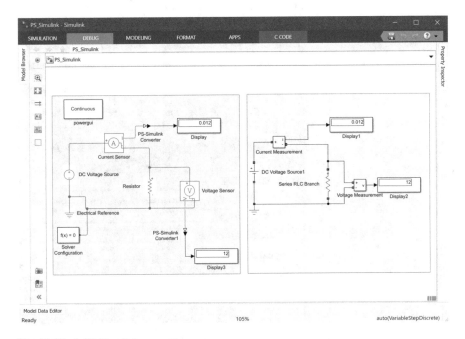

Fig. 12.15 A PS-Simulink converter

Fig. 12.16 Powergui blocks

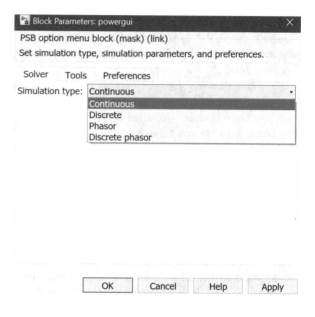

Fig. 12.17 Advanced powergui tools

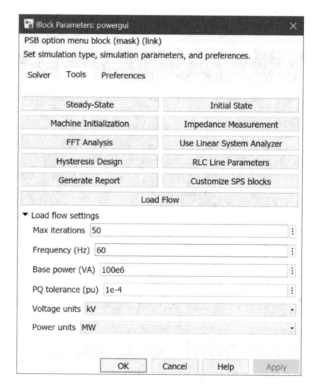

Table 12.1 summarizes the model simulation type of the powergui blocks.

A solver configuration block is another block used in Simscape to specify solver parameters of the model. The block is needed to be present with each physical network. Unlike powergui block, the solver configuration block must be connected at any point of the model. The block validates the model before running the simulation, by providing the solver setting, such as the parameters for starting the simulation from steady state, using different solver, or providing other runtime related constraints. The solver configuration block is available in Simscape→Utilities. Figure 12.18 shows the default parameters of the block. Selecting these options typically enables the user to perform steady-state or transient initialization for the designed electrical model. In-depth analysis of other solver setting parameters is out of the scope of this book.

Table 12.1 Model simulation type of the powergui blocks

Attributes	Simulation type			
	Continuous	Discrete	Phasor	Discrete phasor
Usage	Used for continuous system simulation, where previous states (conditions or results) of a system are taken into consideration	Used for discrete system simulation, where previous states of a system do not matter and a fixed time step is taken into consideration	Used for continuous systems, where only electrical phasors are needed for calculation	Used for discrete systems, where only electrical phasors are needed for calculation
Feature	Considers continuous time steps and integrated system, provides better accuracy and fast runtime for small systems	Considers only a fixed time step, provides better accuracy and fast runtime for large systems	Considers phasors for continuous time steps, calculation is faster in large systems as it uses continuous phasors and converts network differential equations into algebraic equations	Considers phasors for fixed time steps, calculation is more faster in large systems as it uses discrete phasors and converts network differential equations into algebraic equations
Notable applications	Small systems (usually with 20–30 states in a model), such as zero crossing analysis	Systems requiring finite cycle analysis, such as nonlinear power electronic switches	Systems requiring phasor analysis, such as machine transient modeling	Multimachine system design, such as three-phase motor or generator

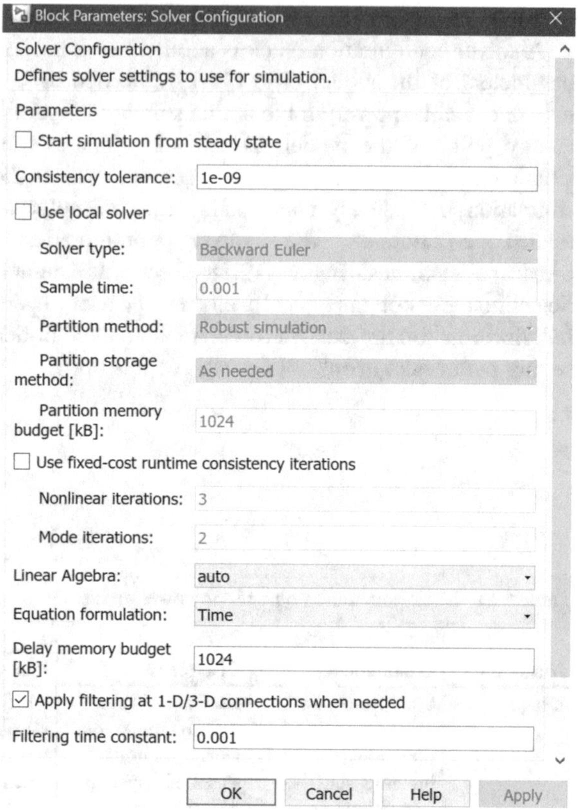

Fig. 12.18 Default
parameters of the solver
configuration block

12.6   Building a Model in Simulink

All the components that are required to design a model are discussed earlier. In this
section, a complete Simulink model will be designed and simulated to get comfortable with the platform. A step-by-step guideline to design a Simulink model to
generate sine wave will be demonstrated in the following example:

Generating Sine Wave
For generating a sine wave model in Simulink, the following blocks are required:

1. Sine wave
2. Scope

In Fig. 12.19, the procedures to pick up these two blocks are shown. From
Simulink Library Browser, any block can also be found out by using the name of
that block, instead of navigating. In Fig. 12.20, two alternated procedures to find out
the Scope block are shown. In Fig. 12.20a, the name of the block "Scope" is typed in
the search option of the Simulink Library Browser. After hitting the "Enter" button
from the keyboard, the "Scope" block will appear. The same searching procedure

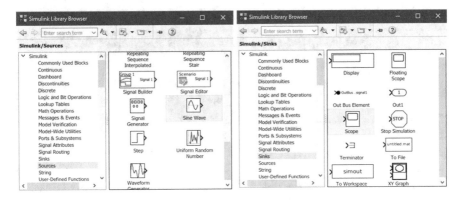

Fig. 12.19 Sine wave and scope from Simulink Library Browser

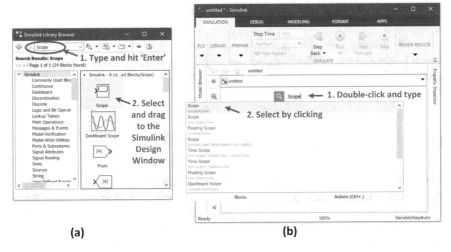

(a)   **(b)**

Fig. 12.20 Two ways of inserting the scope block

can also be performed in the Simulink design window as shown in Fig. 12.20b. By double-clicking on the design window, a blue strip will appear with the search symbol. Type the name of the block, which will create the appearance of a list of available options. Select the desired block from the list by clicking on it. The block will readily appear on the design window.

For generating a sine wave, a model is designed by utilizing the Sine Wave and Scope block. These two blocks are connected with a line as shown in Fig. 12.21. The Sine Wave block can be customized according to users' choice, which can be done by double-clicking on the block. After double-clicking on the Sine Wave block, the parameter window will appear as shown in Fig. 12.22. On the parameter window, there is a short description of the parameters of the Sine Wave block that can be

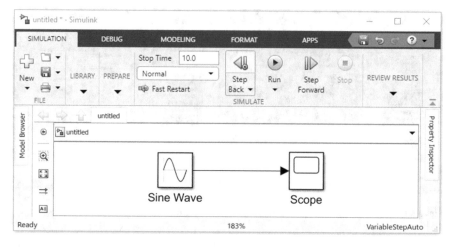

Fig. 12.21 A sine wave can be observed from the scope block

altered. In this example, the amplitude and the frequency of the sine wave are set to 5 and 1, respectively. After clicking the "Apply" button, the Sine Wave block is customized and the model is ready to be simulated.

12.7   Simulate a Model in Simulink

After building a simulation model, the next step is to simulate the design and observe the desired outputs. Multiple parameters can be adjusted and customized before simulating a model, which affects the outputs in different ways. In this section, some of these parameter customizations will be discussed with examples. If the parameters are not customized, the simulation will occur for a default setup. Before discussing the customization of different parameters, the simulation procedure will be covered with the default setup.

12.7.1   "Run" Option

To simulate any model in Simulink, the user need to press the "Run" button from the Simulink header toolstrip, which is shown in Fig. 12.23. The "Run" button is a dropdown box with two options. The first option "Run" implies simulating the model with default speed. The second option "Simulation Pacing" provides an option to simulate the model at a slower pace. The second option is useful to watch the output in a slower mode to visualize the change more accurately.

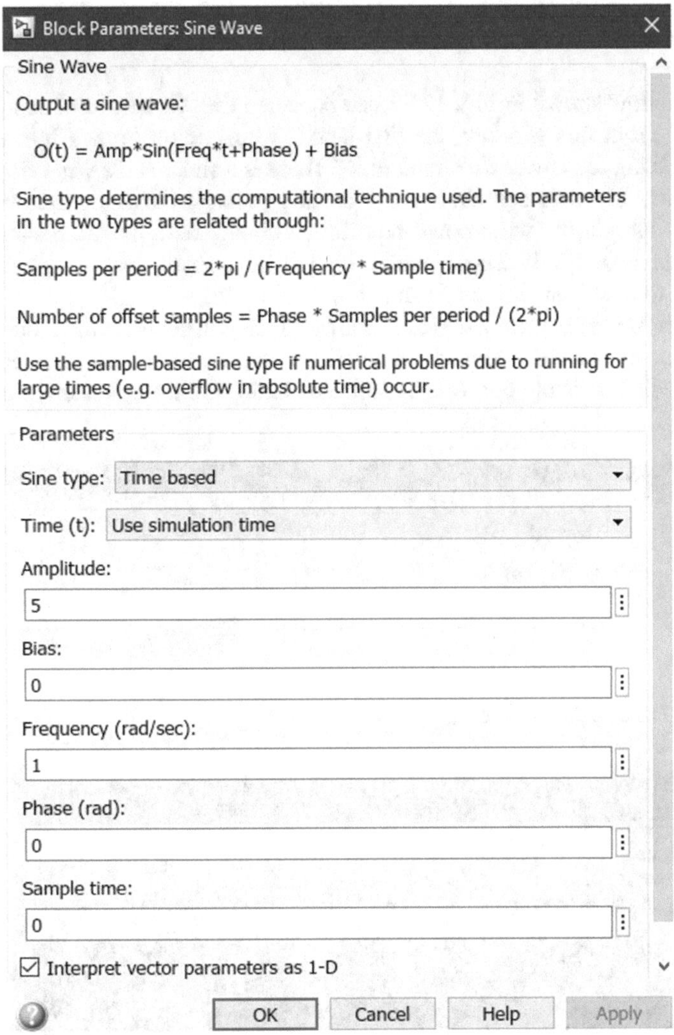

Fig. 12.22 Block parameters of the sine wave

For checking out the difference between these two simulation options, let us consider our previously designed Sine Wave Generator. Double-click on the "Scope" block to observe the output waveform. Initially, before the simulation, a blank scope window will appear. If the first option in the "Run" button is placed, the simulation will occur too fast that the user will have no time to stop or pause the simulation in midway. The scope window will show the final output, i.e., the generated sine wave. It will be generated so fast that the user will not be able to keep a track of the entire generator starting from the initial time to the end time. The following waveform in Fig. 12.24 will appear in the scope window:

However, by selecting the second option "Simulation Pacing," the speed of the simulation can be customized. If the second option is chosen, the following window will appear:

The window shown in Fig. 12.25 can be called the "Simulation Pacing Options window." From this window, the first task is to check the square box that states "Enable pacing to slow down simulation." There is a slider and a write-down box in the window. The user can choose any desired pacing by moving the slider or by writing a value in the write-down box. In the above example, the pacing that has been chosen is 0.45 s. Whenever the pacing will be enabled, a clock sign will appear right next to the "Run" button as shown in Fig. 12.26.

The next step is to press the "Run" button which will simulate the model with the selected pacing. As the model will simulate slower, two more options—"Pause" and "Stop"—will be visible and can be utilized during the time of the simulation, as

Fig. 12.23 Running the sine wave generator

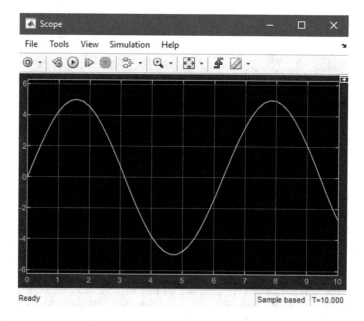

Fig. 12.24 Scope output of the sine wave generator

shown in Fig. 12.27. The lower left part of the window, which is marked as 3, verifies that the pacing is active. The marked 4 part indicates the percentage of the completion of the simulation. From the scope window, the generated wave can be tracked from the start point to the end. Another option that is marked as number 5 in the Simulink header strip is the "Stop Time," which indicates the duration of the simulation. If the "Stop Time" is specified as 20, it will indicate that the simulation will stop after 20 s. It should be noted that the simulation time is not identical to clock time. The simulation time may vary based on the computer speed, the complexity of the designed model, and other aspects. The default value of the "Stop Time" in Simulink is 10 s.

12.7.2 "Step Forward" and "Step Back"

"Step Forward" and "Step Back" options provide opportunities to pause a simulation and step forward and backward with a predefined step size. The first step to enable this feature is to click on the "Step Back" dropdown option, and select the "Configure Simulation Stepping" option as shown in Fig. 12.28.

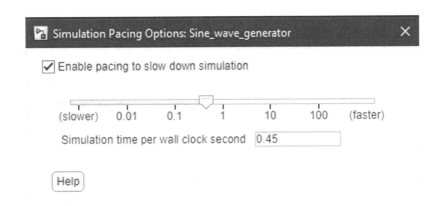

Fig. 12.25 Simulation pacing options of the sine wave generator

Fig. 12.26 Running the sine wave generator with pacing

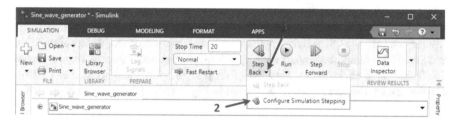

Fig. 12.27 Scope output of the sine wave generator with pacing

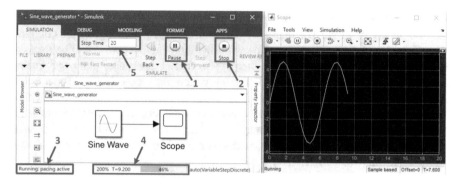

Fig. 12.28 "Step Forward" and "Step Back" options

This step will create the appearance of the following window, from where the user can enable the above-mentioned feature by clicking the checkbox next to the "Enable stepping back" line. In this window, the user can customize the maximum number of saved back steps, and the interval between them. Another option appears, where the user can decide how many steps they want to go forward, or backward for each click on the "Step Forward" and "Step Back" options. In Fig. 12.29, the default values are shown for each of the options:

After customizing the stepping option, the user can click on the "Run" button for simulation. After clicking the "Run" button, the "Pause" and the "Stop" options will appear. To observe each step of the simulation meticulously, the user needs to click on the "Pause" button, which will cause the appearance of three options by replacing the "Pause" option. The three options are "Step Back," "Continue," and "Step Forward." In Fig. 12.30, the Simulink window along with the "Scope" window for the Sine Wave Generator model after clicking the "Run" button is shown, where the "Scope" window is blank initially. In Fig. 12.31, the same appearance is shown right after clicking the "Pause" button after a while. Figures 12.32 and 12.33 illustrate the concept of stepping forward and stepping backward, respectively. While clicking the "Step Forward" button during the pause condition makes the simulation go one step forward (Fig. 12.32) and after clicking the "Step Back" option, the simulation steps one step backward (Fig. 12.33). If the "Continue" button is pressed, the simulation

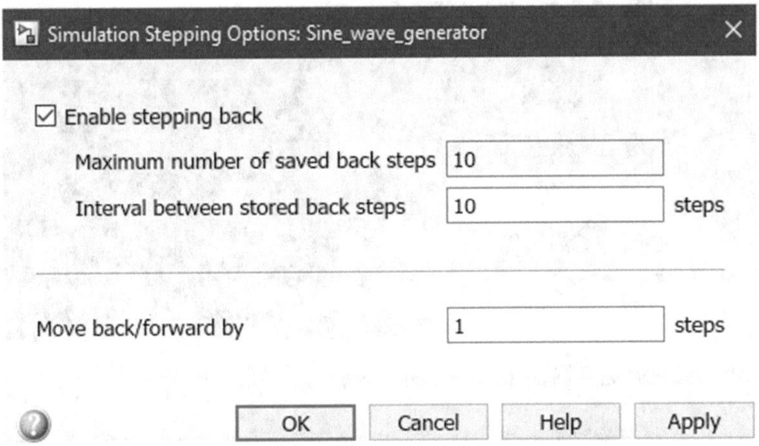

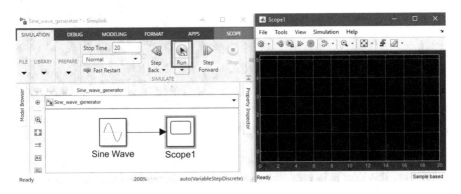

Fig. 12.29 Simulation stepping options of the sine wave generator

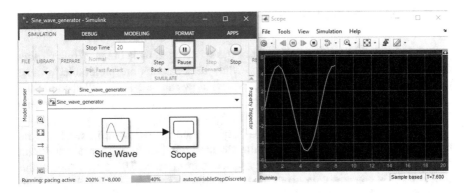

Fig. 12.30 Initial blank scope window in the sine wave generator

Fig. 12.31 Paused sine wave generator

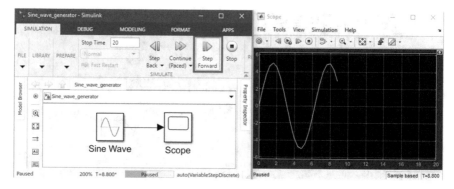

Fig. 12.32 The sine wave generator with step forward

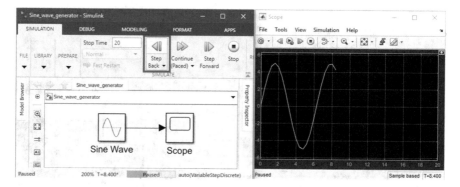

Fig. 12.33 The sine wave generator with step back

will start to continue from the paused position. Thus, for a detailed analysis of a simulation, this feature is very fruitful and effective in the engineering domain.

12.7.3 Customizing the Style of the "Scope" Figure

In default mode, the background color of the scope window is black. While writing scientific papers, or publishing books, white backgrounds are preferable in most cases. In Simulink, the style of the "Scope" window can be customized according to users' choices. The first step is to navigate to the "View→Style" option from the "Scope" window as shown in Fig. 12.34.

After clicking the "Style" option, a "Style" window will appear, which can be changed as shown in Fig. 12.35. In the changed window, the "Figure color" and "Axes background color" are changed into white from black. The "Ticks, labels, and grid colors" is changed into black from gray color. The Line color is changed into black from yellow, and finally, the line width is set as 1 in place of 0.75. After

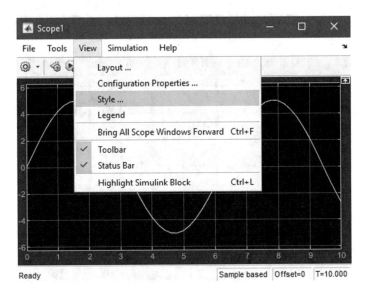

Fig. 12.34 Custom scope styling

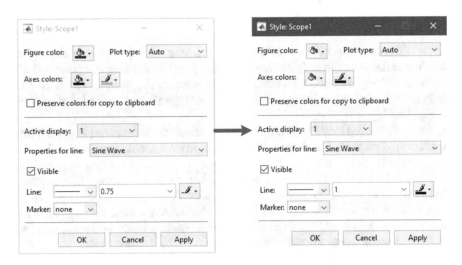

Fig. 12.35 Scope style window

changing all these options from the "Style" window, the final step is to click "Apply," which will create the output figure of the "Scope" as shown in Fig. 12.36. Although some certain changes have been made from the "Style" window, the users may change any options from this window to customize the style of the figure of the "Scope" window.

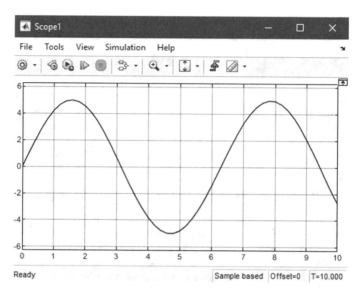

Fig. 12.36 New scope style

12.7.4 "Solver" Option

While simulating a design "Solver" option decides the solving algorithm that works for a particular simulation. In default, Simulink provides the "Auto" option, which selects a solving algorithm automatically for a particular design. The user can observe which solving algorithm is used by looking at the red marked zone shown in Fig. 12.37.

As the following model is simulated in the default mode of solving option, it can be observed that the Solver name in this simulation is "Variable StepDiscreet" in the "Auto" Mode. If the user wants to customize the solver information, it can be performed by first clicking on the setting option from the "Solver Information" feature window, as marked in blue in Fig. 12.38.

After clicking the "Solver Setting" option, the "Configuration Parameters" window will appear (Fig. 12.39), where parameters relevant to solver information can be configured according to the users' choice.

The selection of a particular solver depends on the type of solver that will be appropriate for a certain simulation. The types of solver can be categorized into two parts—Variable-step and Fixed-step type. Within each of these types, multiple solving algorithms are available in Simulink that can be utilized according to the users' choice. In Table 12.2, all the available solver is categorized in the following table based on the two different types of solvers as mentioned earlier:

The users have the liberty to pick any of the solver algorithms from the above-mentioned table for a particular simulation. In default, the setup remains in "Auto" mode, which automatically decides a particular algorithm for a certain Simulink

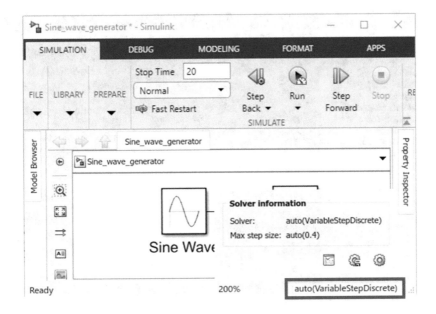

Fig. 12.37   Solver information

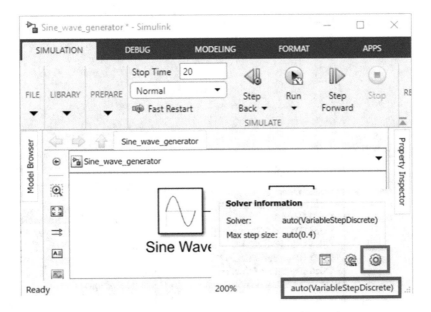

Fig. 12.38   Settings in solver information

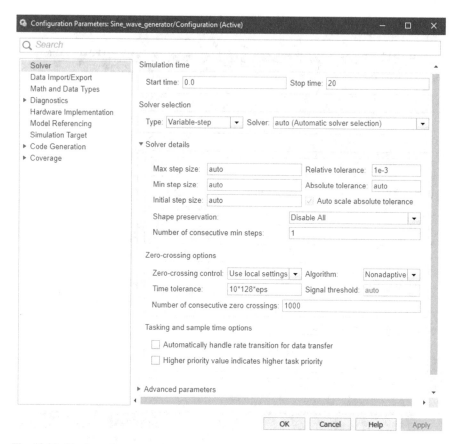

Fig. 12.39 Configuration parameters

Table 12.2 Different types of solver algorithms in Simulink

Solver type	Fixed-step solver	Variable-step solver
Continuous	• Ode1 (Euler) • Ode1be (Backward Euler) • Ode2 (Heun) • Ode3 (Bogacki-Shampine) • Ode4 (Runge-Kutta) • Ode5 (Dormand-Prince) • Ode8 (Dormand Prince) • Ode14x (Extrapolation)	• Ode15s (Stiff/NDF) • Ode23 (Bogacki-Shampine) • Ode23s (Stiff/Modified Rosenbrock) • Ode23t (Modified stiff/Trapezoidal) • Ode23tb (Stiff/TR-BDF2) • Ode45 (Dormand-Prince) • Ode113 (Adams) • OdeN (Nonadaptive) • Daessc (DAE solver for Simcape)
Discrete	• Fixed-step discrete solver	• Variable-step discrete solver

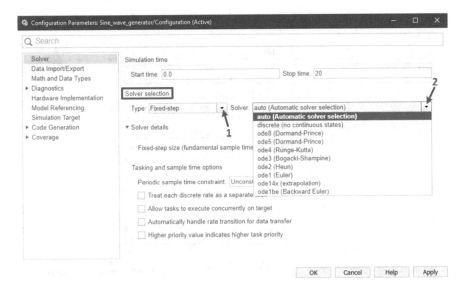

Fig. 12.40 Solver selection in configuration parameters

design. For configuring any particular solver, the first step is to navigate to the "Solver" section of the "Configuration Parameters" window, and then, select the type of the solver by clicking the dropdown option named "Type" under the "Solver Selection" section. Later, the second step is to click on the "Solver" dropdown option to choose any particular solver algorithm that falls under the previously selected types of the solver. These two steps are highlighted in Fig. 12.40 by 1 and 2 with red marks:

For each type of solver, Simulink also provides the option for configuring other parameters to customize solver options with more details. The available parameters that can be configured for each type of solver are listed in Tables 12.3 and 12.4 with some details:

12.7.5 Data Import and Export

In the "Configuration Parameters" window, a "Data Import/Export" section can be found, which can be utilized to link Simulink with the Workspace of MATLAB. The input data can be loaded from the Workspace by using the option "Connect Input" under the subsection "Load from workspace." The initial state can also be assigned externally as well.

Another feature of this section is to save data in the Workspace from the Simulink. The users may pick the data which they want to export in the Workspace. Thus, the parameters of this window can be configured to make collaboration between Simulink and MATLAB by importing and exporting data. The "Data Import/Export" window is shown in Fig. 12.41.

Table 12.3 Different types of parameters for fixed-step solver algorithm in Simulink

Fixed-step solver	
Parameters	Details

1. Periodic sample time constraint	• The constraint of the periodic sample time can be selected
	• There are three options to select for this parameter:
	    a. Unconstrained
	    b. Ensure sample time-independent
	    c. Specified
	• In "Unconstrained" mode selection, a "Fixed-step size" option will appear, which will be needed to specify. The "Fixed-step size" option will be explained later
	• In "Ensure sample time-independent" mode selection, the Simulink model needs to have a sample time inherited by itself
	• In "Specified" mode selection, the "Sample time properties" option will appear, where the priorities of sample time need to be assigned. "Sample time properties" will be explained later

2. Fixed-step size (fundamental sample time)	• The step size of the numerical solver, which is a fixed value • Set "Auto" or type any valid number • It will only appear if the "Periodic sample time constraint" is assigned in the "Unconstrained" mode

3. Sample time properties	• It mentions and specifies the priorities of the sample time
	• The type of input of this parameter is a $n \times 3$ size matrix, where n indicates the number of sample times
	• It is ordered from the fastest to the slowest of the sample times
	• For each sample time, three values are assigned, which are period (sample rate), offset, and priorities, respectively, in order.
	• Sample format:
	    [Period_1 Offset_1 Priority_1 (1st sample time)
	    Period_2 offset_2 priority_2 (2nd sample time)
	    Period_n Offset_n Priority_n (nth sample time)]

4. Treat each discrete rate as a separate task	• A checkbox option, which has two modes "On" and "Off" • By clicking on the check box, "On" mode can be initiated • In "On" mode, multitasking is enabled, where the Simulink model can be executed at different sample rates • In "Off" mode, all the blocks will be executed for a single sample rate
5. Higher priority value indicates higher task priority	• A checkbox option, which has two modes "On" and "Off" • By clicking on the check box, "On" mode can be initiated • In the "Sample time properties," the priority value can be assigned. However, which priority value will indicate higher or lower task priority can be assigned by this option • In "On" mode, a higher priority value indicates a higher priority task • In the "Off" mode, a higher priority value signifies a lower priority task

Table 12.4 Different types of parameters for variable-step solver algorithm in Simulink

Variable-step solver	
Parameters	Details
1. Max step size	• The maximum step size of the numerical solver • Set "Auto" or type any valid number • In default, it will be assigned as "Auto"
2. Min step size	• The minimum step size of the numerical solver • Set "Auto" or type any valid number • In default, it will be assigned as "Auto"
3. Initial step size	• The initial step size of the numerical solver • Set "Auto" or type any valid number • In default, it will be assigned as "Auto"
4. Relative tolerance	• It specifies the maximum acceptable error of the solver relative to the size of every state during every time step • Default value: $1e^{-3}$ • It can be set any valid number
5. Absolute tolerance	• It specifies the maximum acceptable error when the measured state value approaches zero • Default value: "auto" • It can be set any valid number
6. Shape preservation	• It has two options—disable all and enable all • In the "enable all" option, shape preservation is applied to all the signals. In this mode, the integration accuracy in each state is improved by using derivative information in each time step • In the "disable all" option, shape preservation is not applied • Default value: "disable all"

7. Zero-crossing control	• It is used to enable or disable zero-crossing detection
	• It has three options to choose:
	    a. Use local settings
	    b. Enable all
	    c. Disable all
	• In the "use local settings" option, zero-crossing detection can be enabled for each block manually. For enabling the zero-crossing detection of a particular block, navigate to the parameter box of that block, and check on the "enable zero-crossing detection" option
	• By selecting the "enable all" option, zero-crossing detection will be enabled for all the blocks in a Simulink model
	• By selecting the "disable all" option, zero-crossing detection will be disabled for all the blocks in a Simulink model

8. Algorithm	• It falls under the category of the zero-crossing option • It is used to specify the algorithm for zero-crossing detection • Simulink provides two algorithms for such task—adaptive and nonadaptive algorithm • Adaptive algorithm is more suitable for models that have strong zero behavior • The nonadaptive algorithm can provide accurate detection; however, it may take a longer simulation time • Default value: nonadaptive algorithm

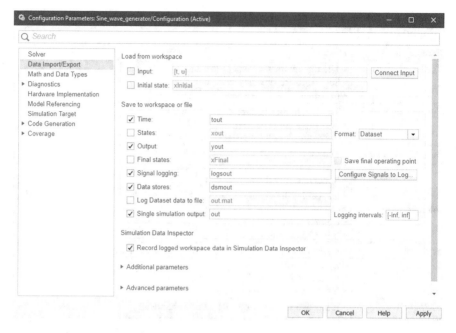

Fig. 12.41 Data Import and Export in configuration parameters

12.7.6 Math and Data Types

"Math and Data Types" is another section in the "Configuration Parameters" window. In the Math subsection, the users can configure two parameters. Firstly, the choice regarding how to handle denormal numbers during simulation. A denormal number can be defined as a number, which is nonzero and smaller than the lowest floating normalized number. In some aspects, denormal values are flushed to zero. Therefore, under the "Math" subsection, there is a dropdown option named "Simulation behavior for denormal numbers," which has two options to configure: Gradual Underflow, which does not make any changes for the denormal numbers, and Flush to Zero (FTZ), which converts any denormal number to zero.

Secondly, there is a checkbox option named "Use algorithms optimized for row-major array layout" under the "Math" subsection. If the checkbox is checked to enable this option, such configuration will allow Simulink to enable an efficient algorithm that will traverse the data in row-major order. For selecting such an option, the generated algorithm will be efficient for a row-major array layout, however will not be fruitful for a column-major array layout.

Unchecking, or disabling, the above option indicates that an efficient algorithm will be enabled, which will traverse the data in column-major order. Therefore, the generated code will be efficient, if the array layout is column-major.

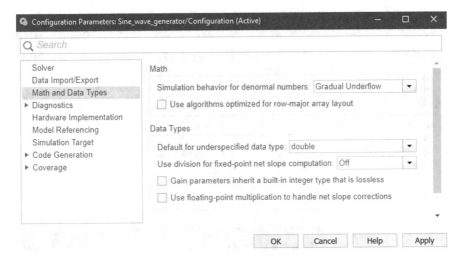

Fig. 12.42 Math and Data Types in configuration parameters

In the Data Types subsection, any unspecified data can be configured as either double or single. There are some other parameters, such as the options of computing fixed-point net slope computation, handling net slope corrections, etc. All of these parameters can be configured by using the window as shown in Fig. 12.42.

12.7.7 Diagnostics

A Diagnostics window appears after a simulation model is run to simulate, which shows errors, or any kind of warnings that need to be fixed. If there is no error, or associated warning, the "Diagnostics" window will not appear after the simulation. Simulink provides the users' flexibility to choose for which conditions a simulated model should show warnings, errors, or none. The "Diagnostics" window is shown in Fig. 12.43, which is a section of the "Configuration Parameters" window.

12.7.8 Other Parameters

In the "Configuration Parameters" window, there are five more sections under which different parameters can be customized. For example, the "Hardware Implementation" section can be configured to specify the types and characteristics of a computer-based model, to be more specific, an embedded system designed using Simulink. In this chapter, all these parameters will not be explained, as the book aims to work as a crash course for the readers. To learn more about these parameters, select a section from the "Configuration Parameters" window, and click on the

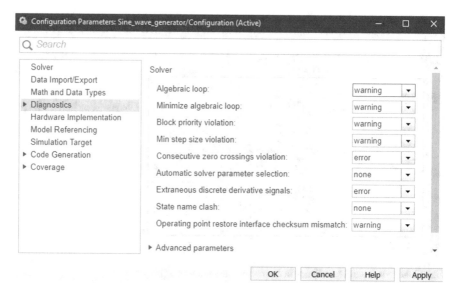

Fig. 12.43 Diagnostics in configuration parameters

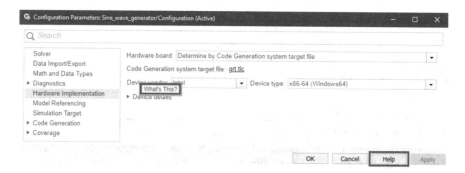

Fig. 12.44 Hardware implementation in configuration parameters

"Help" button from the lower-left portion of the window. A sample is shown in Fig. 12.44, where the "Hardware Implementation" section is chosen, and after clicking the "Help" button, the help page appears as shown in Fig. 12.45. From this page, a concise overview of this section can be learned. To know more about the individual parameter configuration of this section, move the mouse to a certain parameter and right-click on it. An option box named "What's This?" will appear. By clicking on it, a help page will appear describing all the relevant information on when and how to configure this particular parameter. In Fig. 12.44, a sample is shown where after right-clicking on the "Device vendor" parameter, the "What's This?" option appears. After clicking on this option, a short guideline to configure this parameter will appear, which is shown in Fig. 12.45. For learning

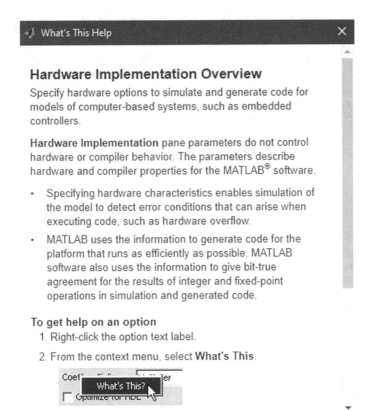

Fig. 12.45   Hardware implementation overview

more details about the configuration of this parameter, click on the "Show more information" option red marked in Fig. 12.46.

12.8   User-Defined Block in Simulink

In Simulink, multiple user-defined blocks are available, such as C caller, C Function, MATLAB Function, MATLAB System, etc. The algorithms of the block can be written in C, C++, MATLAB, or Fortran depending on a particular user-defined block. From the Simulink Library Browser, navigate to the "Simulink→User-Defined Functions," where all the available blocks are listed. In the following example, the MATLAB Function block is utilized to illustrate how to create a user-defined block in Simulink.

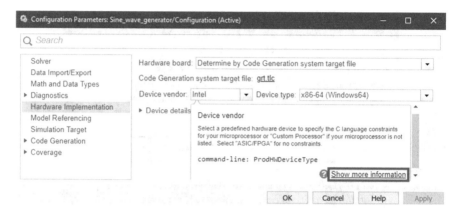

Fig. 12.46 Getting more information about device vendor

```
function [Magnitude, Angle] = Cartesian_to_polar(Real, Imaginary)
Magnitude = sqrt(Real^2 + Imaginary^2);
Angle = atan(Imaginary/Real)*(180/pi);
```

Fig. 12.47 Code—Function for conversion from rectangular to polar coordinates

Example 12.1 A rectangular to polar converter block is created by utilizing the MATLAB Function block in this example. A step-by-step guideline is provided below to create such a user-defined block:

Step 1: Drag the "MATLAB Function" block from the "Simulink Library Browser→Simulink→User-Defined Functions→MATLAB Function" to the Simulink design window.

Step 2: Double-click on the "MATLAB Function," which will open a MATLAB script file in the Editor.

Step 3: Write a user-defined function by specifying input and output parameters to define the task. In this example, the input parameters are the real and imaginary parts of the number represented in rectangular form. The outputs are the magnitude and angle in polar form. A user-defined function for rectangular to polar conversion is written as in Fig. 12.47 in the MATLAB script:

Step 4: Save the MATLAB script (.m file) in the same working directory as the Simulink file (.slx file). It will create two input and output ports in the MATLAB Function block that was dragged before in the Simulink design window.

Step 5: Drag two "Constant" blocks from the "Simulink Library Browser→Simulink→Sink" to provide inputs of the real and imaginary number. Connect both of them with the input port of the "MATLAB Functions" block.

Step 6: Get two "Display" blocks from the "Simulink Library Browser→Simulink→Sink" to display the two output parameters—Magnitude and Angle. Connect these displays with the output ports of the "MATLAB Functions."

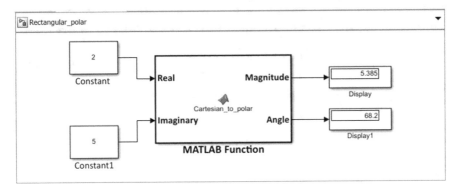

Fig. 12.48 Simulink diagram of the function for conversion from rectangular to polar coordinates

```
function [Mean_value,Standard_deviation,Variance] = fcn(input)
Mean_value = mean(input);
Standard_deviation = std(input);
Variance = var(input);
```

Fig. 12.49 Code—Function for computing mean, standard deviation, and variance

Step 7: Double-click on the "Constant" blocks to set two different values—real and imaginary parts—in the blocks.

Step 8: Click on the "Run" button to simulate the model. The overall Simulink design after the simulation is shown in Fig. 12.48, where the "MATLAB Function" acts as a user-defined block:

Example 12.2 A user-defined Simulink block that can compute mean, standard deviation, and variance. To create this Simulink block, the MATLAB Function block is utilized. After double-clicking the "MATLAB Function" block, a MATLAB script appears. The MATLAB code to define the task of the "MATLAB Function" block is shown in Fig. 12.49.

In the above code, a user-defined function *fcn*() is created, where the outputs are Mean_value, Standard_deviation, and Variance. Three "Display" blocks are connected with the output ports to display the results. In the "input" port, a "Constant" block is connected, where a vector array [5 2 10 2 8 7 9] is assigned from its parameter window. The overall design of the model is shown in Fig. 12.50 after the simulation.

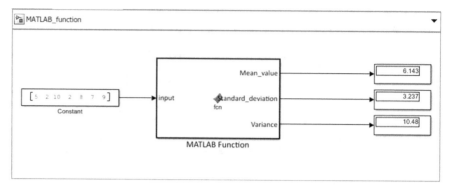

Fig. 12.50 Simulink diagram of the function for computing mean, standard deviation, and variance

12.9   Using MATLAB in Simulink

From Simulink, it is possible to shift any simulated output, or data, to the MATLAB Workspace, from where it can be used to perform any task using MATLAB coding. One of the options to perform such a task is to utilize a Simulink block "To Workspace." An example is shown below to show the usage of this block to establish the usage of MATLAB in Simulink:

Example 12.3 In this example, a solar cell is simulated to demonstrate the PV characteristic curve for different solar irradiance values with the usage of both Simulink and MATLAB. The utilized blocks and the navigation routes from them are summarized in the following Table 12.5:

The designed model by utilizing the above-mentioned blocks is shown in Fig. 12.51. The value of the PS Constant block indicates the solar irradiance value, which is set as 800 initially. Except for three "To Workspace" blocks, all other blocks are kept with their default parameter values. The "To Workspace" block is customized by changing the "Variable name" and the "Save format." In Fig. 12.52, the customization of the "To Workspace" block is shown, where the "Variable name" is changed to "I1" and the "Save format" is changed to "array." For "To Workspace1" block, the "Variable name" and the "Save format" are changed into "P1" and "array," respectively. Finally, the "To Workspace2" block is modified by setting the "Variable name" and "Save format" as "V1" and "array." It is to be noted that the "To Workspace" and "Display" blocks indicate the current; the "To Workspace1" and "Display1" blocks refer to the power; and the "To Workspace2" and "Display2" blocks indicate the voltage from the solar cell circuit.

In this example, we are interested in plotting the PV characteristic curve for different solar irradiance values by utilizing MATLAB. Therefore, in the first simulation, the value of the PS Constant block is set to 800 initially which represents the value of solar irradiance. To shift the data of voltage, current, and power to the

Table 12.5 Blocks and navigation routes utilized in Example 12.3

Name of the block	Navigation route
Solar cell	Simscape→Electrical→Sources→Solar cell
PS constant	Simscape→Foundation library→Physical signals→Sources→PS constant
Solver configuration	Simscape→Utilities→Solver configuration
PS-simulink converter	Simscape→Utilities→PS-Simulink converter
Simulink-PS converter	Simscape→Utilities→Simulink-PS converter
Current sensor	Simscape→Foundation library→Electrical→Electrical sensors→Current sensor
Voltage sensor	Simscape→Foundation library→Electrical→Electrical sensors→Voltage sensor
Variable resistor	Simscape→Foundation library→Electrical→Electrical elements→Variable resistor
Ramp	Simulink→Sources→Ramp
To workspace	Simulink→Sinks→To workspace
Product	Simulink→Math operations→Product
Display	Simulink→Sinks→Display

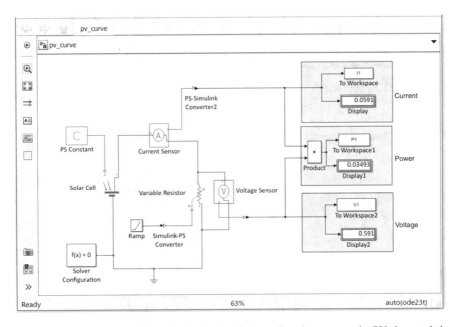

Fig. 12.51 Simulink diagram for the simulation of solar cell to demonstrate the PV characteristic curve for different solar irradiance values

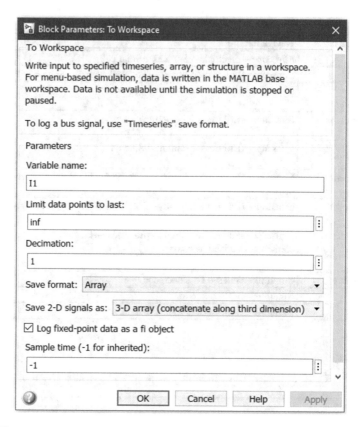

Fig. 12.52 Block parameters

Fig. 12.53 Variables in the workspace

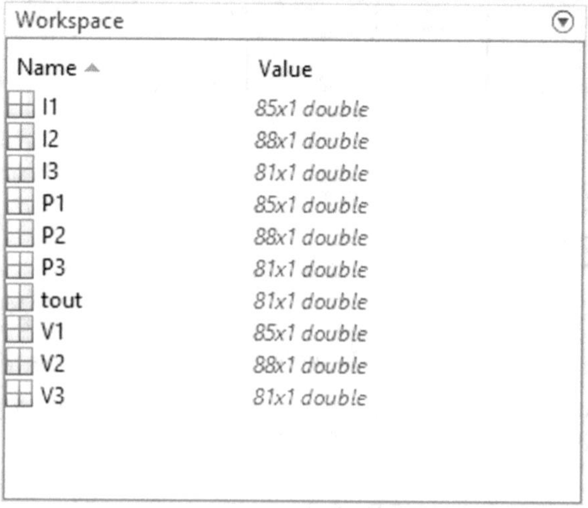

```
clc;
figure(1);
plot(V1,P1,'LineWidth',2);
hold on;
plot(V2,P2,'LineWidth',2);
hold on;
plot(V3,P3,'LineWidth',2);
grid on;
xlabel('Voltage, V (Volt)');
ylabel('Power, P (Watt)');
title('PV characteristic curve of photovoltaic cell');
labels={'Solar Irradiance: 800 W/m^2','Solar Irradiance: 1000 W/m^2',...
    'Solar Irradiance: 1200 W/m^2'};
legend(labels,'Location','Northwest');
```

Fig. 12.54 Code—Plotting the PV characteristic curve for different solar irradiance values

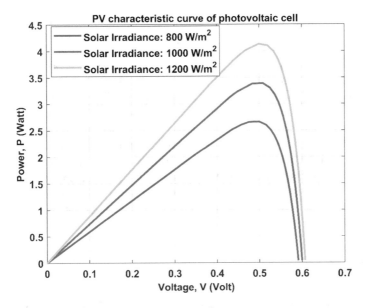

Fig. 12.55 PV characteristic curve of photovoltaic cell

MATLAB Workspace, the "To Workspace" blocks are used. For the solar irradiance value 800, the variable names of the three "To Workspace" values are set as "I1," "P1," and "V1" as mentioned earlier. Afterward, click the "Run" button to simulate the model, which will shift the values of I1, P1, and V1 to the MATLAB Workspace directory. Later, change the value of the "PS Constant" block to 1000 and change the variable names of the three "To Workspace" blocks to "I2," "P2," and "V2." Later, simulate the model again by clicking the "Run" button. Finally, simulate the model for the new "PS Constant" value 1200 and the changed variable name of the three

"To Workspace" blocks "I3," "P3," and "V3." To check whether the values of the variables are shifted to the MATLAB Workspace, go to the MATLAB starting page, and in the "Workspace" directory, all the variables will be available with the assigned name in the "To Workspace" blocks as shown in Fig. 12.53.

As the variables are shifted to the MATLAB Workspace, we can use MATLAB Editor to write code using the available variables. As we are interested in plotting the PV characteristic curves for different solar irradiance values, the following code (Fig. 12.54) in the MATLAB editor can do this task.

After running the above-mentioned code in the MATLAB Editor, the PV characteristic plots appear as shown in Fig. 12.55. Here, we have utilized only the values of powers and voltages for plotting the PV curve. By utilizing the values of voltages and currents, the VI curve can also be plotted in the same manner. Thus, using MATLAB, any simulated model can be analyzed further in more detail.

12.10   Conclusion

This chapter acts as introductory guidelines for the readers to proceed further with Simulink. By going through the contents of this chapter, the readers will be able to understand the basic features of the Simulink along with different elements that will be necessary to realize the contents of the later chapters. The basic procedures to build any model in the Simulink design window, usage of the Simulink Library Browser, and simulating a model are illustrated step-by-step with examples in this chapter. In addition, this chapter also covers how to create a user-defined block and configure it according to the users' interests. Finally, the interaction between MATLAB and Simulink is shown with examples in the last section of this chapter to realize their usability and applications in the engineering domain.

Exercise 12

1. How can you describe the importance of Simulink in the engineering domain?
2. (a) What are the ways to open Simulink from the MATLAB window?
   (b) Use one of the ways and create a blank Simulink file titled "First_Exercise. slx."
   (c) Drag the following blocks in the Simulink design window of the "First_Exercise.slx" file:
       (i) Add
       (ii) Constant
       (iii) Display
   (d) Design a model that can perform subtraction using the previous blocks.

(e) Simulate the model and verify the result by performing it in the MATLAB command window.

3. (a) What is the purpose user-defined blocks?

(b) Create a user-defined block that can convert any polar coordinates to rectangular coordinates.

(c) Using the user-defined block mentioned in (b), design a model that can perform polar to rectangular conversion

(d) Convert the following two numbers in their rectangular forms by using the previous model:

    (i) 10 ∠ 45 *rad*

    (ii) 20 ∠ 30 *rad*

4. (a) Recreate the model shown in Example 12.3 and simulate the model for solar irradiance 1000.

(b) Generate the graph of P and V with respect to time using a single scope. Customize the following parameters of the style of the graph in the Scope window:

    (i) Figure color: White

    (ii) Axes background color: White

    (iii) Ticks, labels, and grid colors: Black

    (iv) Linewidth: 1.0

    (v) Line color of the P graph: Blue

    (vi) Line color of the V graph: Red

(c) Generate the PV characteristic curve of the photovoltaic cell using an XY Graph block in the design. (The XY Graph is available in Simulink→Sinks→XY Graph). Change the maximum and minimum value of the x-axis into [0 1], and y-axis into [0 3] by double-clicking on it.

(d) Use MATLAB "plot" command to generate VI characteristic curve of photovoltaic cell for solar irradiances 800, 1000, 1200, and 1600.

Chapter 13
Commonly Used Simulink Blocks

13.1 Sink

In Simulink, it is essential to display signal values or signal graphs. Hence, sinks are one of the most commonly used blocks in the Simulink model to visualize the output results in the form of values or graphs. There are different forms of sinks available in the Simulink platform. In this chapter, some of the basic blocks will be explained, such as display, scope, floating scope, to workspace, XY graph, etc.

13.1.1 Display

"Display" block is used to observe values from a particular signal line. To drag the "Display" block in the Simulink design window, navigate to the path—Simulink Library Browser→Simulink→Sinks→Display.

Navigation route:
   Simulink Library Browser→ Simulink→ Sinks→ Display

After dragging the block into the design window, double-click on the block, which will cause the appearance of the window named "Block Parameters: Display." From this parameter window, the users have the opportunity to customize some of the parameters of the block, such as the Numeric display format. There are nine available formats available for the "Display" block. Another parameter of this block is called "Decimation," which indicates the frequency of displaying data on this block. The last customizable parameter is the checking box option named "Floating display." By checking this box, a "Display" block can be made floating, which means the block does not require to be connected with the signal line. By selecting a

© The Author(s), under exclusive license to Springer Nature Switzerland AG 2022   361
E. Hossain, *MATLAB and Simulink Crash Course for Engineers*,
https://doi.org/10.1007/978-3-030-89762-8_13

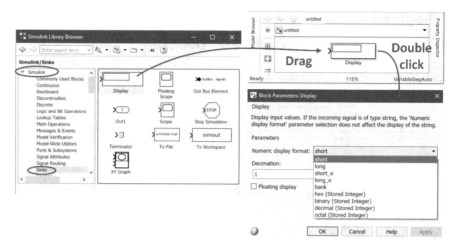

Fig. 13.1 The display block

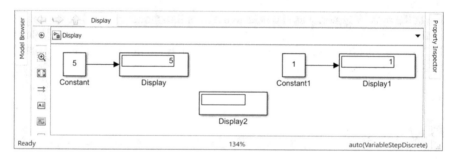

Fig. 13.2 Three display blocks

signal line before simulation, it is enough to show the data in a Floating display. In Fig. 13.1, the "Display" block finding, dragging, and parameter window appearance are shown graphically.

An example of the usage of the "Display" block is shown in Fig. 13.2, where the "Constant" block is used as a source. In the following figure, three "Display" blocks can be found, where "Display 2" is in the floating mode. In "Display" and "Display 1," the values of the "Constant" blocks can be observed after the simulation happens; however, in "Display 2," it is blank.

In Fig. 13.3, the signal line between the "Constant" and "Display" blocks is selected, and simulated afterward by clicking the "Run" button. It causes the "Display 2" block, which is on the floating mode, to show the output of the "Constant" block. On the other hand, in Fig. 13.4, the signal line between the "Constant 1" and "Display 1" blocks is selected before the simulation. As a result, this time "Display 2" presents the signal value of "Constant 1." It is to be noted that

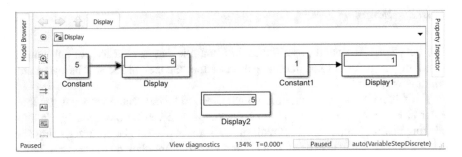

Fig. 13.3 Display2 shows the same output as Display block

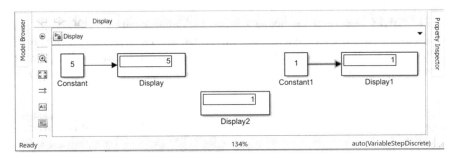

Fig. 13.4 Display2 shows the same output as Display1 block

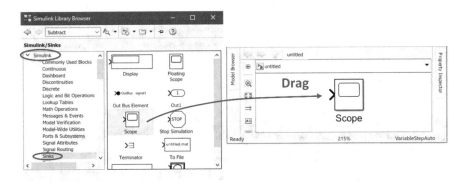

Fig. 13.5 The scope block

when the checkbox option of the "Floating display" is checked, the port sign disappears from the "Display" block.

13.1.2 Scope

The Scope is a frequently used block in Simulink that is used to plot any output signal graph. In Fig. 13.5, the navigation of the "Scope" block and its parameter window is shown:

The "Scope" has one input port, which needs to be connected with a signal line. After simulation, the output graph of the signal line can be observed by double-clicking on the "Scope" block. If multiple line output needs to be observed using one "Scope" block, it can also be attained. For doing so, right-click on the "Scope" block and choose the option "Signal and Ports→Number" of input ports. Under "Number of input ports," there are options to select a number to define the desired number of input ports for a scope.

The "Style" feature of the Scope window is demonstrated in Sect. 12.7.3. The Scope has some other features as well. One of the most widely used features is its scaling feature. If a graph in the Scope window does not fit quite well, it can be configured to be scaled either along the X-axis or Y-axis. Even scaling along both

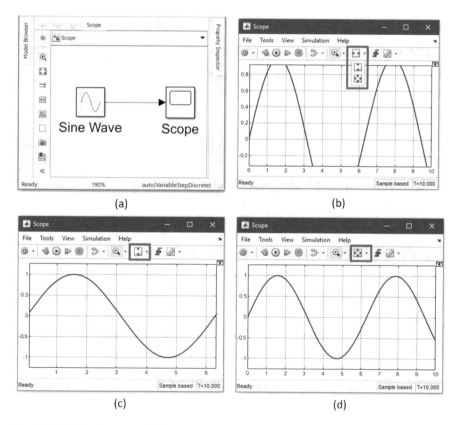

Fig. 13.6 Scaling the scope window

the X- and Y-axis can also be performed for a graph to be fitted in the Scope window perfectly. An example is provided below to show how the scaling feature works in the Scope:

In Fig. 13.6a, a Sine Wave Generator is connected with a Scope block. This simple model is generated to demonstrate the feature of the Scope window more easily. In Fig. 13.6b, a red marked box is placed to encircle a particular option of the Scope window from its upper tool strip. In this red marked box, three symbols are available in a dropdown option list. In that figure, the first symbol indicates "Scale X-Axis Limits," while the second refers to "Scale Y-Axis Limits." Finally, the third one represents "Scale X- and Y-Axes Limits." In Fig. 13.6b, the first option is chosen that indicates the "Scale X-Axis Limits" option. The model is simulated for 10 s. Therefore, the highest limit of the X-axis is 10. While the Scope window is scaled for the X-axis limit, the entire graph along the X-Axis is observable; however, the Y-axis limit is not scaled in this case. Hence, some of the upper and lower portions of the graph are invisible in Fig. 13.6b. In Fig. 13.6c, another phenomenon is shown where the Y-axis scaling option is chosen. In this case, the graph is scaled over the Y-axis but does not scale over the X-axis. Hence, some of the portions of the graph along the X-axis are not observable. Finally, in Fig. 13.6d, the scaling of both X- and Y-Axes is chosen, which causes the scaling of the graph over both axes. Hence, the overall graph is visible. Thus, the user has the preference to choose how they want to scale the graph according to their requirement.

Another important feature of the Scope window is the Layout feature. If multiple graphs are plotted in the same Scope window, they can be separated by using the principle of Subplot. The layout can be defined based on which the graphs will be plotted in separate subwindows of the original Scope window. An example is provided in Fig. 13.7 to explain the Layout feature of the Scope window.

In Fig. 13.7a, four Sine Wave blocks are connected with the four input ports of a Scope window. The method of customizing the number of input ports of a Scope block is explained earlier. After simulating the model, the graph of Fig. 13.7b can be visible, from which it can be observed that all of the four sine wave graphs are plotted in the same Scope window. This layout can be changed by clicking on the "Configuration Properties" option from the upper tool strip as marked by "1" in Fig. 13.7b. After clicking on this dropdown option, select the Layout feature, which is marked by "2" in the figure. This will cause the appearance of a small window as shown in Fig. 13.7c. This window represents the layout in a matrix format. The user has the preference to create subwindows by selecting the number of box arrays from this small window. In Fig. 13.7c, four-column box arrays are chosen, which form four-column subwindows in the original Scope window. The four figures shift to one of these four subwindows as shown in Fig. 13.7d. The Layout can be changed in another format as well. For example, in Fig. 13.7e, another layout option is chosen. In this case, four box arrays are chosen in such a format that it creates a 2 by 2 matrix format, i.e., four boxes in two rows and two columns formation. By selecting this layout, the subwindows are created exactly in the same manner as shown in Fig. 13.7f. Thus, based on the preference of the layout, the Scope window can be

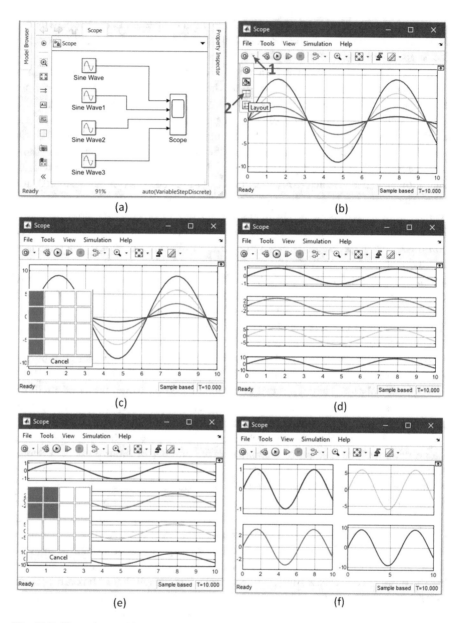

Fig. 13.7 Illustrating multiple plots on the scope window

formatted to create as many subwindows as possible to make the graphs more presentable.

13.1.3 *Floating Scope*

The "Floating Scope" block allows observing output signal without any line connection. To observe the signals of different lines, the "Floating Scope" block provides more flexibility as it can work as a replacement for multiple "Scope" blocks and direct line connections. An example is given in Fig. 13.8, where three separate "Sine Wave" blocks with different amplitudes are connected with three separate "Out" blocks. To observe the output of these three signal lines, three "Scope" blocks can be utilized by connecting them with the signal lines. However, another alternative way is to use the "Floating Scope," which does not require any direct line connection and only one is sufficient to observe the multiple signal outputs. The navigation route of the Floating Scope is given below:

Navigation route:
  Simulink library browser→Simulink→Sinks→Floating scope

In Fig. 13.8, the objective is to observe the output signals of Sine Wave 1, Sine Wave 2, and Sine Wave 3. These three sources are connected with three Out 1 blocks, which are available on Simulink Library Browser→Ports and Subsystems→Out 1. In the figure, a Floating Scope can be seen, which can be used to either plot one of the sine wave signals or all of them according to users' preference. The steps for using floating scope are mentioned below:

1. Double-click on the Floating Scope, which will create the appearance of the following Scope window in Fig. 13.9. Select the "Signal Selector" option from this window.

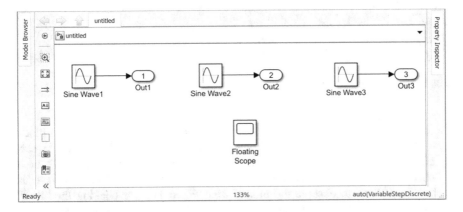

Fig. 13.8 Floating scope

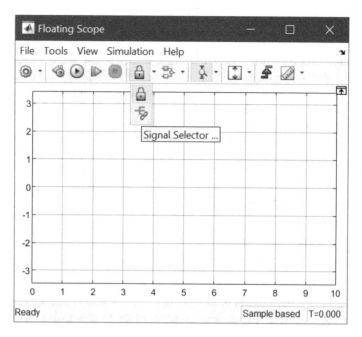

Fig. 13.9 Signal selector from scope window

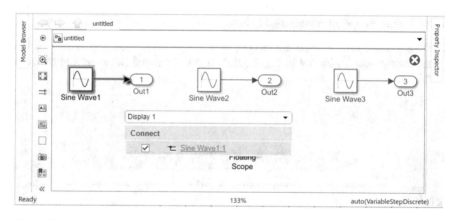

Fig. 13.10 Observing the plot of Sine Wave1

2. Later get back to the Simulink Design Window and select the Sine Wave 1 block and its associated signal line as shown in Fig. 13.10. This will make the appearance of the Display 1 window, where under Connect option, there will be a checkbox option named Sine Wave 1: 1. Click on the Checkbox, if the user is only interested to observe the plot of Sine Wave 1. After clicking the "Run" option, the desired output signal of the Sine Wave 1 block will appear on the

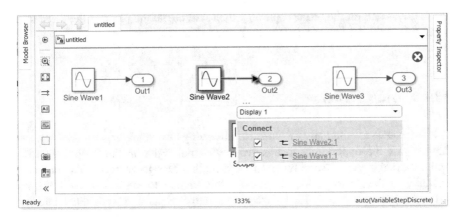

Fig. 13.11 Observing the plot of Sine Wave1 and Sine Wave2

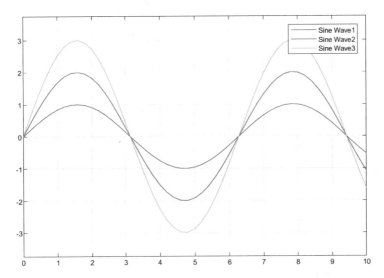

Fig. 13.12 Three plots of Sine Wave1, Sine Wave2, and Sine Wave3 in the same window

Floating Scope window. However, in this example, we are interested in observing three of the output signals of the Sine Wave blocks together in the Floating Scope window. Therefore, we will skip the clicking of the "Run" option for now.

3. Afterward, select the Sine Wave 2 block along with its signal line as shown in Fig. 13.11, which will add another checkbox option in the Display 1 window named Sine Wave 2: 1. Continue the same process for the Sine Wave 3 block as well.

Finally, click on the "Run" option, which will create the plots of the three sine wave signals in the Scope window as shown in Fig. 13.12.

13.1.4   Add Viewer

Add Viewer is an alternative feature for observing any output signal. Add Viewer can provide the opportunity to observe any output signal in the Scope Window, without dragging any Scope block in the Simulink Design Window. The previous example of the floating scope is used in this section to implement Add Viewer feature as shown in Fig. 13.13. To observe the output signal of a particular signal line using Add Viewer, the first step is to select that signal line. The second step is to select the "Add Viewer" option available on the header Strip under Simulation mode. By clicking on the "Add Viewer" option, a small window will appear from where multiple options can be chosen. In this example, we are interested to utilize the "Scope" option; hence, the next step is to select the "Scope" from the Add Viewer window. These steps are illustrated graphically in the following figure for more comprehensibility of the reader:

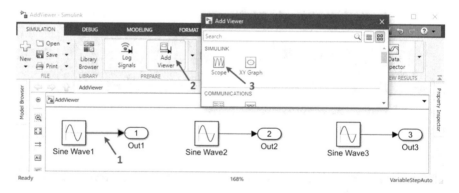

Fig. 13.13 The Add Viewer feature

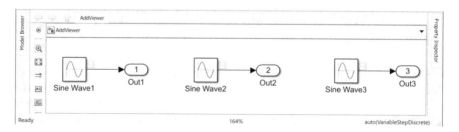

Fig. 13.14 Small fade scope icons next to the Sine Wave blocks

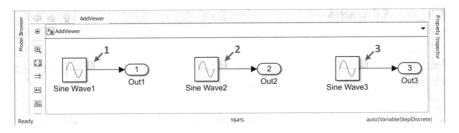

Fig. 13.15 Clicking on the scope icons will show the scope window

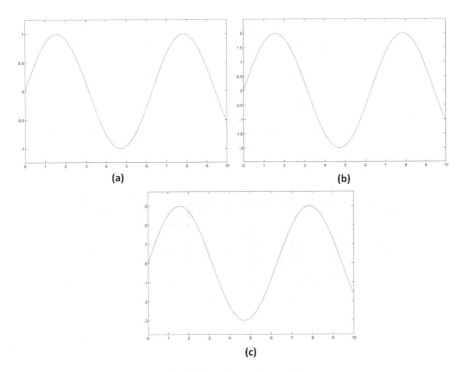

Fig. 13.16 The scope windows after clicking the small scope icons

After selecting the three signal lines associated with Sine Wave 1, Sine Wave 2, and Sine Wave 3 one by one and performing the above-mentioned steps separately for each line, we will create the appearance of the design as shown in Fig. 13.14. In this figure, it can be observed that a small fade scope window appears right next to the Sine Wave blocks, or at the start point of each output signal. After simulating the model by clicking on the "Run" button, click on any of the small scope icons (Fig. 13.15). This will create the appearance of the scope window with the associated output signal line plot as shown in Fig. 13.16. The Add Viewer makes the overall design simple by avoiding the usage of extra blocks in the design.

13.1.5   XY Graph

XY Graph can be utilized to plot a simple 2D graph. It is a graphical tool available in Simulink that performs almost the same task as the "plot" function of MATLAB. The navigation route of the XY Graph is shown below:

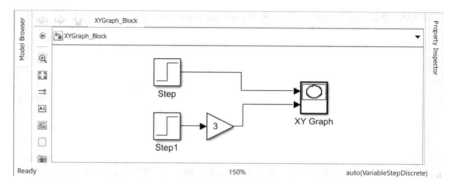

Fig. 13.17   Simulink block for XY graph

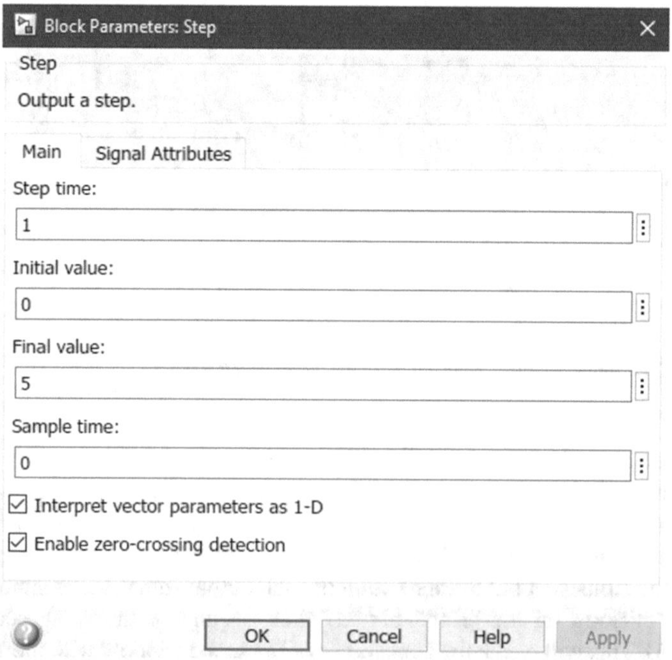

Fig. 13.18   Block parameters of step block

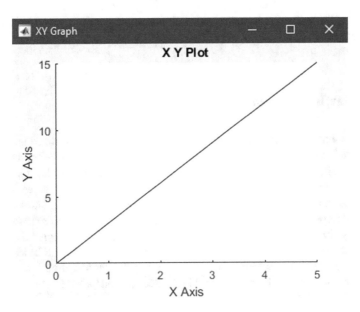

Fig. 13.19 XY plot

Navigation route:
  Simulink Library Browser→Simulink→Sinks→XY Graph

In Fig. 13.17, an example of the usage of the XY Graph block is shown, where two inputs are provided by using two "Step" blocks. For the second input, a "Gain" block is used which multiplied the step signal three times as assigned in the Gain block parameter window. It is to be noted that the "Gain" block is available on Simulink Library Browser→Simulink→Math Operations→Gain.

In Fig. 13.18, the parameter window of the "Step" block is shown. After simulating the model, double-click on the XY Graph block, which will create the appearance of the XY plot as shown in Fig. 13.19. This block can also be used from Add Viewer by following the same procedures described in the previous section.

13.2 Source

For any Simulink model, the Source plays one of the most important parts of the design. For an electrical circuit, a DC or AC source is necessary for the circuit to become active. Similarly, for any physical system, a source is the working factor. Therefore, different blocks that may act as a source in the Simulink are described in this section. In Simulink, different source blocks are available; however, in this section, only the most commonly used Source blocks are covered to make this section concise.

13.2.1 Pulse Generator

The Pulse Generator is a source block that can generate any customized pulse signal. This block is available on Simulink Library Browser→Simulink→Sources→Pulse Generator. An example is shown in Fig. 13.20, where the Pulse Generator is connected with a "Scope" block to observe the nature of a pulse signal. In Fig. 13.20, the parameter window of the Pulse Generator is also shown, through which the pulse signal can be customized according to users' preferences.

The amplitude of the pulse signal is set as 5, the period of the signal as 2, and the pulse width as 10. To observe the generated pulse signal, a Scope block is added with the Pulse Generator block. After simulating the model, double-click on the Scope signal to observe the pulse signal in the Scope window as shown in Fig. 13.21.

13.2.2 Ramp

Ramp block can generate ramp signals, which is an important source for many physical models. The navigation path of the Ramp block is given below:

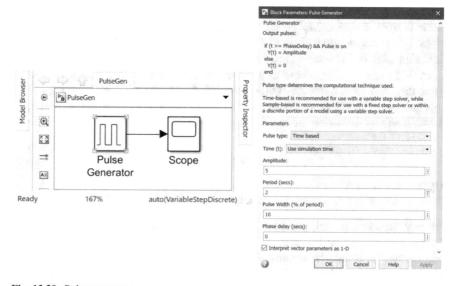

Fig. 13.20 Pulse generator

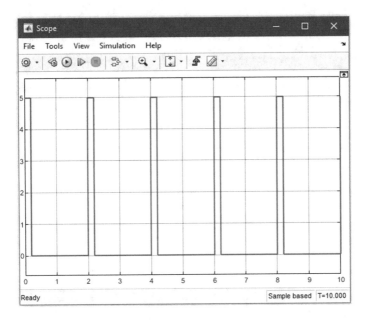

Fig. 13.21 Pulse signal in the scope window

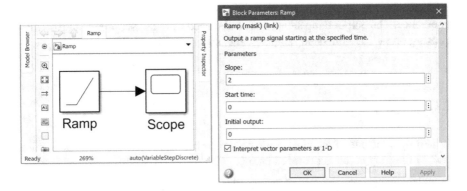

Fig. 13.22 Ramp block and its parameters

Navigation route:
Simulink Library Browser→ Simulink→ Sources→ Ramp

In Fig. 13.22, a Ramp block is connected with a Scope block. The slope of the Ramp block is assigned as 2. The start time and the initial output of the ramp signal are both set as zero. The user can customize these parameters according to their

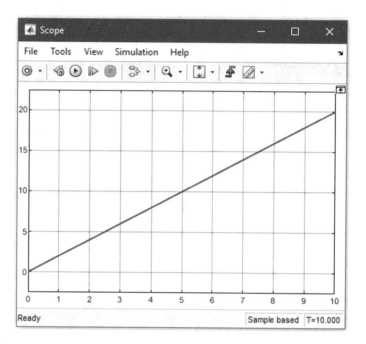

Fig. 13.23 Ramp signal in the scope window

preference. The output ramp signal can be observed from the Scope block by double-clicking on it. The generated ramp signal is shown in Fig. 13.23.

13.2.3 Step

A Step signal is one of the most commonly used blocks in Simulink. The navigation path of the Step block is provided below:

Navigation route:
   Simulink Library Browser→Simulink→Sources→Step

In Fig. 13.24, a Step block is dragged into the Simulink design window along with a Scope block. The parameters of the Step block that can be customized in the Simulink can also be observed in Fig. 13.24. In the following example, the Step time is selected as 2. The initial and the final values of the Step signal are assigned as 0 and 10, respectively. The sample time is considered as zero for this example. According to this particular customization, the output Step signal can be observable via the Scope block as shown in Fig. 13.25.

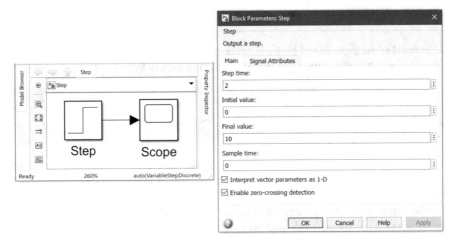

Fig. 13.24 Step signal block and its parameters

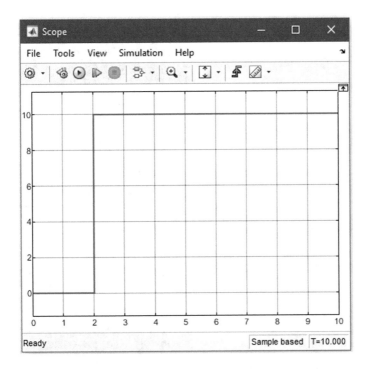

Fig. 13.25 Step signal in the scope window

13.2.4   Sine Wave

In Simulink, the Sine Wave block is available that can generate a sine wave signal. The destination path of the Sine Wave block is shown below:

Navigation route:
    Simulink Library Browser→Simulink→Sources→Sine Wave

For any sine wave, it is important to have the flexibility to customize different parameters such as amplitude, frequency, phase, etc. In Simulink, these parameters can be customized as shown in Fig. 13.26. The sine wave function in the Simulink can be observed in the parameter window of the Sine Wave block (Fig. 13.26), where the parameters are explained initially. In the following example, the amplitude and the frequency of the sine wave signal are selected as 5 and 1. The rest of the parameters are assigned as zero. The output sine wave signal can be seen through the Scope block, which is demonstrated in Fig. 13.27.

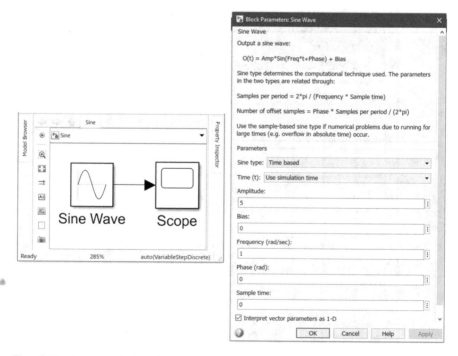

Fig. 13.26   Sine wave block and its parameters

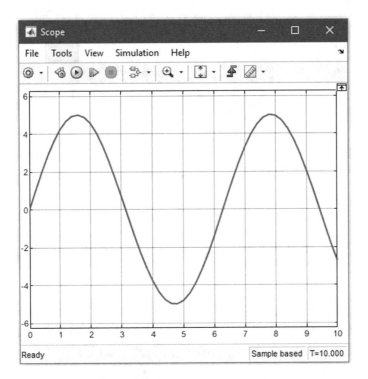

Fig. 13.27 Sine wave in the scope window

13.2.5 Constant

Constant is another source available in Simulink, capable of providing any numeric values or array of numbers as input. The navigation path of the Constant block is given below:

Navigation route:
Simulink Library Browser→Simulink→Sources→Constant

In Fig. 13.28, two examples are provided, where in the first example, a single numeric value is given as input. In the other example, an array of numbers is provided as input. In the Display blocks, the outputs can be observed.

In Fig. 13.29, the parameter customization of the Constant blocks for the two different examples is shown. It is to be noted that, for providing an array as an input, the third bracket needs to be used along with commas among the different numbers.

The Constant block can also be used to provide a matrix as the input. An example is given in Fig. 13.30 to show the procedure of assigning a matrix input via the Constant block:

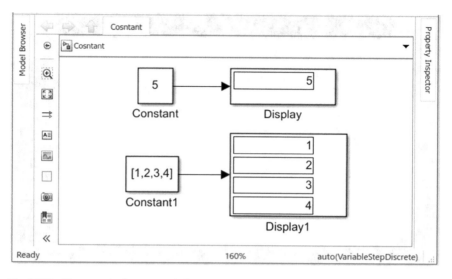

Fig. 13.28 Sine wave in the scope window

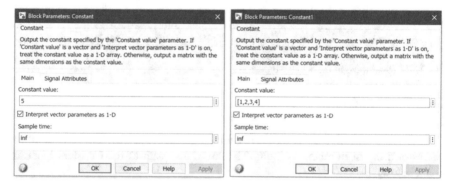

Fig. 13.29 Block parameters of the constant blocks

13.3   Math Operators

One of the important features of Simulink is its vast library of Math Operators. By using several blocks that can perform mathematical operations, any mathematical model can be simulated in the Simulink. Out of several Math Operators blocks, some of the widely used blocks will be explained in this section.

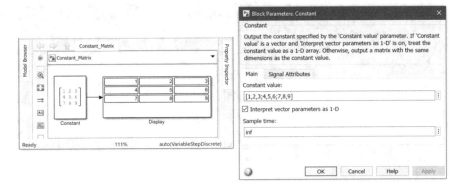

Fig. 13.30 Constant matrix

13.3.1   Abs and MinMax

Abs and MinMax are two mathematical operators that can be utilized in the Simulink. These two blocks can be found in the Simulink Browser by following the steps mentioned in Fig. 13.31.

The Abs block can determine the absolute value of any number including a complex number as well. In Fig. 13.32, two examples are given, where in the first example, a negative number -2 is given as input. The Abs block can determine the absolute value of -2, which is 2, and is displayed in the Display block. In the second example, a complex number $-3 + 4i$ is given as input. The absolute value of this number should be $\sqrt{(-3)^2 + 4^2} = 5$, which can be observed in Display 2.

In Fig. 13.33, two examples of MinMax block are demonstrated. MinMax block is capable of finding out either the maximum or the minimum value from an array of numbers. In the first example of Fig. 13.33, the MinMax block is configured to determine the minimum value from an array input, which is given by a Constant block. From the array of numbers, it can be seen that the minimum value is -2.5, the same value displayed in the Display 3. In the second example, for the same given input array, the MinMax block is configured in its maximum mode to determine the maximum value from the array. In Display 4, it can be observed that the block can determine the maximum value from the array, which is 10.

It is to be noted that, to configure the MinMax block either in minimum or maximum mode, the parameter window needs to be customized by double-clicking on the block. In Fig. 13.34, the parameter window of the MinMax block is shown, where the Function has a dropdown options list that incorporates the min and max options. While the user needs to configure the MinMax block in minimum mode, the Function option needs to be selected as min. For configuring the block in the maximum mode, the max option needs to be selected. Thus, the MinMax block can be utilized for determining either maximum or minimum values from an input array.

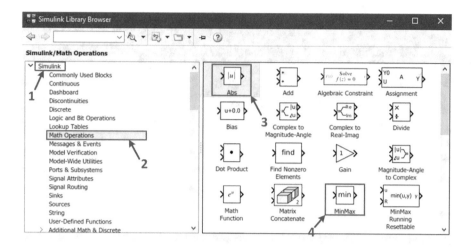

Fig. 13.31 Math operations in the Simulink library browser

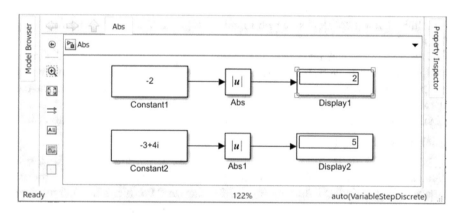

Fig. 13.32 Math operations in the Simulink library browser

13.3.2   Add, Subtract, and Sum of Elements

Some of the basic operations of mathematics are addition, subtraction, and the sum of elements. In Simulink, these three operations can be performed graphically using separate blocks—Add, Subtract, Sum of Elements—which are shown in Fig. 13.35.

The navigation routes of these three blocks are listed below:

Navigation route:

(continued)

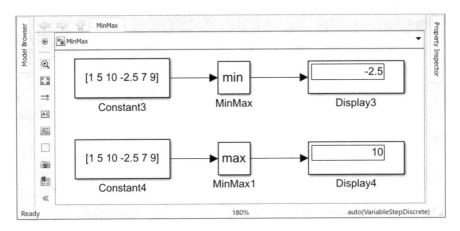

Fig. 13.33 Use of the MinMax block

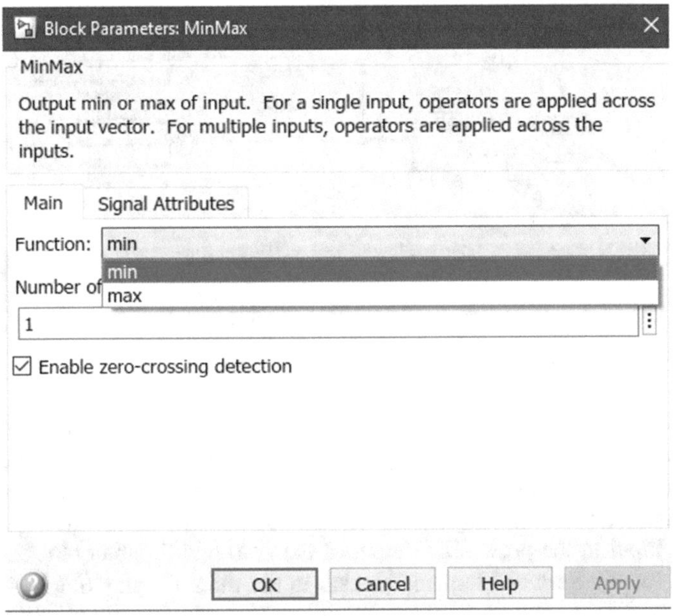

Fig. 13.34 Block parameters of the MinMax block

Add: Simulink Library Browser→Simulink→Math Operations→Add
    Subtract: Simulink   Library   Browser→Simulink→Math
Operations→Subtract
    Sum of Elements: Simulink Library Browser→Simulink→Math
Operations→Sum of Elements

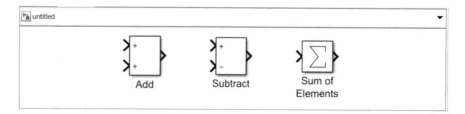

Fig. 13.35 Blocks for Add, Subtract, and Sum of Elements

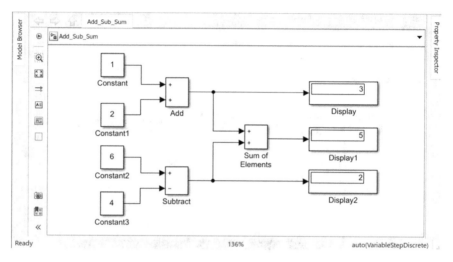

Fig. 13.36 Use of the Add, Subtract, and Sum of Elements blocks

An example is given in Fig. 13.36, where all three of these blocks are utilized in the same model. Two Constant blocks are connected with the two input ports of the Add block. Similarly, two other Constant blocks are also connected with the input ports of the Subtract block as well. The Constant blocks are used to provide numerical input to the ports. The output of the Add and Subtract blocks are set as the inputs for the Sum of Elements blocks. In the three Display blocks, "Display" indicates the output of the Add block; "Display 2" shows the output of the Subtract block; and "Display 1" refers to the output of the Sum of Elements block.

13.3.3 Product and Divide

Product and Divide blocks in the Simulink perform the multiplication and division operation. The navigation routes of these two blocks are provided below:

Navigation route:
  Product: Simulink   Library   Browser→Simulink→Math
Operations→Product
    Divide: Simulink Library Browser→Simulink→Math Operations→Divide

Product

In Fig. 13.37, two examples of the implementation of the Product block are shown. In the first example, two numeric numbers are provided as input via two Constant blocks to perform the multiplication of these two numbers. In the second example, instead of a single numerical value, two matrices are provided as input for the Product block. The Constant 3 block is used to give the first matrix input of size 2×2, and the Constant 4 block provides a 2×3 size matrix as the second input. The Product 1 block is customized to perform Matrix Multiplication from the parameter window of the Product 1 block as shown in Fig. 13.38. By selecting the Multiplication option "Matrix(*)," the Product 1 block can be used to perform matrix multiplication. From the output shown in Display 1 (Fig. 13.37), it can be observed that the size of the output matrix is 2×3, which proves that the matrix multiplication has occurred. If the user wants to perform element-wise multiplication, the size of the two input matrices needs to be of the same size, and the Multiplication option from the Product 1 parameter window needs to be assigned as "Element-wise(.*)."

Divide

Divide block can be used to perform Division operations. In Fig. 13.39, three separate examples are provided for three different inputs. In the first example, two

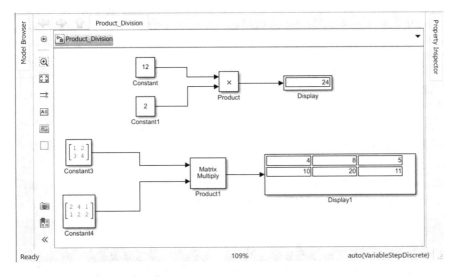

Fig. 13.37 Use of the product block

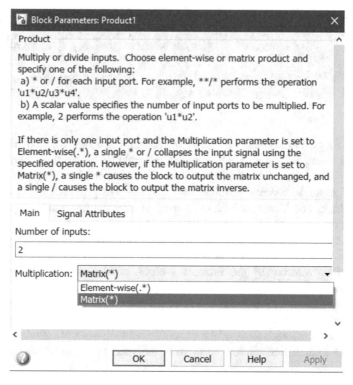

Fig. 13.38 Block parameters of the product block

numeric numbers are provided to perform simple division among them. In the second example, two matrices of the same size are provided as input to perform element-wise division. Finally, in the last example, a single matrix is provided as input, and the Divide 2 block is customized from its parameter window to perform Matrix operation instead of Element-wise operation (Fig. 13.40). By selecting the "Matrix(*)" option, the Divide 2 block is utilized to determine the inverse matrix of the given input matrix in the last example.

13.3.4   Sum and Sqrt

Sum and Sqrt are two widely used mathematical operators of Simulink. The navigation routes of these two blocks are listed below:

Navigation route:
  Sum: Simulink Library Browser→Simulink→Math Operations→Sum
  Sqrt: Simulink Library Browser→Simulink→Math Operations→Sqrt

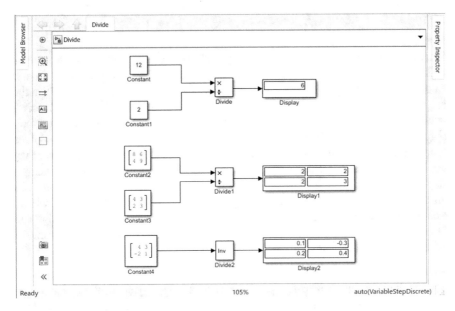

Fig. 13.39 Use of the divide block

Fig. 13.40 Block parameters of the divide block

Sum

The Sum block can be utilized to perform multiple addition, subtraction, or combination of both of them as shown in Fig. 13.41. In the first example, two numeric values are summed up. Therefore, in the Sum block, two additions are placed via the customization made from its parameter window. In Fig. 13.42, the parameter window of the Sum block is shown on the right side. On that window, the blank option named "List of signs" is filled with "|++." Here, the "|" sign is used to place space among the two signs mentioned later by "++." In the Sum block, the user may use multiple signs to perform multiple addition or subtraction operations using a single block. To place the signs in the block in a more presentable way, the user has the flexibility to control the space among the signs by using the "|" sign. A second example is shown in Fig. 13.41, where the two "+" signs and one "-" sign are incorporated in the Sum 1 block. The customization of the block from the parameter window can be observed from the left figure of Fig. 13.42.

Sqrt

Sqrt block in the Simulink can perform three operations—square root, reverse square root, and signed square root. A Simulink model example demonstrating these three operations is provided in Fig. 13.43. In the model, three examples performing these three operations are shown. In the first example, a positive numeric number, 36, is given as input of the Sqrt block. Sqrt block can be customized from its parameter window, which has been shown in Fig. 13.44. In the parameter window, there is a

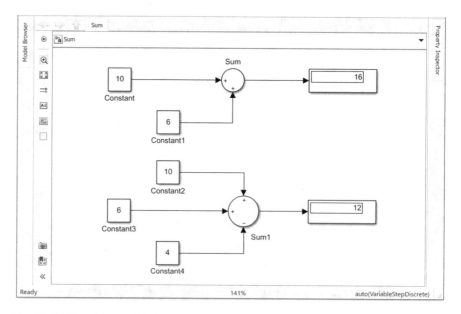

Fig. 13.41 Use of the sum block

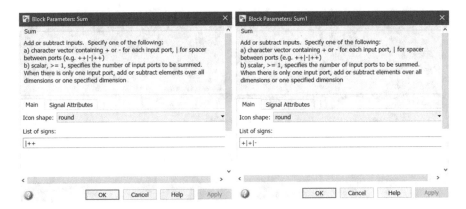

Fig. 13.42 Block parameters of the sum block

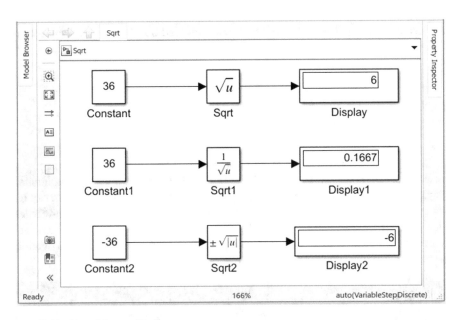

Fig. 13.43 Use of the sqrt block

dropdown option box under the name Function. This dropdown box has three options—sqrt, signed Sqrt, and rSqrt. In the first example shown in Fig. 13.43, the sqrt option is chosen for the Sqrt block. For the Sqrt 1 block, the rSqrt option is selected, which indicates the reverse square root operation. Right after selecting the rSqrt option from the dropdown option, it can be seen that the symbol within the Sqrt 1 block changes from $\sqrt{u}$ to $\frac{1}{\sqrt{u}}$.

Here, u indicates the given input. Similarly, for the Sqrt 2 block, the Function option is set as SignedSqrt, which indicates the operation that will be performed by

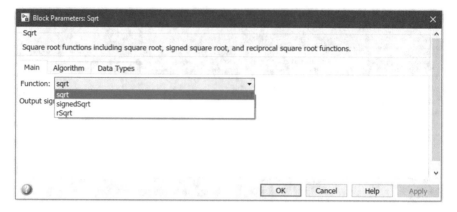

Fig. 13.44 Block parameters of the sqrt block

the block is $\pm\sqrt{|u|}$. The results of the three configured Sqrt blocks are displayed in three Display blocks.

13.3.5   Complex to Magnitude-Angle and Complex to Real-Imag

In Chap. 4, the representation of any complex number in two different forms—rectangular and polar form—is explained. In Simulink, the Complex to Magnitude-Angle block is available, where any complex number in its rectangular form can be given as input, and the block can determine the magnitude and the angle. Another block named Complex to Real-Imag block in the Simulink can perform the determination of real and imaginary values from a complex number. The navigation routes of these two blocks are given below:

Navigation route:
Complex   to   Magnitude-Angle: Simulink   Library
Browser→Simulink→Math Operations→Complex to Magnitude-Angle
Complex to Real-Imag: Simulink Library Browser→Simulink→Math
Operations→Complex to Real-Imag

A model given in Fig. 13.45 comprises two examples where these blocks are implemented. For both of these two blocks, a complex number, $2 + 5i$, is given as input via the Constant blocks. In the first example, the Complex to Magnitude-Angle

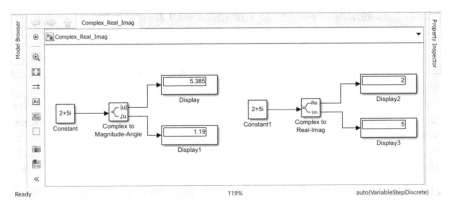

Fig. 13.45   Use of the Complex to Magnitude-Angle and Complex to Real-Imag blocks

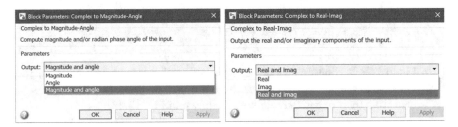

Fig. 13.46   Block parameters of the Complex to Magnitude-Angle and Complex to Real-Imag blocks

block can calculate the magnitude and angle of the complex number in its polar form. From the Display and Display 1 blocks, the magnitude and the angle can be observed. It is to be noted that the user has the flexibility to choose the output of the block either as only magnitude, or only angle, or both of them. It can be customized from the parameter window of the Complex to Magnitude-Angle block as shown in Fig. 13.46. Similarly, in the second example, the same complex number is given as input to the Complex to Real-Imag block, which determines the real and imaginary values from the complex number that are displayed in the Display 2 and Display 3 blocks. Similar to the Complex to Magnitude-Angle block, this block also provides the opportunity to choose the output as only real value, only imaginary value, or both of them from its parameter window (Fig. 13.46).

13.3.6 Magnitude-Angle to Complex and Real-Imag to Complex

Magnitude-Angle to Complex and Real-Imag to Complex blocks perform just the opposite operations of the previously mentioned two blocks. Given the input in terms of magnitude and angle, the Magnitude-Angle to Complex block can determine the complex number in rectangular form. For the Real-Imag to Complex block, the real and imaginary numbers are given as inputs, and the block can determine the complex number in its rectangular form. The navigation routes of these two blocks are mentioned below:

Navigation route:
    Magnitude-Angle   to   Complex: Simulink   Library
Browser→Simulink→Math Operations→Magnitude-Angle to Complex
    Real-Imag to Complex: Simulink Library Browser→Simulink→Math
Operations→Real-Imag to Complex

In Fig. 13.47, a Simulink model comprising the implementation of these two blocks is given. The customization of the input port of these two blocks can be performed from their parameter window, as shown in Fig. 13.48.

13.3.7 Math Function

The Math Function block is a very useful block to perform mathematical operations in the Simulink. The Math Function block is available in the following path of the Simulink Library Browser:

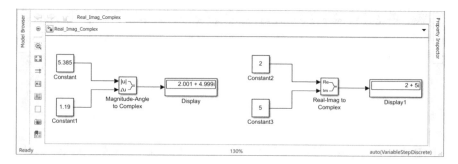

Fig. 13.47 Use of the Magnitude-Angle to Complex and Real-Imag to Complex blocks

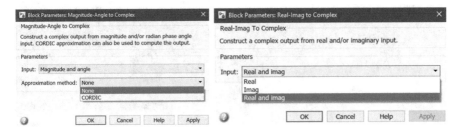

Fig. 13.48 Block parameters of the Magnitude-Angle to Complex and Real-Imag to Complex blocks

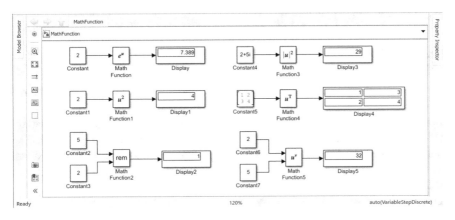

Fig. 13.49 Use of the math function blocks

Navigation route:
   Math   Function: Simulink   Library   Browser→Simulink→Math
Operations→Math Function

The Math Function block can be utilized to perform 14 different mathematical operations. To choose a particular mathematical operation using the Math Function block, it needs to be customized first from its parameter window. In Fig. 13.49, six operations out of the available 14 operations are shown to grasp the idea of utilizing the Math Function block. In Fig. 13.50, the parameter window of the Math Function block is shown, where under the Function option, a dropdown options list is available. In the list, the name of the 14 mathematical functions is enlisted. The user can choose any one of the function names from this list to make the Math Function block perform that particular operation. For example, in Fig. 13.49, the first Math Function block is customized for a function name "exp." Hence, it can be observed that the symbol within the Math Function block is changed into e^u, where u is the input of the block. In the second example, the function name of the Math

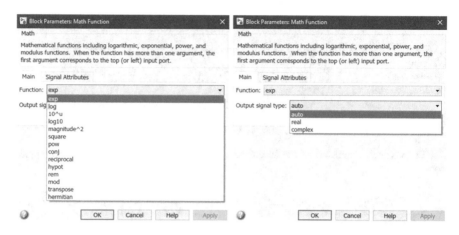

Fig. 13.50 Block parameters of the math function blocks

Function 1 block is selected as "square." Similarly in the third example, "rem" option is chosen for the Math Function 2 block, which automatically creates two input ports in the block, due to its necessity to perform the remainder operation. Three more examples are also provided in the model shown in Fig. 13.49. The parameter window of the Math Function has another parameter called "Output signal type," which can be customized as auto, real, or complex. The user may choose any of these types to display the output results in the Display blocks.

13.3.8 Trigonometric Function

The Trigonometric Function in the Simulink block can perform trigonometric operations. The navigation route of this block is listed below:

Navigation route:
Trigonometric Function: Simulink Library Browser→Simulink→Math Operations→Trigonometric Function

The Trigonometry Function block can perform 16 operations in the Simulink. A model is provided in Fig. 13.51, where three examples are given with three different operations. To select a particular trigonometric operation, the parameter window of the Trigonometry Function block needs to be customized, which is shown in Fig. 13.52. In the parameter window, 16 operations are available, out of which three operations are shown in the model of Fig. 13.51. In this model, the first block is customized for "sin" operation, the second block for "acos" operation, and the third block for "tanh" operation. In the second example, a "Gain" block is used in the

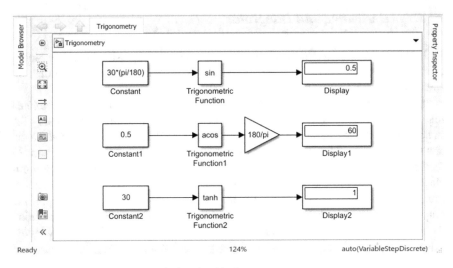

Fig. 13.51 Use of the trigonometric function blocks

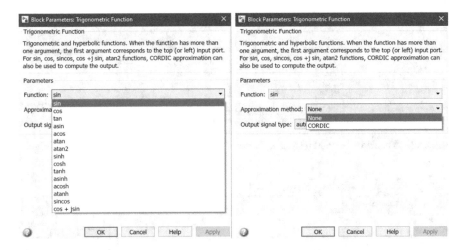

Fig. 13.52 Block parameters of the trigonometric function blocks

middle to multiple the answer with "180/pi" value to convert the output from radian to degree value. Therefore, the Display 1 indicates the value 60 in degree unit instead of radian unit.

13.3.9 Derivative and Integrator

In Simulink, derivation and integration can be performed by using graphical blocks
Derivative and Integrator. The navigation routes of these two blocks are demonstrated in Fig. 13.53.

In Fig. 13.54, a Simulink model is shown where a ramp signal is differentiated
first by using the Derivative block, and later the output is integrated by using the
Integrator block. The input ramp signal, the output signal from the Derivative block,
and the output signal from the Integrator block are connected with the input ports of
the Scope block. The number of input ports of the Scope block can be set by right-
clicking on the Scope block and navigating to "Signal and Ports→Number of Input
Ports→3." As integration is the opposite of differentiation, in this example, the
output of the Integrator block should be identical to the input signal. After simulating
the model, the input and the output signals can be observed from the Scope window
as shown in Fig. 13.55.

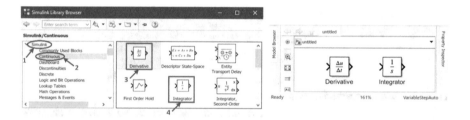

Fig. 13.53 The derivative and integrator blocks

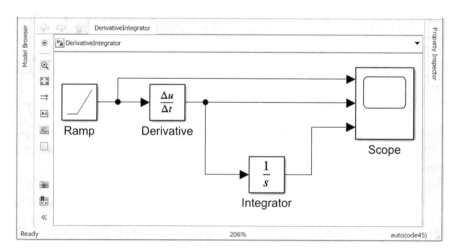

Fig. 13.54 Use of the derivative and integrator blocks

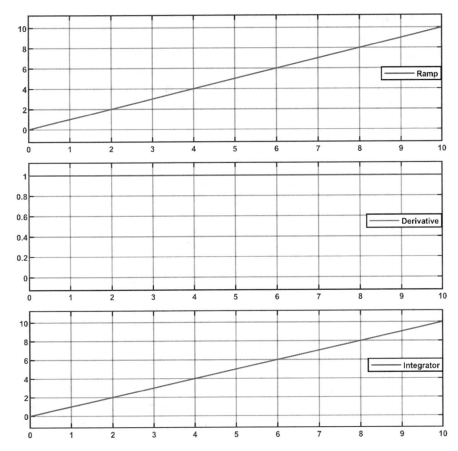

Fig. 13.55   Scope window output of the ramp, derivative, and integrator blocks

13.4   Port and Subsystem

One of the interesting features of the Simulink is the subsystem. A subsystem is a group of blocks that constitutes a part of the entire system. The blocks that comprise a subsystem reside in another layer; therefore, a subsystem can make the original design model simple to represent. It can also be helpful when a particular part of a system is required multiple times in the same design. In that case, a subsystem can be very useful, as it works almost like a user-defined block. The difference between a subsystem and a user-defined block is that a subsystem does not require to be programmed by MATLAB. While creating a subsystem, two types of ports—In1 and Out1—become very necessary to establish the connection between the subsystem and the original system. Apart from that, there are also some other types of ports, which are necessary while designing a model in the Simulink. Mux and Demux are two widely used port blocks in the Simulink. The procedures of creating a subsystem and the usage of the other ports are described in the following subsections.

13.4.1   Subsystem, In1, and Out1

To explain the procedures of creating a subsystem, an example is considered where a subsystem will be created that incorporates a Derivative and an Integrator block. The subsystem will have two inputs and two outputs. The first input will represent the Derivative signal of the given first input. On the other hand, the second output of the subsystem will indicate the integration of the second input signal. To complete this objective, the following steps need to be followed:

Step 1: Drag a Derivative and an Integrator block to the Simulink design window following the navigation route described previously in Sect. 13.3.9.

Step 2: Select both of the blocks, which will create the appearance of three blue dots as indicated in the first figure of Fig. 13.56. Drag the mouse over these three blue dots, which will create a small blue ribbon with multiple options represented by small icons. Move the mouse over these options one by one, which will show the name of each option. Find out the option called "Create Subsystem" as shown in the second figure of Fig. 13.56, and click on it.

Step 3: Another alternative method of creating a subsystem is to skip the second step and follow the step that will be mentioned in this step. Step 3 is to select the Derivative and the Integrator block as before, and right-click on the mouse. This will create the appearance of a small window from where choose the option "Create Subsystem from selection" as shown in Fig. 13.57. Another shortcut method of creating a subsystem is to select the two blocks and type "ctrl+G."

Step 4: By following either Step 2 or Step 3, a subsystem can be created. When a subsystem is created, it will appear as the figure shown in Fig. 13.58 with two input and two output ports. Right-click on the Subsystem, which will make the appearance of another layer named Subsystem that incorporates the original two blocks as shown in Fig. 13.59.

In Fig. 13.59, it can be observed that In1 and In2 blocks appear in the input ports, while Out1 and Out2 ports appear in the output ports automatically. The name of these ports can be changed from their parameter window.

Fig. 13.56 Creating subsystem

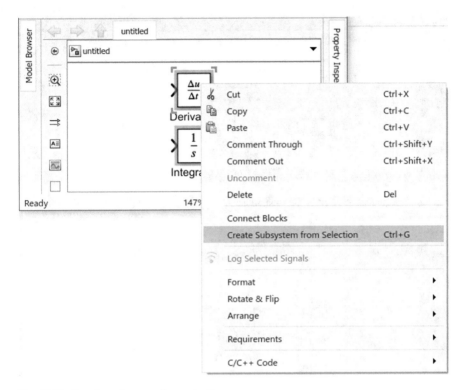

Fig. 13.57 Creating subsystem from selection

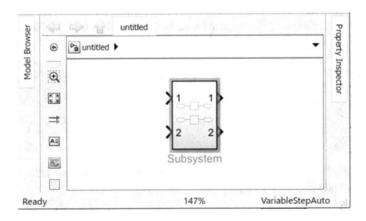

Fig. 13.58 Subsystem

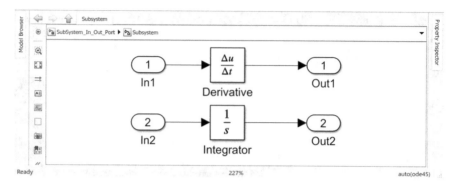

Fig. 13.59 Simulink diagram of the subsystems

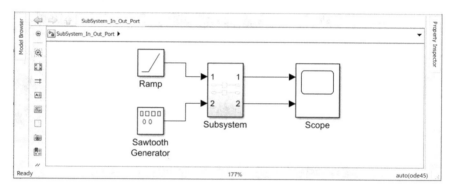

Fig. 13.60 Complete Simulink diagram of the subsystems

The subsystem is created, and the next step is to complete the original model by providing input signals to the two input ports and connecting a scope to the output side. The complete model is shown in Fig. 13.60, where a Ramp block is connected to port 1 and a Sawtooth Generator block is connected to port 2. The navigation routes of both of these sources are given below:

Navigation route:
   Ramp: Simulink Library Browser→Simulink→Sources→Ramp
   Sawtooth Generator: Simulink Library Browser→Simulink→Quick Insert→Sources→Sawtooth Generator

After simulating the model, the output signals can be observed from the Scope window as shown in Fig. 13.61.

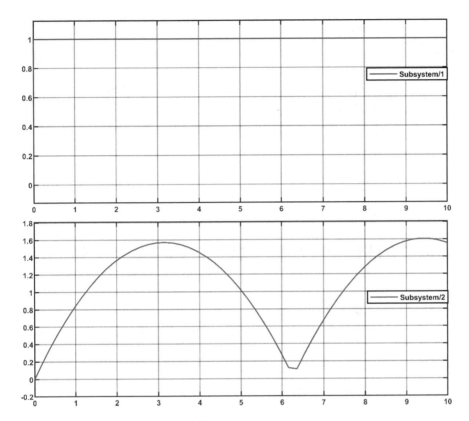

Fig. 13.61 Output signals of the two subsystems

13.4.2 *Mux and Demux*

Mux block is used to combine multiple signal lines into a signal one, while the Demux block is used to do the opposite. The navigation routes of these two blocks are listed below:

Navigation route:
  Mux: Simulink Library Browser→Simulink→Signal Routing→Mux
  Demux: Simulink Library Browser→Simulink→Signal Routing→Demux

A Simulink model is shown in Fig. 13.62, where the usage of both of these blocks is shown in two separate examples. In the first example, the two signal lines from two constant blocks are connected with a Mux block, which converts them into a single output line. However, when this output line is connected with the Display block, it can be observed that the Display block shows the two original values

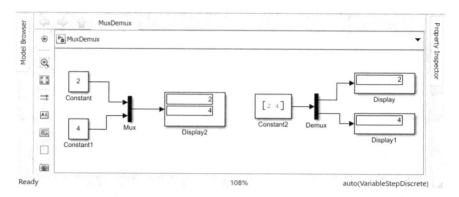

Fig. 13.62 Simulink diagram of Mux and Demux

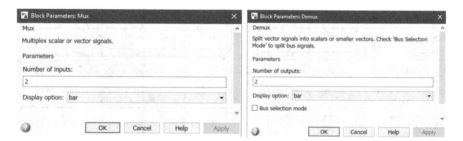

Fig. 13.63 Block parameters of Mux and Demux

obtained from the two Constant blocks. It proves that the Mux block does not create any data loss in the process. In the second example, one Constant block is connected with the input of Demux. The Constant block provides an array of two values as input. The Demux creates two output lines, one of which displays the first value of the array and the second output line displays the other value of the array.

The number of inputs of the Mux block and the number of outputs of the Demux block can be customized from their parameter windows as shown in Fig. 13.63.

13.5   Logical Operator, Relational Operator, Programs, and Lookup Table

Logical operators and relational operators can be useful in some applications in the engineering domain. Although they can be programmed in MATLAB, Simulink provides the opportunity to implement the same logic via graphical blocks. In this section, the procedures to implement both logical and relational operators will be explained with examples. In Simulink, programs such as "If" and "Switch Case" can also be implemented graphically, which will be demonstrated via examples. Another important block of Simulink is the Lookup Table, which is useful while performing graphical approximation. All of these topics will be covered in this section with the necessary illustrations.

13.5.1   Logical Operator

In Simulink, the Logical Operator block can be found in the following path of the Simulink Library Browser:

Navigation route:
   Logical Operator: Simulink Library Browser→Simulink→Logic and Bit Operations→Logical Operator

By using the Logical Operator block, seven logical operations can be performed. To select a particular logical operator, double-click on the block for opening the parameter window. In the parameter window of the Logical Operator, a dropdown options list will be available under the Operator option as shown in Fig. 13.64. In this dropdown list, seven logical operators such as AND, OR, NAND, NOR, XOR, NXOR, and NOT are available to implement.

In the Simulink models shown in Fig. 13.65, two examples are provided with two different logical operators—AND and OR.

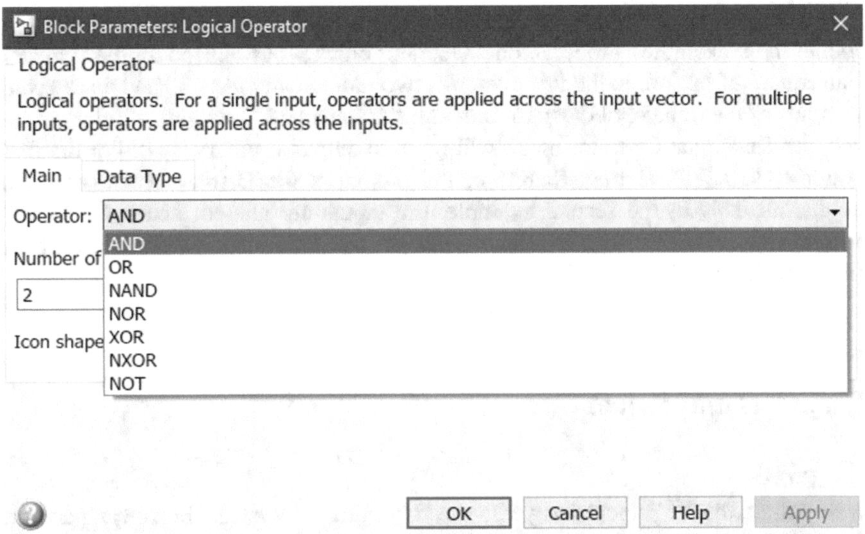

Fig. 13.64 Block parameters of logical operator

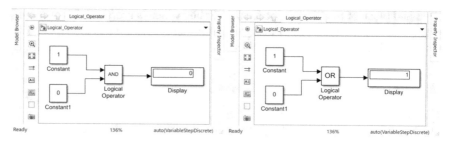

Fig. 13.65 Logical AND and OR operators

13.5.2   Relational Operator

In Simulink, the Relational Operator block is available in the following navigation path of the Simulink Library Browser:

> Navigation route:
>   Relational Operator: Simulink Library Browser→Simulink→Logic and Bit Operations→Relational Operator

By double-clicking on the Relational Operator block, the parameter window of the block can be opened. From this parameter window, ten relational operators are available in a dropdown options list under the Relation Operator option as shown in Fig. 13.66.

In Fig. 13.67, two examples are provided side by side with two different inputs. In both of these examples, the Relational Operator block is customized for the "Greater than equal, $\geq$" option. In the first example, two numeric numbers 10 and 5 are given as inputs of the Relational Operator block. If the statement "1st input $\geq$2nd input" is true, the Relational Operator block will provide output 1, otherwise 0. For the first example, "10 $\geq$5" is true; therefore, the output in the Display block after the simulation is 1. In the second example, the inputs are altered; hence, "5 $\geq$ 10" becomes false. Consequently, the output of the Relational Operator block as displayed in the Display 1 block is 0.

13.5.3   If and Switch Case

"If" Block

In Simulink, the "If" block along with the "If Action Subsystem" block can perform the "If-else" program. The "If" block is used to provide the conditions, and the "If Action Subsystem" is utilized to define the tasks for different conditions. The navigation path of both of these blocks is given below:

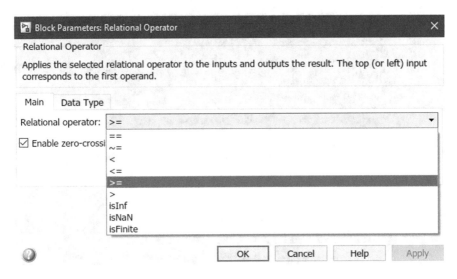

Fig. 13.66 Block parameters of relational operator

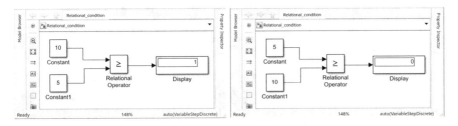

Fig. 13.67 Demonstration of the operation of the relational operator block

Navigation route:
   If: Simulink Library Browser→Simulink→Ports and Subsystems→If
   If Action Subsystem: Simulink Library Browser→Simulink→Ports and Subsystems→If Action Subsystem

The implementation of the above-mentioned blocks is demonstrated in the following figures. In Fig. 13.68, the parameter window of the "If" block is shown. At the beginning of the parameter window, the general format of the "If" expression is described. For each "If" and "Elseif" statement, a condition (expression) and action are required to be defined. The action will only be fulfilled if the input satisfies the associated condition or expression. The input can be a single variable, or multiple variables, depending on the application. From the parameter window, the number of inputs can be specified. For this example, the number of inputs parameter is set as 1. The second step of customization is to provide the sets of expressions for both "If"

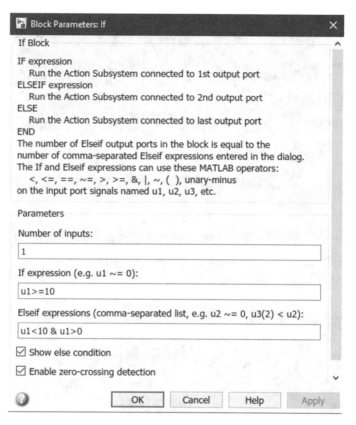

Fig. 13.68 Block parameters of the if block

and "Elseif." In this example, the expression of the "If" statement is defined as $u1 \geq 10$, and for the "Elseif" statement, the assigned expression is $u1 < 10 \,\&\, u1 > 0$. It is to be noted that the operators used to define the expression are explained previously in Sect. 13.3.4. At the end of the parameter window, there are two checkbox options, one of which is named "Show else statement." This is an optional condition that depends on the users' preference. For this example, "Else" is considered to be included; hence, the checkbox option is checked, as shown in Fig. 13.68.

To summarize, the number of inputs is set as 1; hence, the "If" block has one input port in the model shown in Fig. 13.69. "If," "Elseif," and "Else"—all three of these statements—are selected, which have created three output ports for the "If" block. For each of these three statements, an "If Action Subsystem" block is connected. These three connection lines can be seen as dotted lines from the model designed in Fig. 13.69. The reason for such dotted lines can be understood from the logic of operations of these two types of blocks. When an input is provided via the Constant block to the input port of the "If" block, the block checks the defined expressions one by one from its parameter window by considering the input value. If the first

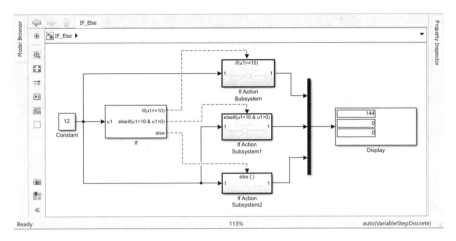

Fig. 13.69 Simulink diagram of the if…else block for the constant 12

expression under the "If" statement satisfies the expression for the given input, the line connection between the first output port to the "If Action Subsystem" block becomes active, and the other two output ports remain inactive. Therefore, only the task defined in the first "If Action Subsystem" block is executed in such a scenario. If the expression of the "If" statement does not satisfy the input value, the "If" block will shift to the next "Elseif" expression for the verification. If this expression is satisfied by the input value, the second output port will be activated, and only the task of the "If Action Subsystem 1" will be executed. Similarly, if the "Elseif" expression turns out to be false for the input value, the third output port will be activated. Hence, only the task defined in the "If Action Subsystem 2" block will be performed. This is the general overview of the working procedure of the "If" and "If Action Subsystem" blocks.

The input of Fig. 13.69 is set in such a way that it satisfies the "If" expression. Therefore, the task of the "If Action Subsystem" block is executed and the result can be observed in the Display block. As only the task of the first "If Action Subsystem" block is executed, the output of the other "If Action Subsystem" blocks can be seen as zero from the Display block.

For the second model shown in Fig. 13.70, the input is changed that does not satisfy the "If" expression but satisfies the expression of the "Elseif." Therefore, the task of the "If Action Subsystem 1" block is executed, and the result can be seen from the second output result of the Display block. The rest of the outputs of the Display block becomes zero, in this case.

In Fig. 13.71, the input value does not satisfy any of the "If" and "Elseif" expressions, thereby the task of the "If Action Subsystem 2" connected with the "Else" port becomes active. From the Display block, the third output refers to the result, while the remaining two outputs become zero due to inactivity.

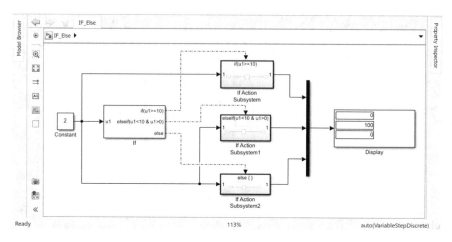

Fig. 13.70 Simulink diagram of the if...else block for the constant 2

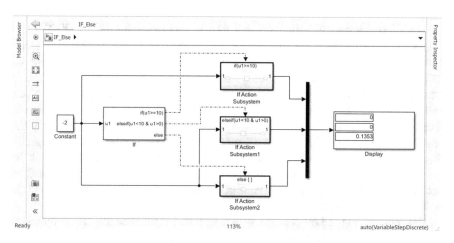

Fig. 13.71 Simulink diagram of the if...else block for the constant −2

The tasks performed by each "If Action Subsystem" blocks need to be defined before the simulation. This can be performed by customizing their parameter windows. The defined tasks for this particular example are demonstrated in Fig. 13.72. The results can be verified from the output of the Display block.

"Switch Case"

The "Switch Case" block works almost as similar as the "If" block. Instead of "If," "Elseif," and "Else" statements, in the "Switch Case" block, the expressions are defined under "Case [1]," "Case [2]," ..., "default" statements. The user has the preference to create as many cases as necessary. To define the number of cases, and the associated condition, the parameter window of the "Switch Case" block needs to be customized as shown in Fig. 13.73.

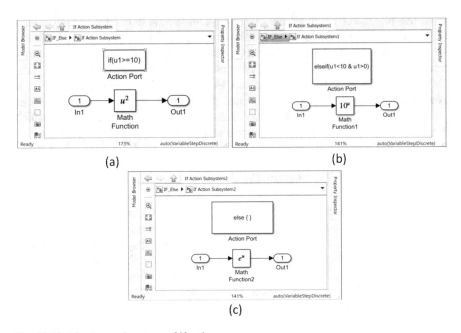

Fig. 13.72   The three subsystems of if action

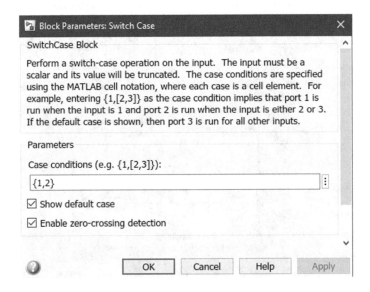

Fig. 13.73   Block parameters of the switch case

In the parameter window, the "Case Conditions" is defined as {1,2}, which implies if the input is 1, the output port 1 will run, and for input 2, the port 2 will become active. Inside the curly bracket, each numeric value separated by commas creates an individual case. The numeric value defines the input value for which the port associated with that case will be run. For example, for a condition {[1,2], 3}, two separate cases will be created as inside the curly bracket there are two types of data separated by one comma. For the first data, an array of [1,2] can be seen, which indicates that port 1 of Case 1 will become active when the input will be either 1 or 2. If the input becomes 3, port 2 associated with Case 2 will become active. If the "Show default case" option is checked, another port named "default" will be created as an output port of the "Switch Case" block. The "Switch Case" block only determines which case will run based on the given input. For each case, the task that needs to be performed can be defined by the "Switch Case Action Subsystem" block, which works almost similar to the "If Action Subsystem" block. When a particular case becomes active, the "Switch Case Action Subsystem" associated with that case becomes active, and only the task defined under that particular block is executed. In our example, two cases are considered along with an additional "default" case. Therefore, the "Switch Case" block has three output ports that are connected with three separate "Switch Case Action Subsystem" blocks. The tasks of these three blocks are defined as shown in Fig. 13.74.

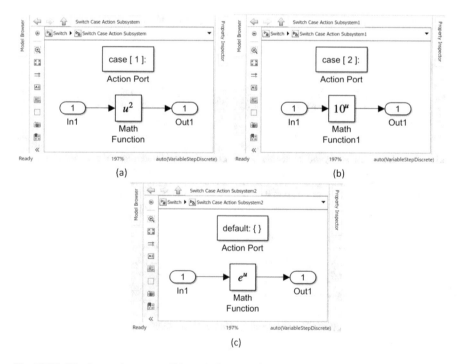

Fig. 13.74 The three subsystems of the switch case action

In Fig. 13.75, the complete model is shown with an input set as 1. The "Case Conditions" is selected as {1,2} as mentioned earlier for this example. As the input is 1, Case [1] becomes activated and the task of the "Switch Case Action Subsystem" is executed. The task is defined in Fig. 13.74a, and it can be observed that the task is to provide the square output of its input. The inputs of the three "Switch Case Action Subsystem" blocks are set as the same value, which is 4. Therefore, the output result of the first "Switch Case Action Subsystem" should be the square of 4. From the display, the same result can be found for the first output, while the remaining two outputs are zero, as they are inactive. In Fig. 13.76, the input of the "Switch Case" block is set as 2, which activates the port of the Case [2]. Hence, the task of the "Switch Case Action Subsystem 1" block is executed and displayed in the Display block. Similarly, from Fig. 13.77, it can be observed that the input of the "Switch Case" block activates the "default" case port, and thus, the task of the "Switch Case Action Subsystem 2" is executed only and displayed in the Display block.

It is to be noted that both the "Switch Case" and the "Switch Case Action Subsystem" blocks are available in the destination of the Simulink Library Browser→Simulink→Ports and Subsystems.

13.5.4 Lookup Tables

In Simulink, different Lookup Tables are available. In this section, the Lookup Table Dynamic block will be explained in detail due to its wide usage in the engineering domain. Lookup Table Dynamic block can be useful to approximate missing values based on a given data. It can produce data for interpolation or extrapolation, and also for other applications. The navigation route of this block in the Simulink Library Browser is listed below:

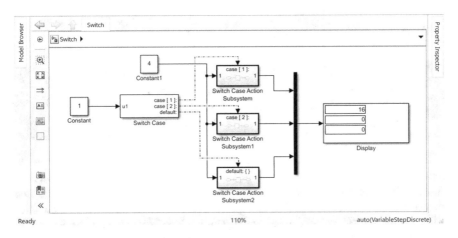

Fig. 13.75 Switch case action subsystems with constant 1

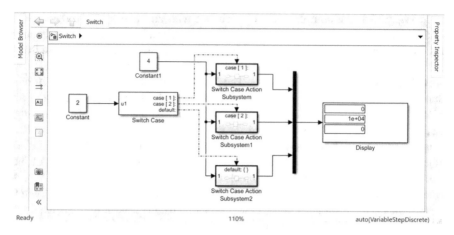

Fig. 13.76 Switch case action subsystems with constant 2

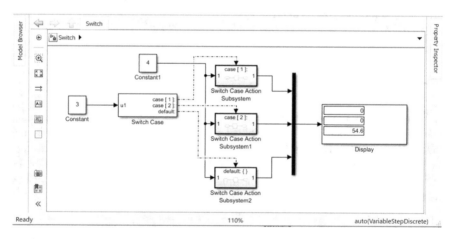

Fig. 13.77 Switch case action subsystems with constant 3

Navigation route:
    Simulink   Library   Browser→Simulink→Lookup   Tables→Lookup
    Table Dynamic

In the following example, the Lookup Table Dynamic block is utilized to approximate unknown values of a given dataset base. In the block shown in Fig. 13.78, there are three input ports—x, xdat, and ydat. "xdat" and "ydat" are the given dataset of x and y pairs. "x" represents the values of x for which the values of corresponding "y" values are unknown. The Lookup Table Dynamic in Fig. 13.79 can approximate the values of "y" for the corresponding "x" values based on the given dataset, as shown in the Display block.

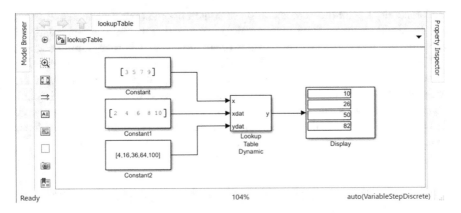

Fig. 13.78 Simulink diagram of the Lookup Table

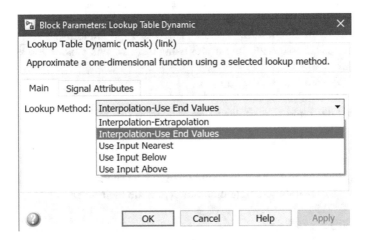

Fig. 13.79 Block parameters of the Lookup Table Dynamic

13.6   Conclusion

A Simulink model is comprised of blocks and lines; hence, the readers need to learn
about as many blocks as possible that will be useful while designing any physical
model. With that intent in mind, this chapter presents some of the most commonly
used blocks which are categorized among different types for the convenience of
keeping track of them. The chapter initiates by introducing some of the common
blocks of sink and source type. The source type blocks are important to provide input
to any physical model, while the sink acts for displaying the output. In Simulink,
almost all possible mathematical operations can be performed by utilizing different
math operator blocks, some of which are incorporated in this chapter with necessary
illustrations. The subsystem is another important feature of Simulink, which is

explained along with the demonstration of some important ports such as In1, Out1, etc. In Simulink, apart from mathematical operators, many logical and relational operators are also available in terms of different blocks. In the programming context, "if" and "switch" are two important programs, which can also be implemented in Simulink with graphical tools. This chapter introduces all of these aspects with the usage of relevant blocks. Finally, the chapter ends by demonstrating the Lookup Table considering its importance in the engineering domain. The main objective of this chapter is to provide the readers some practical insights regarding several blocks to make them comfortable with the Simulink environment more rigidly.

Exercise 13

1. Name some of the types of commonly used Simulink blocks, and provide two examples from each type.
2. (a) What are the importance of source type blocks in Simulink?
   (b) Generate the following sine waves using the Sine Wave block and show them in the two Scope blocks:
   (i) $5 \sin (2\pi \times 60 \times t)$
   (ii) $2 \sin (50 \times t)$
   (c) Show the same sine waves as mentioned in (b), by utilizing single floating scope.
3. (a) Write down the formulas to convert a coordinate from its polar form to rectangular form and vice versa.
   (b) Consider the following two complex numbers in their rectangular form. Use Complex to Magnitude-Angle block to determine their polar form:
   (i) $0.5 + 3i$
   (ii) $2 + 0.5i$
   (c) Perform the previous task in (b), by using blocks in the mathematical operations except for the Complex to Magnitude-Angle block. Verify the result with the previous one.
4. (a) What is a subsystem? Write down some of the advantages of creating a subsystem.
   (b) Create a subsystem that can take two inputs (input1, input2), and provide the following output:

$$\text{Output} = 2 \sin (\text{input1}) + \frac{d}{dt}(\cos (\text{input2})) + \sqrt{\text{input1}^2 + \text{input2}^2}$$

   (c) Show the result using any relevant blocks from the sink type.

5. (a) What are the general format of the "if" and "switch" blocks?

(b) What is the navigation route of the Lookup Dynamic Table block in the Simulink Library Browser?

(c) Given any number input, create a model that can determine whether it is even or not. If it is even, display 1, and if it is odd, display 0. Use any convenient Programs blocks for the model.

Chapter 14
Control System in Simulink

14.1 Control System

Control system is an important part of the engineering field, which makes it crucial to incorporate some fundamental ideas regarding this topic with the usage of Simulink. Simulink is widely used as a simulation platform to implement in numerous applications related to the control system. A control system signifies the system that can regulate, or control, the output of a system for a given input. The concept of the control system is very important to ensure the stability of a system and govern the output according to a predetermined reference. A control system can be categorized into two types—open-loop control system and closed-loop control system—based on the structure of a control system. These two individual types will be explained with necessary illustrations utilizing Simulink in the following subsections:

14.2 Open-Loop Control System

The open-loop control system refers to such systems whose output is controlled with respect to the input. In such systems, the effect of feedback is not considered; hence, the output of such systems is never measured. The design of an open-loop system is not complicated; hence, it is cheaper to implement. Open-loop systems have many applications in the real world, such as microwave, washing machine, etc. The basic structure of an open-loop system is demonstrated through a block diagram in Fig. 14.1, where the control system incorporates both the controller and the fundamental process blocks to produce the desired output response subjected to a given input.

© The Author(s), under exclusive license to Springer Nature Switzerland AG 2022
E. Hossain, *MATLAB and Simulink Crash Course for Engineers*,
https://doi.org/10.1007/978-3-030-89762-8_14

Fig. 14.1 Input and output
of a control system

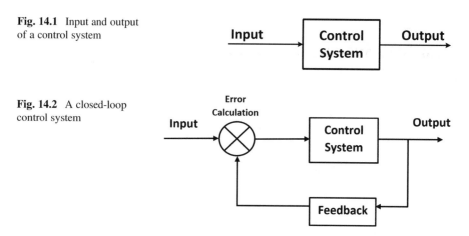

Fig. 14.2 A closed-loop
control system

14.3   Closed-Loop Control System

In a closed-loop control system, the output of the system is considered for making adjustments to the input signal for controlling the output with more accuracy. To consider the effect of output in such a manner is regarded as feedback in the control system domain. The feedback creates an interconnection between the output and input; hence, error calculation is a necessary part of such systems to make the proper adjustments. A basic layout of a closed-loop control system is provided in Fig. 14.2. A closed-loop control system can be made more accurate and stable than an open-loop system. However, due to the incorporation of feedback in such systems, the closed-loop systems are comparatively more complex and costly to implement in real-time applications.

14.4   Open-Loop vs Closed-Loop Control System

The differences between the open-loop and closed-loop control systems are tabulated in Table 14.1.

14.5   Simulink Model Design

This section describes the design of an open-loop and a closed-loop control system.

Table 14.1   Various aspects of open-loop and closed-loop control systems

Aspects	Closed-loop control system	Open-loop control system
1. Stability	A closed-loop system may become unstable, as the controlled parameters may change for different aspects	An open-loop system will never become unstable unless the controller is unstable
2. Sensitivity	The sensitivity of a closed-loop control system is comparatively better	The open-loop control system has poor sensitivity
3. Gain	The gain of a closed-loop system is comparatively low	An open-loop system possesses high gain
4. Accuracy and complexity	A closed-loop system is comparatively more accurate due to its high capability of noise cancellation, but complex in design	An open-loop system is simple in design but has less accuracy as such a system is affected significantly by unknown noises or disturbances

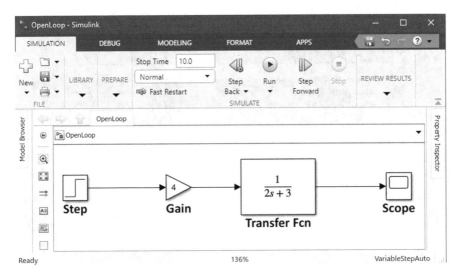

Fig. 14.3   Simulink diagram of an open-loop control system

14.5.1   Open-Loop Control System

In this section, an open-loop control system will be designed using Simulink to realize the concepts via simulation. The characteristics of a system are defined by the transfer function of that system. The concept of the transfer function is explained in detail previously in Sect. 9.2.2. In an open-loop system, the output is not connected via any feedback. In Fig. 14.3, a Simulink model demonstrating an open-loop system is shown. The blocks that are used for the simulation are summarized in the following Table 14.2 with their navigation routes:

Table 14.2 The blocks and navigation routes used in Fig. 14.3

Blocks	Navigation routes
Step	Simulink Library Browser→Simulink→Sources→Step
Gain	Simulink Library Browser→Simulink→Sources→Gain
Transfer Fcn	Simulink Library Browser→Continuous→Transfer Fcn
Scope	Simulink Library Browser→Sinks→Scope

Fig. 14.4 Block parameters of the transfer function block for the open-loop control system

The parameters of the Transfer Fcn block are customized from its parameter window as shown in Fig. 14.4. The considered transfer function of the system in this example is $\frac{1}{2s+3}$. To customize this exact function in the Transfer Fcn block from its parameter window, the numerator and the denominator of the function need to be defined in terms of an array. In the parameter window shown in Fig. 14.4, the first parameter is named "Numerator coefficients," which is set as [1]. For the "Denominator coefficients," an array [2 3] is assigned, where 2 represents the coefficient of "s" and 3 represents the constant. The concept of definition is almost similar to the built-in function *tf()*, which was used in the MATLAB section. By providing the coefficients of the numerator and the denominator, the Transfer Fcn block can figure out the transfer function of the system, and it will appear automatically within the Transfer Fcn block as shown in Fig. 14.3. The Gain block is customized by setting

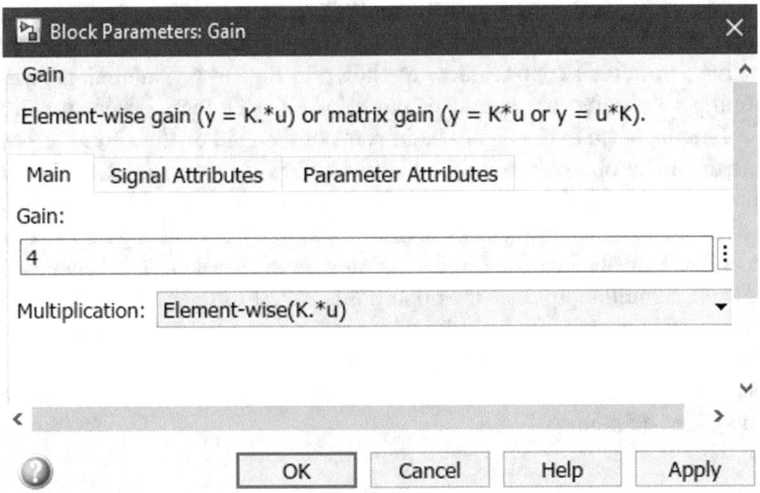

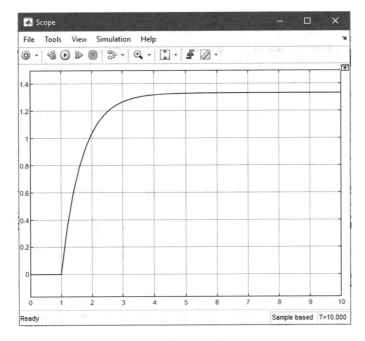

Fig. 14.5   Block parameters of the gain block of the open-loop control system

Fig. 14.6   The output in the scope window of the open-loop control system the value 4 (Fig. 14.5), which will be multiplied with the transfer function of the system. After simulating the model, the output response of the open-loop system can be observed from the Scope window as shown in Fig. 14.6. Simulink provides the opportunity to observe the output response of an open-loop system for different transfer functions and gain value. The impacts of the gain on the output response of the system can be observed from the Scope window by changing the value of gain from the parameter window of the Gain block. Thus, the transfer function can also be changed to simulate a completely different system, by only changing the customization of the Transfer Fcn. Before designing a control system in practice with real instruments, Simulink provides the opportunity to test different systems for varying parameters with an objective to build up a prototype on the Simulation platform. It reduces the cost of performing testing procedures with real instruments, and also the efforts. This is one of the most important benefits of utilizing Simulink for the control system domain.

14.5.2   Closed-Loop Control System

A Simulink model of a closed-loop control system is illustrated in Fig. 14.7. The considered transfer function for such a system is $\frac{1}{2s^2+3s+5}$. As a controller, PID is considered for the following system. As the system is a closed-loop system, there needs to be a feedback gain among the output and the input. In this example, unity feedback gain is considered. If the feedback gain is not unity, a Gain block needs to be inserted into the line between the output and the input. The input of the system is provided by using the Step block. The feedback gain connects with the input via the

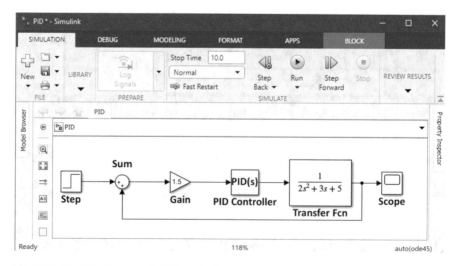

Fig. 14.7 Simulink diagram of a PID controller

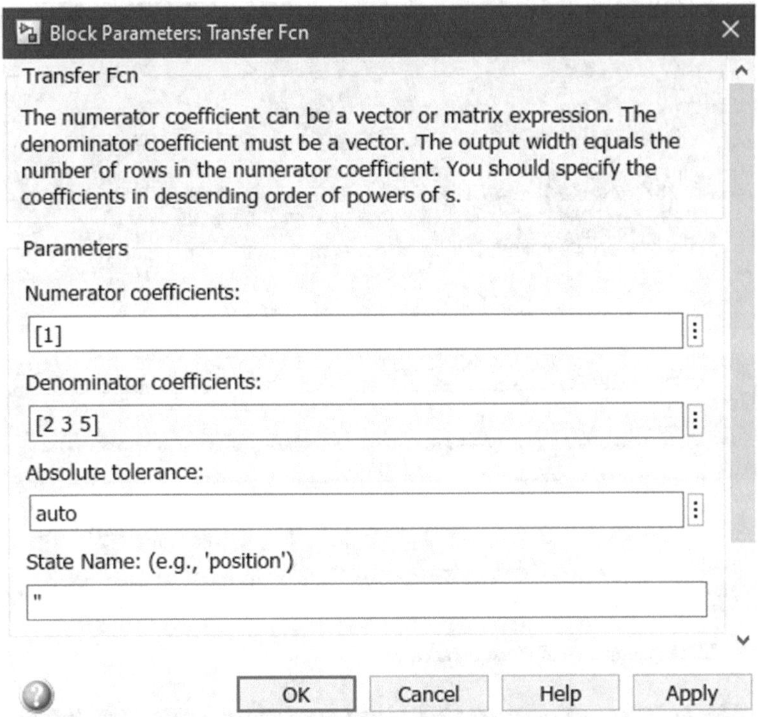

Fig. 14.8 Block parameters of the transfer function block for the PID controller

Sum block. The Sum block can be customized according to the specification of the system. For this example, the feedback gain is added to the input; therefore, two "+" signs are used in the Sum block customization. It can be altered from the parameter window of the Sum block according to the users' preference for the system.

The customization of the Transfer Fcn block is shown in Fig. 14.8, where the numerator coefficients and the denominator coefficients are provided. The customization of the Sum block is demonstrated in Fig. 14.9. Another block named the PID Controller can also be customized from its parameter window. The PID Controller has three parameters—Proportional (P), Integral (I), and Derivative (D) constants. The block can be utilized for P, I, PI, PD, or PID controller by choosing any of the controller types from the dropdown option list of its parameter window as shown in Fig. 14.10. For this example, the PID controller is chosen. The values of the three constants can be defined from the parameter window as well. For ensuring the stability of the system, the values of these three constants need to be tuned by varying the values. By selecting appropriate values, the complete design of the closed-loop control system using the PID controller is ready for the simulation. After simulating the model, the response can be observed from the Scope window as shown in Fig. 14.11.

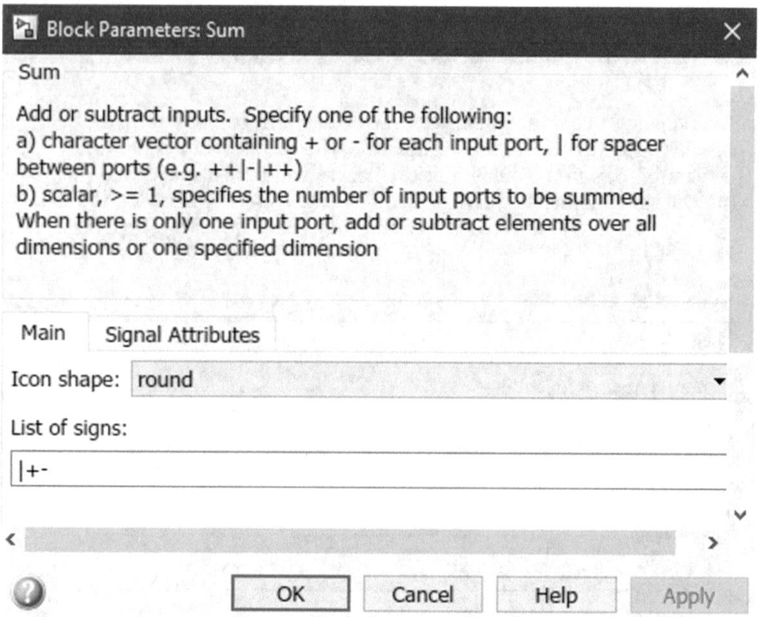

Fig. 14.9 Block parameters of the sum block

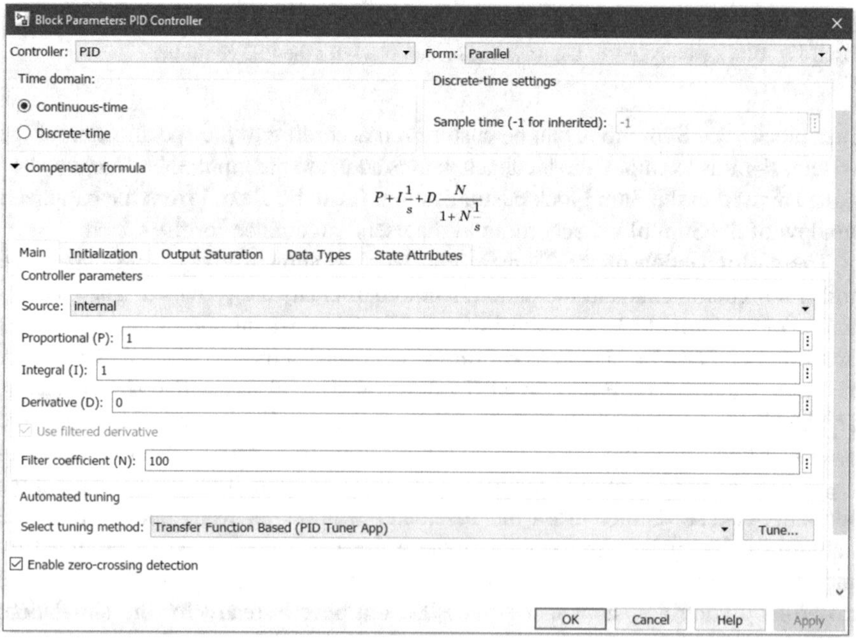

Fig. 14.10 Block parameters of the PID controller

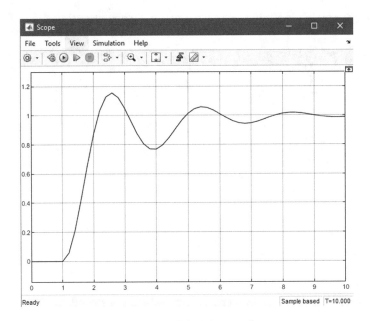

Fig. 14.11 The output in the scope window of the PID controller

The Simulation result shows that the response becomes stable after 9 s approximately. The characteristics of this system can be further analyzed by experimenting with different parameter values. The parameter value can be swept or changed by using the MATLAB command window as well. An example of how to change the value of a parameter or variable using MATLAB command window is shown in the following example, where the same closed-loop model is used for better comprehensibility.

Customizing Variables Utilizing MATLAB Command Window
For this example, the value of the Gain block is considered as the variable, which will be defined by using the MATLAB command window. For doing so, the first step is to set the value of the gain as "K" or other letters from the parameter window of the Gain block as shown in Fig. 14.12.

The next step is to define the name of the variable in the MATLAB command window by assigning a value in there, as shown in Fig. 14.13.

The complete design of the Simulink model is demonstrated in Fig. 14.14, where the only change is made in the Gain block.

After simulating the model again, the model will now be simulated for the new value of the gain, $K = 6.5$. Thus, instead of changing the value from the Simulink, the value of a particular parameter of a block can be changed from the MATLAB command window as well. The output response of the system obtained from the Scope window is shown in Fig. 14.15.

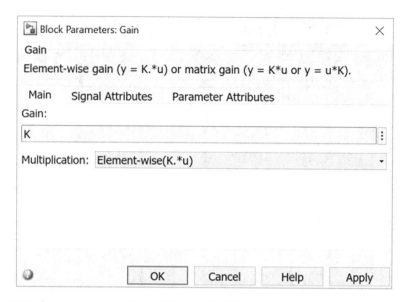

Fig. 14.12   Block parameters of the gain block of the PID controller

Fig. 14.13   Defining the
variable for the gain

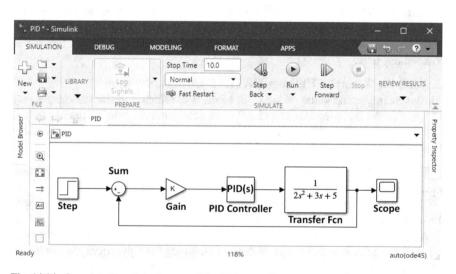

Fig. 14.14   Complete Simulink diagram of the PID controller

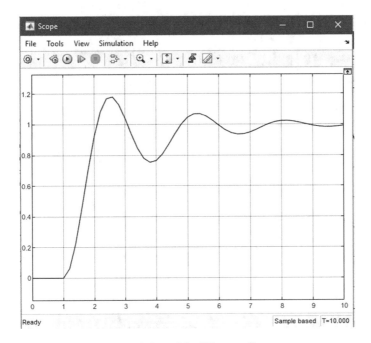

Fig. 14.15 The output in the scope window of the PID controller

Fig. 14.16 Defining another variable for the gain

For another simulation, the value of the gain K is set as 9 from the MATLAB command window (Fig. 14.16), and the model is simulated to observe the output. The response of the system is shown in Fig. 14.17, from where the effects of gain on the system can be realized.

Due to the increase of gain, it can be observed that the value of the early oscillation is increased for the response shown in Fig. 14.17. Thus, the characteristics of any control system can be analyzed by utilizing the Simulink model.

14.6   Stability Analysis

Stability analysis is a key part of designing any control system. The stability of a control system can be determined by following multiple procedures such as the Pole-Zero Map, Bode Plot, Nyquist Plot, etc. The procedures to comment on the stability

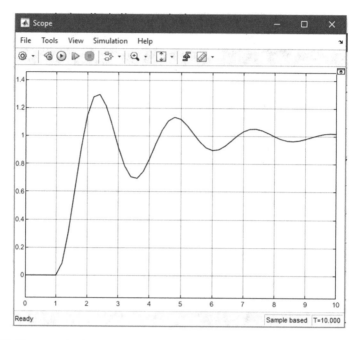

Fig. 14.17 The output in the scope window of the PID controller for the new gain

of a system—stable or unstable—based on these plots of a system are explained in Sect. 9.6 in detail. In this section, a stable and an unstable system will be designed utilizing Simulink to demonstrate the procedures of generating Zero-Pole plotting, Bode plotting, and Nyquist plotting.

14.6.1 Stable System

A control system that incorporates a transfer function, $G(s) = \frac{50}{s^2+12s+1}$ with the Step input, is designed and presented in Fig. 14.18. The blocks utilized to design this model are enlisted with their navigation routes in the following Table 14.3:

The output response of the system can be observed from the Scope window, and the stability of such systems can be determined by generating different plots such as Pole-Zero Plot, Bode Plot, and Nyquist Plot.

The customizations of the Step and Transfer Fcn blocks are illustrated in Fig. 14.19. The numerator of the transfer function is assigned to be [50], whereas the denominator is provided as an array of [1 12 1], to represent the transfer function, $G(s) = \frac{50}{s^2+12s+1}$.

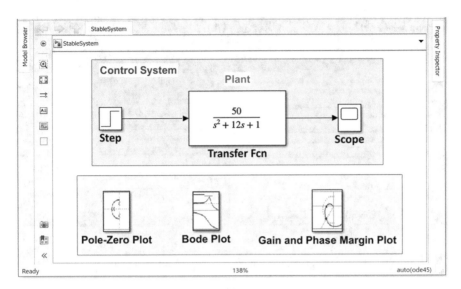

Fig. 14.18   A stable control system with a step input

Table 14.3   Navigation routes of the blocks used in the design of Fig. 14.18

Blocks	Navigation path on the Simulink library browser
Step	Simulink→sources→step
Transfer Fcn	Simulink library browser→continuous→transfer Fcn
Scope	Simulink library browser→sinks→scope
Pole-zero plot	Simulink→control design→linear analysis plots→pole-zero plot
Bode plot	Simulink→ control design→linear analysis plots→bode plot
Gain and phase margin plot	Simulink→control design→linear analysis plots→gain and phase margin plot

Pole-Zero Plot

The Pole-Zero Plot of a control system can be used to determine the stability of a system. If all the poles of a system lie in the left-half plane of the coordinate system, the system can be regarded as a stable system. In Simulink, the Pole-Zero Plot block can be utilized to generate such plots. The configuration of this block to produce the Pole-Zero Plot is summarized below:

Step 1: Double-click on the Pole-Zero block which will create the appearance of a parameter window as shown in Fig. 14.20a. Click on the "+" button as red marked by "1," which will create a right-side box in the parameter window (Fig. 14.20a).

Step 2: Select the input signal line as shown in Fig. 14.21a that will automatically enlist the name of the signal on the right-side box of the parameter window of the Pole-Zero Plot block (Fig. 14.21b). Select the name and click on the "<<" button as red marked by "3," to shift the signal to the left box. From the "Configuration" tab of the left-side box, select "Open-loop Input," as shown in Fig. 14.21c.

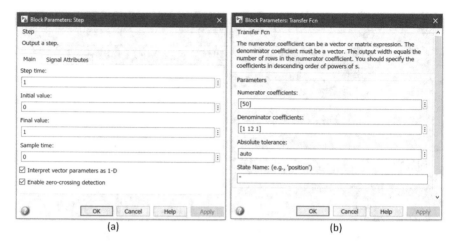

Fig. 14.19 Block parameters of the step and transfer function blocks

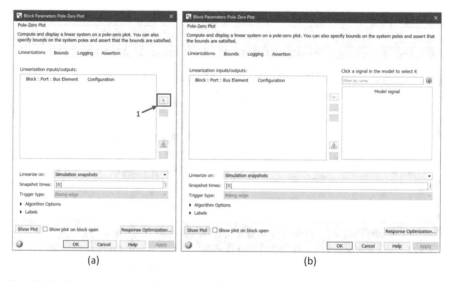

Fig. 14.20 Block parameters of the pole zero plot

Step 3: Select the output signal line that is connected with the Scope, and follow the same procedures mentioned in Step 2. From the "Configuration" tab, choose the "Open-loop Output" option this time. Make sure to check the "Show plot on block open" option in the end. This concludes the customization of the Pole-Zero Plot, which looked like Fig. 14.22 in its final stage.

Bode Plot

The Bode Plot is also used to determine the stability of a system. More details on the Bode Plot can be found in Sect. 9.6.3. To generate the Bode Plot diagram of the same

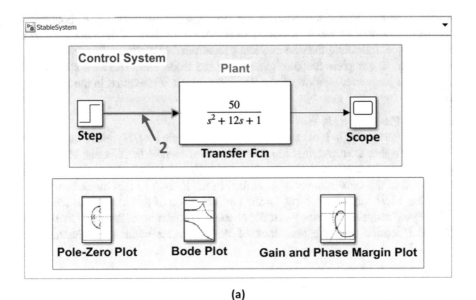

(a)

(b)

Fig. 14.21 Open-loop input of the control system

system, drag the "Bode Plot" block in the Simulink design window. The configuration of the Bode Plot block is similar to the customization of the Pole-Zero Plot block. Hence, following the previous steps mentioned in the Pole-Zero Plot section is sufficient to complete the configuration of the Bode Plot. The final customized look of the parameter window of the Bode Plot block is presented in the following figure:

Gain and Phase Margin Plot

Gain and Phase Margin Plot can be utilized to plot Bode Plot, Nyquist Plot, Nichols Plot, etc. In this example, this block will be used to plot the Nyquist Plot of the system. The Customization of the Gain and Phase Margin Plot block can be performed in the same manner as described in the Pole-Zero Plot block customization. One additional requirement for the customization of this block is to select the "Plot Type" from four options—Bode, Nichols, Nyquist, and Tabular. Choose the "Nyquist" option, and the final look of the parameter window is illustrated in Fig. 14.24.

Simulation

After simulating the model for 100 s, the output response of the system can be seen by double-clicking on the Scope block. The output response of this system is shown in Fig. 14.25, from where it can be observed that the response becomes stable after some time.

(c)

Fig. 14.21 (continued)

Fig. 14.22 Block parameters of the Bode plot

The Pole-Zero Plot of the system can be produced by double-clicking on the Pole-Zero Plot block. A figure window will appear and click on the "Run" button from the head tool strip of this window. It will generate the Pole-Zero Map of the system, which is shown in Fig. 14.26. From the figure, it can be observed that all the poles lie on the left-half plane, hence proving the stability of the system.

For generating the Bode Plot of the system, double-click on the Bode Plot block, which will generate a figure window. From that figure window, click on the "Run" button from the header tool strip. The Bode Plot of the system will generate, which is demonstrated in Fig. 14.27. From the figure, it can be observed that both of the margins are positive; therefore, the system is stable. More details regarding the procedures to comment on the stability of a system from its Bode Plot can be found in Sect. 9.6.3.

Finally, to produce the Nyquist Plot of the system, double-click on the Gain and Phase Margin block, and click on the "Run" button from its figure window. It is seen that the contour does not enclose $(-1,0)$ point, and there is no pole in the right-half plane; therefore, the system is stable. More details on the plot are available in Sect. 9. 6.4. The Nyquist Plot of the system will appear as follows:

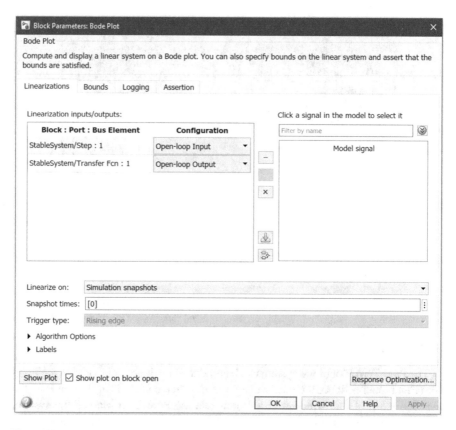

Fig. 14.23 Customization of the Bode plot parameters

14.6.2   Unstable System

An example of an unstable system is demonstrated in this section utilizing Simulink. Like the previous stable system, different plots such as Pole-Zero Map, Bode Plot, and Nyquist Plot of the system are produced to perform stability analysis. The entire Simulink design of the unstable system is provided in Fig. 14.29. The blocks that are used in this design are the same as the previous model as listed in Table 14.3.

The customizations of the Step and the Transfer Fcn blocks are shown in Fig. 14.30. The transfer function of this unstable system is set as $G(s) = \frac{s+1}{s^3 - 20s^2 - 10s + 1}$.

The Pole-Zero Plot, Bode Plot, and the Gain and Phase Margin Plot are customized in the same manner as demonstrated for the stable system. Hence, it will be skipped for this example to reduce redundancy. After simulating the model for 10 s, the output response of the unstable system can be observed from the Scope window, which is shown in Fig. 14.31. From the response, it can be seen that the response is increasing steadily, and becoming unstable.

Fig. 14.24   Block parameters of the Gain and Phase Margin Plots

The Pole-Zero Plot, the Bode Plot, and the Nyquist Plot of the unstable system are illustrated in Fig. 14.32a–c, respectively, to realize the instability of the system (see Sect. 9.6 for more details).

14.7   Conclusion

This chapter introduces the reader to the two different types of control systems—open-loop and closed-loop control systems. Before diving into the Simulink model, the basic structures of these two types of control systems along with their distinctions are explained for building up the fundamental knowledge regarding the topic. Later, the Simulink model designs of these two types are illustrated with step-by-step guidelines to understand the characteristics via simulation. For both of these types, separate example models are designed and simulated. The utilization of the MATLAB command window for parameter definition of any Simulink block is

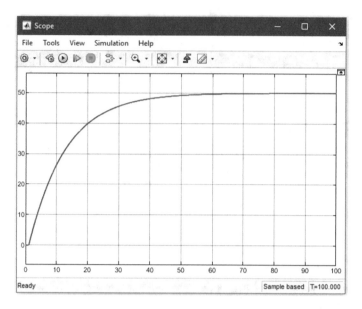

Fig. 14.25 The output response in the scope window

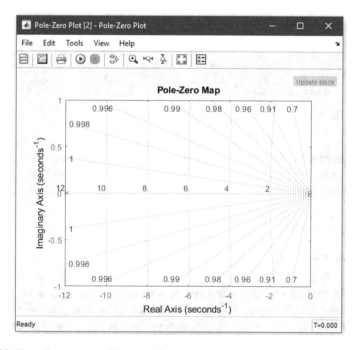

Fig. 14.26 The pole zero map of the control system

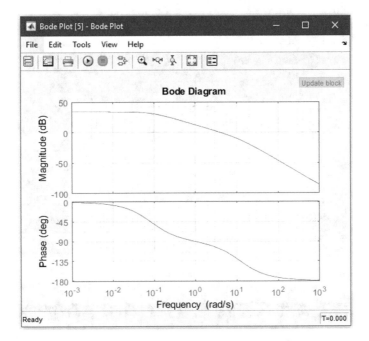

Fig. 14.27 The Bode plot of the control system

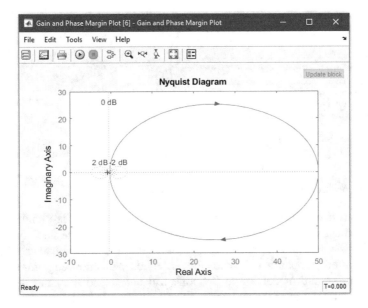

Fig. 14.28 The Gain and Phase Margin Plot of the control system

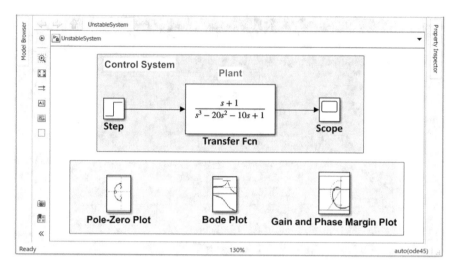

Fig. 14.29 An unstable control system with a step input

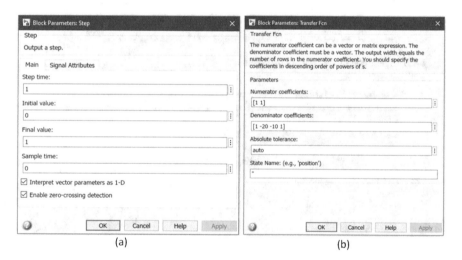

Fig. 14.30 Block parameters of the step and transfer function blocks

also demonstrated in this chapter. Finally, the stability analysis of any control system is demonstrated in this chapter with necessary illustrations. For the stability analysis, Pole-Zero Map, Bode Plot, and Nyquist Plot are covered in this chapter. The main objective of this chapter is to provide the reader adequate knowledge for designing any control system model in the Simulink and understand the characteristics of different control systems via simulation.

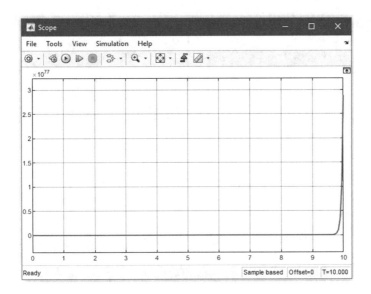

Fig. 14.31 The output response of the unstable system in the scope window

Exercise 14

1. Define control system and write down the name of the two types of control systems based on structure.
2. (a) List the advantages of an open-loop control system.
   (b) Consider an open-loop system with a transfer function, $G(s) = \frac{s+2}{2S^2+1}$, and gain $= 3$. Design the model in the Simulink and show the response in a Scope window.
   (c) Simulate the previous model for the gain of $K = 5$, $K = 8$, and $K = 12$. Write down the effects of the gain increments for an open-loop system based on the responses obtained here.
3. (a) List the advantages of a closed-loop system.
   (b) Consider a closed-loop system with a transfer function, $G(s) = \frac{3s+5}{2S^2+5s+2}$, and the gain of $K = 2$. For unity feedback and PID controller, design the model in the Simulink and show the response.
   (c) Simulate the same model by changing the value of the gain to $K = 6$ by utilizing the MATLAB command window. Demonstrate the response of the system.
4. (a) What are the key controller parameters of a PID controller?
   (b) Design the following model (Fig. 14.33) in Simulink and show the response of the system in the Scope window. Set the value of P and I of the PID controller as 1.
   (c) Simulate the same model for a PD controller using the value of P and D as 1. Show the response of the system using Scope.

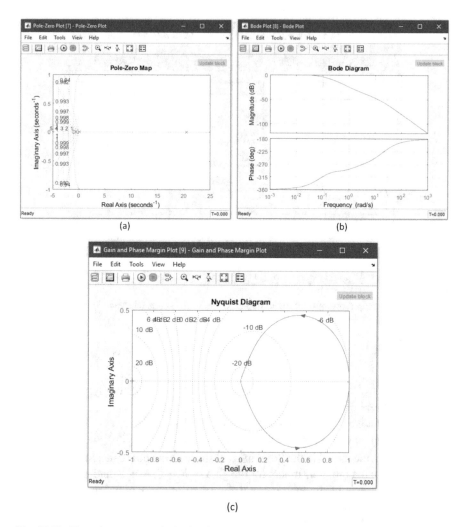

Fig. 14.32 The pole zero map, the Bode plot, and the Nyquist diagram of the unstable system

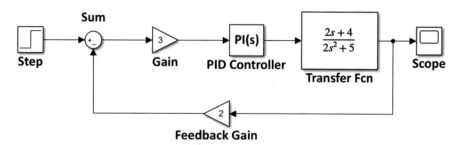

Fig. 14.33 A PID control system with feedback gain

(d) Evaluate the effects of the two controllers—PI and PD—on the response of the above system.

5. (a) State some of the methods to determine the stability of a control system
   (b) Consider an open-loop control system with a transfer function, $G(s) = \frac{400}{s^3 - 4s^2 + 50s + 45}$. Design the model with a Step input to produce the following plots:
       (i) Pole-Zero Map
       (ii) Nyquist Plot
   (c) Comment on the stability of the system based on individual plots generated on (b).

Chapter 15
Electrical Circuit Analysis in Simulink

15.1 Measure Voltage, Current, and Power of a Circuit

In electrical circuit analysis, the parameters that are emphasized most are voltage, current, and power. In this section, the simulation will be performed on both DC and AC circuits with an aim of measuring these parameters.

15.1.1 DC Circuit Analysis

For electrical circuit design, different electrical blocks are available in the Simulink. For this section, we are going to utilize the blocks from Simscape→Electrical→Specialized Power Systems. A DC circuit where a DC voltage source is connected with two resistances in series is modeled in the Simulink design window in Fig. 15.1. After the simulation, the current, voltage, and power of the circuit across one of the resistances are displayed. This is a simple circuit configuration aimed to show how to measure the voltage, current, and power of a DC circuit. The used blocks in Fig. 15.1 are tabulated in Table 15.1, to help the reader navigate easily to find the exact blocks.

In the model shown in Fig. 15.1, the blocks are customized as shown in Fig. 15.2.

The amplitude of the voltage in the DC voltage source block is set at 60 V. The Series RLC Branch block provides the users' flexibility to utilize R, L, C, or any combinations of these three passive elements with a series connection. For this example, the Branch type is selected as "R" from the dropdown option list for both the "Series RLC Branch" and "Series RLC Branch1." The values of the resistances for these blocks are set as 10 Ω and 20 Ω, respectively. To measure the current of the series circuit, the current measurement block is connected in series, while the voltage measurement block is connected in parallel across the resistance— series RLC branch 1—to measure the voltage across it. Two displays are connected

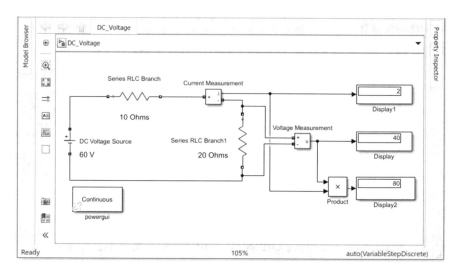

Fig. 15.1 Simulink diagram of a DC electrical circuit with a voltage source and two RLC branches

Table 15.1 Blocks and navigation routes used for the DC circuit analysis in Fig. 15.1

Blocks	Navigation path on the Simulink Library Browser
DC voltage source 1	Simscape→electrical→specialized power systems→fundamental blocks→electrical sources→DC voltage source
Series RLC branch, series RLC branch 1	Simscape→electrical→specialized power systems→fundamental blocks→elements→series RLC branch
Current measurement	Simscape→electrical→specialized power systems→fundamental blocks→measurements→current measurement
Voltage measurement	Simscape→electrical→specialized power systems→fundamental blocks→measurements→voltage measurement
Product	Simulink→math operations→product
Display, display 1, display 2	Simulink→sinks→display
Powergui	Simscape→electrical→specialized power systems→fundamental blocks→powergui

with the output ports of these two measurement blocks to display the results. In a DC circuit, the power across a resistance is the product of the current through the resistance and the voltage across it. Therefore, the product block is utilized where the two inputs are from the output lines of the voltage measurement and the current measurement blocks. The Product block simply multiplies the voltage and current to determine the power which has been displayed in the Display block connected with the output port of the product block.

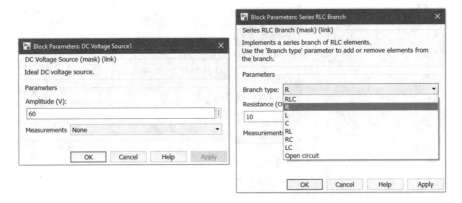

Fig. 15.2   Block parameters of the voltage source and the series RLC branch

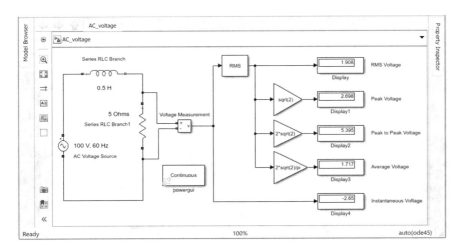

Fig. 15.3   The voltage parameters of the AC circuit are shown in the display blocks

15.1.2   AC Circuit Analysis

In AC circuit analysis, an AC voltage source is required. In Fig. 15.3, a simple series AC circuit simulation is shown, where different voltages associated with the AC voltage source are displayed. In the series circuit, an inductor and a resistor are connected in series with an AC voltage source. The voltage across the resistance is displayed in the "Display" block. The amplitude of the voltage in the AC voltage source is set as 100 V and the frequency is given as 60 Hz (Fig. 15.4). For an AC voltage source, different voltage and current terminologies are available, such as RMS, peak, peak to peak, average, and instantaneous value. In this simulation, all of the values of voltage are displayed. The values of the same terminologies related to current are shown in a later simulation in Fig. 15.5. For determining the RMS value,

Fig. 15.4 Block parameters
of the AC voltage source

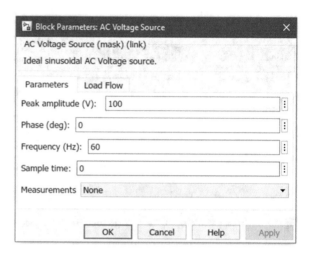

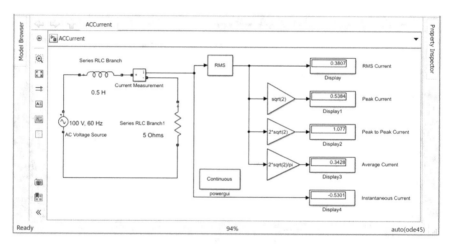

Fig. 15.5 The current parameters of the AC circuit are shown in the display blocks

the RMS block is used. The output signal line from the Voltage Measurement block
indicates the instantaneous value. The other values can be determined from the RMS
value, by applying the following formulas in Eqs. (15.1), (15.2), and (15.3) via the
Gain block:

$$\text{Peak value} = \sqrt{2} \times \text{RMS value} \qquad (15.1)$$

$$\text{Peak to peak value} = 2 \times \sqrt{2} \times \text{RMS value} \qquad (15.2)$$

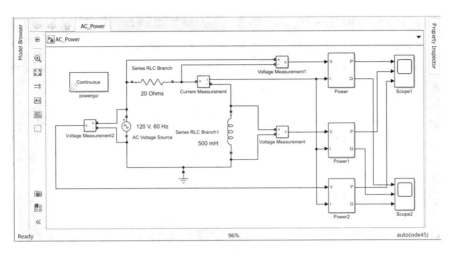

Fig. 15.6 Simulink diagram of a series RL AC circuit

$$\text{Average value} = \frac{2}{\pi} \times \sqrt{2} \times \text{RMS value} \qquad (15.3)$$

In Fig. 15.3, all of these values for the voltage across the resistance are displayed in the "Display" blocks. The same values for the current flowing through the series circuit are displayed in Fig. 15.5.

In an AC circuit, the power measurement can be done by utilizing the "Power" block, where the inputs are voltage and current; and the outputs are real power (P) and reactive power (R). In Fig. 15.6, a series AC RL circuit is simulated. In the circuit, the amplitude of the AC voltage source is selected as 100 V and the frequency is given as 60 Hz (Fig. 15.7). The resistance and the inductance of the circuit are provided as 20 Ω and 500 mH, respectively. In this example, the power across the resistor, the inductor, and the input source or the total power is measured. In an AC circuit, the total power component can be subdivided into two parts—real and reactive power. The vector summation of these two components constitutes the apparent power. In Scope 1, the real powers across the resistor, inductor, and input source are shown (Fig. 15.7). Similarly, in Scope 2, the reactive powers across the resistor, inductor, and input source are demonstrated (Fig. 15.8). It is to be noted that, in the "Power" block, the frequency of the block needs to be identical with the frequency of the input voltage source (Fig. 15.7).

In Fig. 15.8, it can be observed that the total real power or the power of the input source is the summation of the real power across the resistor and the inductor. The same is also true for the reactive power scenario, which can be seen in Fig. 15.9.

Power Factor

The power factor of an AC circuit can be determined by utilizing the real power (P), the reactive power (Q), and the apparent power (S) of the input source. The power factor of a circuit can be calculated by using the power triangle concept as demonstrated below:

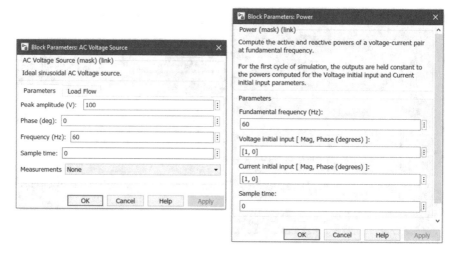

Fig. 15.7 Block parameters of the AC voltage source and the power blocks

In Fig. 15.10, the cosine of the angle Θ is called the power factor. It can also be defined as the ratio of real power to apparent power. Therefore, the power factor of an AC circuit can be formulated as in Eq. (15.4) or (15.5):

$$\text{Power Factor} = \cos \Theta \tag{15.4}$$

$$\text{Power Factor} = \frac{P}{S} = \frac{P}{\sqrt{P^2 + Q^2}} \tag{15.5}$$

In the following example, the power factor of a series RL circuit is determined using the Simulink model by adopting two procedures, where the two above formulas are implemented for the calculation. In Fig. 15.11, the Simulink design model is shown, from where it can be observed that the simulated power factor obtained from the two procedures is the same.

15.2   RLC Circuit Analysis

In any RLC circuit, three components are very necessary—resistance (R), inductance (L), and capacitance (C). Hence, a circuit incorporating three of these elements is named the RLC circuit. In the electrical engineering domain, the most common type of circuit that is implemented in various applications is the RLC circuit. Therefore, learning about RLC circuit simulation in Simulink is essential indeed.

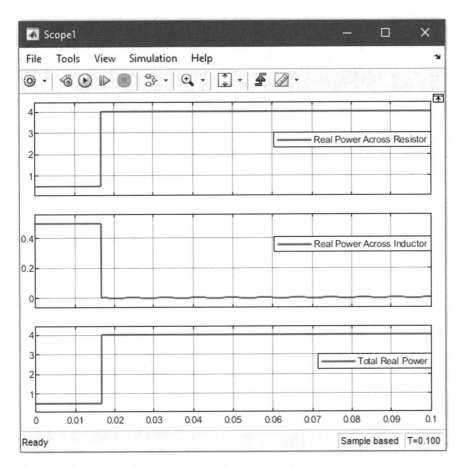

Fig. 15.8 The scope window showing the real power parameters

15.2.1   AC RLC Circuit Analysis

In Simulink, RLC circuit analysis can be performed with less complexity. In Fig. 15.12, an RLC circuit is demonstrated with an AC input source. The voltage and current across the resistance named as the "Series RLC Branch 1" are simulated and displayed.

15.2.2   DC RLC Circuit Analysis

The same example is repeated with the replacement of the AC source with a DC voltage source in Fig. 15.13. For a circuit with a DC source that has zero frequency component, the inductor acts like a short circuit, and the capacitor behaves as an

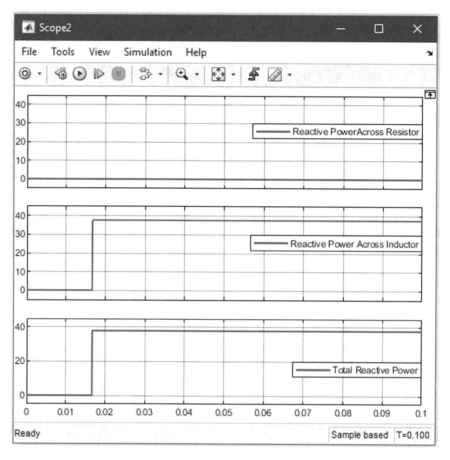

Fig. 15.9 The scope window showing the reactive power parameters

Fig. 15.10 The power triangle

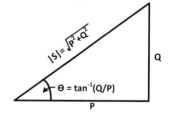

open circuit, which can be observed from the following example as shown in Fig. 15.13. As the capacitor creates an open circuit and the inductor becomes a short circuit with the DC voltage source, the voltage across the resistance becomes the voltage of the DC source. In this simulation, the voltage of the DC source and the resistor are identical, which proves the former concept via simulation. Thus, any kind of RLC circuit simulation can be performed utilizing Simulink.

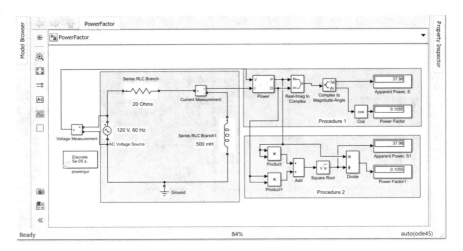

Fig. 15.11 Simulink diagram of a series RL AC circuit

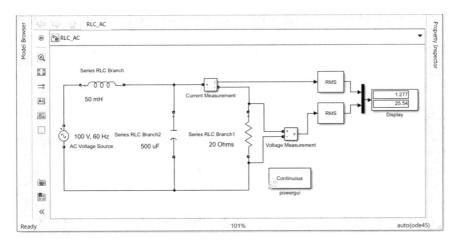

Fig. 15.12 Simulink diagram of a series RLC AC circuit

15.3   Conclusion

This chapter presents the analysis of electrical circuits for both AC and DC with separate examples implemented in the Simulink. In Simulink, the parameters of an electrical circuit can be determined avoiding any complex calculations. The user just needs to utilize different Simulink blocks available in the Simulink Library Browser to create the exact circuit. After running the simulation, the values of different parameters as well as graphical illustrations of input and output signals can also be observed by utilizing Simulink. Before creating any actual hardware testbed, Simulink can be used as a simulation platform to verify the performance of the

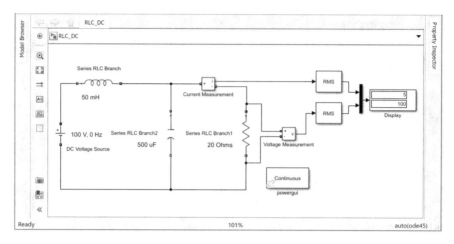

Fig. 15.13 Simulink diagram of a series RLC DC circuit

testbed. The reader can also verify the circuit theories by performing simulations in the Simulink for different circuit analyses. Thus, this chapter can significantly help the readers to understand and analyze the electrical circuit with more depth.

Exercise 15

1. Mention some of the parameters of the electrical circuit.
2. (a) Write some of the differences between DC and AC circuits.
   (b) Consider a series RLC circuit with resistance, $R = 10\ \Omega$; inductance, $L = 0.5$ H; and capacitance, $C = 0.4\ \mu F$, with an AC voltage source. The amplitude of the AC voltage source, $V = 120$ V; and the frequency $= 60$ Hz. Determine the following parameters:
   > (i) RMS voltage across the resistance
   > (ii) Peak-to-peak voltage across the resistance
   > (iii) Average voltage across the resistance
3. (a) What is an RLC circuit?
   (b) Recreate the following Simulink model by using the same parameters:
   (c) Demonstrate the real and reactive power across the resistance of Fig. 15.14 using a Scope block.

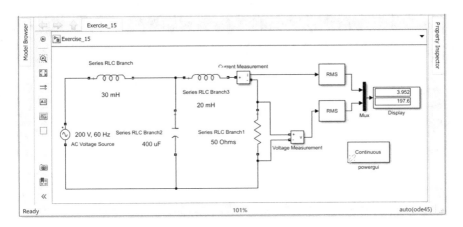

Fig. 15.14 Simulink diagram of a series RLC AC circuit for Question 3b

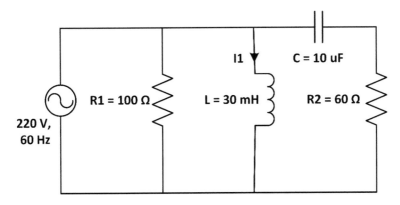

Fig. 15.15 An RLC AC circuit for Question 4b

4. (a) What is the power factor?
   (b) Design a Simulink model for the following circuit, and measure the RMS current (I1) through the inductor, L.
   (c) Simulate the model of Fig. 15.15 to determine the power factor of the source.

Chapter 16
Application of Simulink in Power Systems

16.1 Modeling Single-Phase Power Source in Simulink

In the field of the power system, Simulink is widely used due to its rich collection of different power system blocks. The importance of Simulink in the power system is quite undeniable, as it provides an advanced simulation platform along with versatile facilities. A power source can be categorized into two types—DC power and AC power. For DC sources, the power is unidirectional, while an AC source produces power that changes its direction constantly. In our current age, AC sources are considered as a standard power source due to their efficiency to transfer power over long distances without significant losses.

A single-phase AC power system incorporates one single AC source with two wires. One wire is termed power wire, while the other one is called neutral wire. Electricity flows from the power wire to the neutral wire. Almost all of our household appliances work by using a single-phase AC power source. A Simulink model showing a simple single-phase AC source is shown in Fig. 16.1. The navigation routes of all the utilized blocks of this example are listed in Table 16.1.

The AC Voltage Source block is customized by setting the amplitude of voltage as 120 V and the frequency as 60 Hz (Fig. 16.2). In the USA, the standard frequency for a single-phase AC source is 60 Hz. In some other countries, the standard may become 50 Hz. For this example, 60 Hz frequency is considered.

A Voltage Measurement block is used to observe the voltage output of the source via the Scope window. The output voltage can be seen in Fig. 16.3, which appears after double-clicking the Scope block.

A dual, or split phase, power source is also considered a single-phase power source in some aspects. In such scenarios, two phases may available with one neutral wire. An example is given below (Fig. 16.4), where Phase-A or Phase-B wire with the neutral wire provides 120 V supply individually. However, taking voltage across the Phase-A and Phase-B wires without the neutral wire can be used as a voltage source of 240 V. This type of arrangement can be utilized for multi-functioning

© The Author(s), under exclusive license to Springer Nature Switzerland AG 2022   455
E. Hossain, *MATLAB and Simulink Crash Course for Engineers*,
https://doi.org/10.1007/978-3-030-89762-8_16

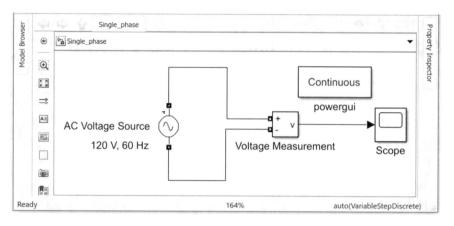

Fig. 16.1 A simple single-phase AC voltage source

Table 16.1 Blocks and navigation routes for modeling a single-phase AC power source

Blocks	Navigation path on the Simulink Library Browser
AC Voltage Source	Simscape → Electrical → Specialized Power Systems → Fundamental Blocks → Electrical Sources → AC Voltage Source
Voltage Measurement	Simscape → Electrical → Specialized Power Systems → Fundamental Blocks → Measurements → Voltage Measurement
Scope	Simulink → Sinks → Scope
powergui	Simscape → Electrical → Specialized Power Systems → Fundamental Blocks → powergui

purposes. For example, low-power loads can be operated using the 120 V arrangements, and for high-power loads, 240 V arrangement can be more feasible.

16.2 Modeling Three-Phase AC Power Source in Simulink

In a three-phase AC power source, three individual AC sources are connected either in delta or wye/star configuration. A concise analysis of the characteristics of the three-phase AC source based on delta and wye configuration including the two sequences is demonstrated in Sect. 8.3.4. Hence, the theoretical concept of this section will be skipped to avoid redundancy. In this section, the aim is to model a three-phase AC power source for both delta and wye configuration by utilizing Simulink and observe the characteristics via simulation graphs. Based on different configurations, the three-phase AC power source can be of five types in general, such as:

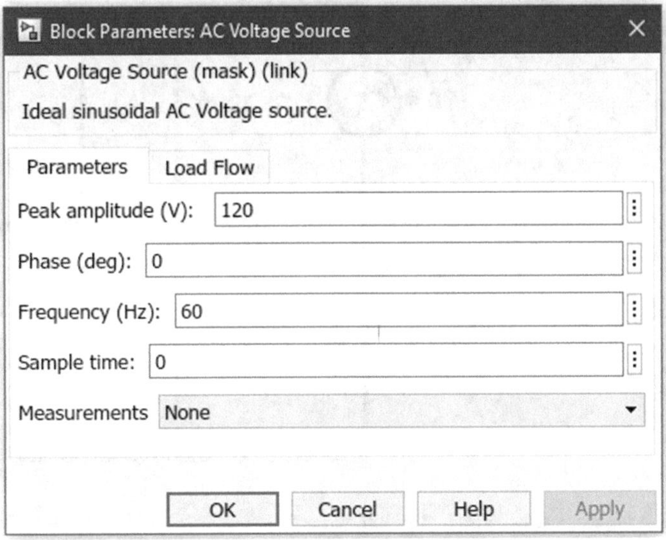

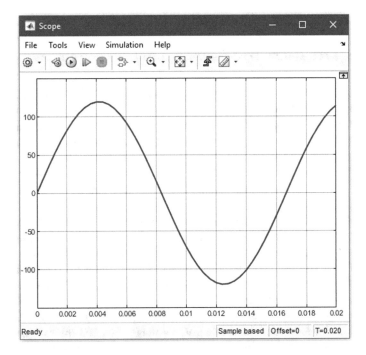

Fig. 16.2 Block parameters of the AC voltage source

Fig. 16.3 The output of the AC voltage source in the scope window

Fig. 16.4 A dual or split
phase power source

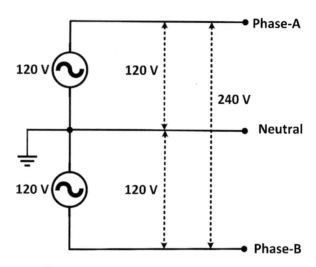

1. Delta-connected three-phase AC source.
2. Wye-connected three-wire three-phase AC source.
3. Wye-connected four-wire three-phase AC source.

16.2.1 Three-Phase Wye-Connected AC Power Source

In a three-phase wye-connected AC power source, three individual AC power sources are connected in a wye configuration. The voltages and currents of the three sources are 120° apart from each other in phase difference. The sequences of the phase differences among the three phases create the categorization of two sequences—"abc" sequence and "acb" sequence—that are described earlier in Sect. 8.3.4. In Fig. 16.5, two examples are provided where a three-phase AC source model in wye configuration is designed and simulated. The blocks utilized in this example are enlisted in Table 16.2.

In Fig. 16.5, two examples are separated by using two Area windows. In the first example, three AC voltage sources are connected in a wye configuration, and the neutral point is grounded. The voltages of the three phases can be observed in the Scope window. This example provides a balanced three-phase AC source; hence, the amplitudes and the frequencies of the three voltage sources are kept similar which are 150 V and 60 Hz, respectively. The customization of the parameter windows of the AC Voltage Sources is shown in Fig. 16.6. The phase degree of the AC Voltage Source, AC Voltage Source1, and AC Voltage Source2 blocks are set as $0°$, $-120°$, and $+120°$, respectively, by maintaining "acb" sequence.

In example two, the same configuration is designed using different blocks. In Simulink, a Three-Phase Voltage Source block is available, through which wye configuration can be made easily without using three separate voltage sources. In the

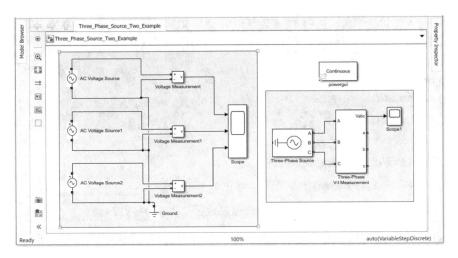

Fig. 16.5 Simulink diagram of a three-phase AC voltage source

Table 16.2 Blocks and navigation routes for modeling a three-phase wye-connected AC power source

Blocks	Navigation path on the Simulink Library Browser
AC Voltage Source AC Voltage Source1 AC Voltage Source2	Simscape → Electrical → Specialized Power Systems → Fundamental Blocks → Electrical Sources → AC Voltage Source
Voltage Measurement Voltage Measurement1 Voltage Measurement2	Simscape → Electrical → Specialized Power Systems → Fundamental Blocks → Measurements → Voltage Measurement
Three-Phase Source	Simscape → Electrical → Specialized Power Systems → Fundamental Blocks → Electrical Sources → Three-Phase Source
Three-Phase V-I Measurement	Simscape → Electrical → Specialized Power Systems → Fundamental Blocks → Measurements → Three-Phase V-I Measurement
Scope, Scope1	Simulink → Sinks → Scope
Ground	Simscape → Electrical → Specialized Power Systems → Fundamental Blocks → Elements → Ground
powergui	Simscape → Electrical → Specialized Power Systems → Fundamental Blocks → powergui

parameter window of the Three-Phase Voltage Source block, three types of configurations are available—Y, Yg, and Yn (Fig. 16.7). Here, the Y configuration indicates that the neutral point is in the floating mode and cannot be accessed. Yg configuration refers that the neutral point is grounded, and for Yn configuration, the neutral point can be accessed externally. In this example, the Yg configuration is chosen to match the configuration of the first example. The line-to-neutral voltages (RMS value) of the three phases are set as $150/\sqrt{2}$. It is to be noted that, in the first example, the peak amplitude of the voltages is set as 150 V. To make the results identical, the RMS voltages of the sources in the second example are assigned as

Block Parameters: Three-Phase Source   ✕

Three-Phase Source (mask) (link)

Three-phase voltage source in series with RL branch.

| Parameters | Load Flow |

Configuration: Yg   ▼

Source

☑ Specify internal voltages for each phase

Line-to-neutral voltages [Va Vb Vc] (Vrms)   [150 150 150]/sqrt(2)   ⋮

Phase angle of line-to-neutral voltages [phia phib phic] (deg)   [0 -120 +120]   ⋮

Frequency (Hz): 60   ⋮

Impedance

☐ Internal   ☐ Specify short-circuit level parameters

Source resistance (Ohms): 0   ⋮

Source inductance (H): 0   ⋮

Base voltage (Vrms ph-ph): 150   ⋮

| OK | Cancel | Help | Apply |

Fig. 16.7 Block parameters of the three-phase AC voltage source

Block Parameters: AC Voltage Source ✕	Block Parameters: AC Voltage Source1 ✕	Block Parameters: AC Voltage Source2 ✕
AC Voltage Source (mask) (link)	AC Voltage Source (mask) (link)	AC Voltage Source (mask) (link)
Ideal sinusoidal AC Voltage source.	Ideal sinusoidal AC Voltage source.	Ideal sinusoidal AC Voltage source.
Parameters Load Flow	Parameters Load Flow	Parameters Load Flow
Peak amplitude (V): 150	Peak amplitude (V): 150	Peak amplitude (V): 150
Phase (deg): 0	Phase (deg): -120	Phase (deg): +120
Frequency (Hz): 60	Frequency (Hz): 60	Frequency (Hz): 60
Sample time: 0	Sample time: 0	Sample time: 0
Measurements Voltage	Measurements Voltage	Measurements Voltage
OK Cancel Help Apply	OK Cancel Help Apply	OK Cancel Help Apply

Fig. 16.6 Block parameters of the three AC voltage sources

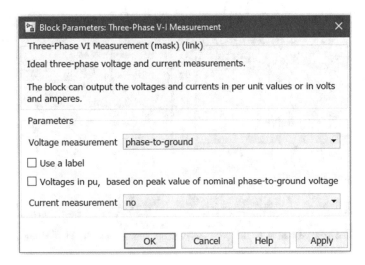

Fig. 16.8 Block parameters of the three-phase V-I measurement

$1/\sqrt{2}$ times of the peak amplitudes of the first example. The Phase angles are set as 0°, -120°, $+120^{\circ}$, for ensuring "acb" sequence as before.

A Three-Phase V-I Measurements block is connected to the Three-Phase Voltage Source. The Three-Phase V-I Measurements block has three inputs, which are connected with the three phases of the Three-Phase Voltage Source block. The output ports of the Three-Phase V-I Measurements blocks are V_{abc}, I_{abc}, and the output terminals for the three phases. This block can be utilized to measure the voltages and currents both in per unit values and in volts and amperes. The parameter window of the Three-Phase V-I Measurements block is shown in Fig. 16.8, from where measurement options for both voltages and currents can be configured. For this example, the phase-to-ground option is chosen for voltage measurement and the per unit value option is kept unchecked. The current measurement option is selected as "no," as in this example, the currents of the phases are not measured. However, the user has the preference to select the current measurement option.

The voltages of the three phases for both of the examples can be observed from the Scope windows as shown in Fig. 16.9.

16.2.2   Three-Phase Delta-Connected AC Power Source

A three-phase delta-connected AC power source is designed in the Simulink, which is shown in Fig. 16.10. In the figure, three AC Voltage Source blocks with three Series RLC Branches are connected in such a way to form a delta connection. The utilized blocks for this example are listed in Table 16.3 with their navigation routes.

The customization of the AC Voltage Source blocks is similar to the customization of the previous wye-connected AC Voltage Source blocks. The peak amplitude

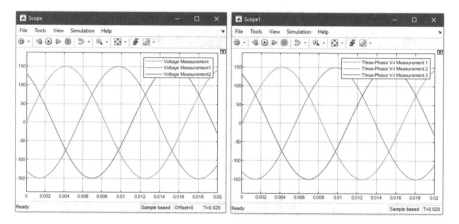

Fig. 16.9 The voltage output of the three AC voltage sources and the three-phase AC voltage source

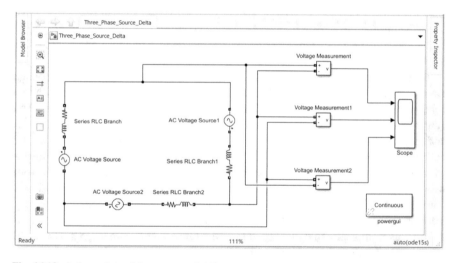

Fig. 16.10 A three-phase delta-connected AC power source

and the frequency are set as 150 V and 60 Hz, respectively. The phase angles of the three sources are set as $0°$, $-120°$, and $+120°$. In Fig. 16.11, the parameter window of one of the AC Voltage Source blocks is shown in the first figure. In the second figure, the parameter window of the Series RLC Branch is shown, where the block is customized for RL branch by setting the resistance value 5000 ohms and the inductance value 100 mH. The other two Series RLC Branch blocks are customized in the same manner.

The voltages of the three phases can be observed from the Scope windows, which are shown in Fig. 16.12.

Table 16.3 Blocks and navigation routes for modeling a three-phase delta connected AC power source

Blocks	Navigation path on the Simulink Library Browser
AC Voltage Source AC Voltage Source1 AC Voltage Source2	Simscape → Electrical → Specialized Power Systems → Fundamental Blocks → Electrical Sources → AC Voltage Source
Voltage Measurement Voltage Measurement1 Voltage Measurement2	Simscape → Electrical → Specialized Power Systems → Fundamental Blocks → Measurements → Voltage Measurement
Series RLC Branch Series RLC Branch1 Series RLC Branch2	Simscape → Electrical → Specialized Power Systems → Fundamental Blocks → Elements → Series RLC Branch
Scope	Simulink → Sinks → Scope
powergui	Simscape → Electrical → Specialized Power Systems → Fundamental Blocks → powergui

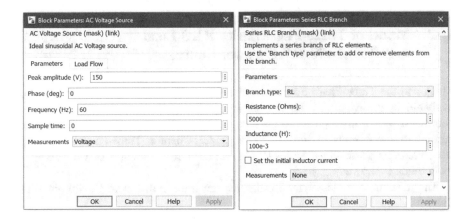

Fig. 16.11 Block parameters of the AC voltage source and the series RLC branch

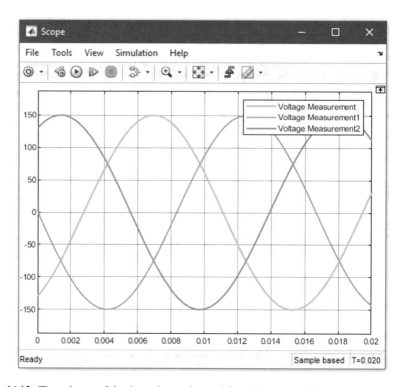

Fig. 16.12 The voltages of the three phases observed from the scope window

16.3 Model of Three-Phase Series RLC Load with Three-Phase AC Power Source

Similar to the Three-Phase Source, a Three-Phase Series RLC Load block is also available in the Simulink. In this section, a Three-Phase Series RLC Load is connected with a three-phase AC source to observe the instantaneous power and current of the circuit. In Fig. 16.13, a model where a Three-Phase Source is connected to a Three-Phase Series RLC Load through a Three-Phase V-I Measurement block is shown.

The navigation routes of the blocks that are used in this example are given in Table 16.4.

To measure the real and reactive power of the system, the voltage and current terminals from the Three-Phase V-I Measurement block are inserted into the input of the Power (3 ph, Instantaneous) block. The outputs of this block are inserted into the input terminals of the Scope block. The V_{abc} and the I_{abc} signal lines are connected to the Scope1 block to observe the voltages and currents of the phases. The parameter windows of the Three-Phase Source and the Three-Phase Series RLC Load blocks

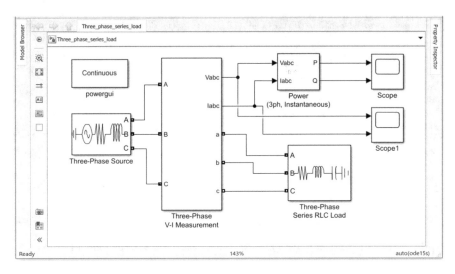

Fig. 16.13 Simulink diagram of a Three-Phase Series RLC Load with Three-Phase AC Power Source

Table 16.4 Blocks and navigation routes for modeling a Three-Phase Series RLC Load with three-phase AC power source

Blocks	Navigation path on the Simulink Library Browser
Three-Phase Source	Simscape → Electrical → Specialized Power Systems → Fundamental Blocks → Electrical Sources → Three-Phase Source
Three-Phase V-I Measurement	Simscape → Electrical → Specialized Power Systems → Fundamental Blocks → Measurements → Three-Phase V-I Measurement
Power (3 ph, Instantaneous)	Simscape → Electrical → Specialized Power Systems → Fundamental Blocks → Power Electronics → Pulse and signal generators/ Measurements
Three-Phase Series RLC Load	Simscape → Electrical → Specialized Power Systems → Fundamental Blocks → Elements → Three-Phase Series RLC Load
Scope, Scope1	Simulink → Sinks → Scope
powergui	Simscape → Electrical → Specialized Power Systems → Fundamental Blocks → powergui

are shown in Fig. 16.14. It can be noted that both the source and the load are customized to act in Yg configuration mode. The graphs of the real and reactive power can be observed from the Scope window, as shown in Fig. 16.15. On the other hand, the voltages and the currents of the system can be seen from the Scope1 window, which is illustrated in Fig. 16.16.

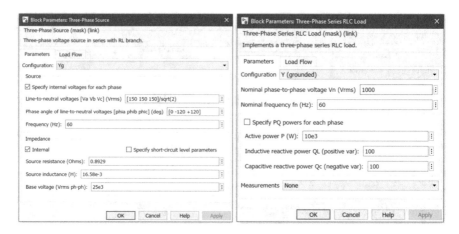

Fig. 16.14 Block parameters of the three-phase source and the three-phase series RLC load

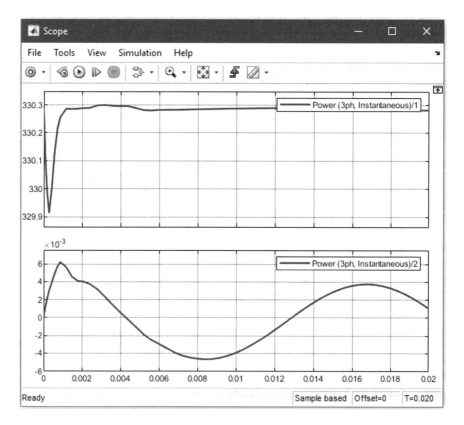

Fig. 16.15 The graphs of the real and reactive power from the scope window

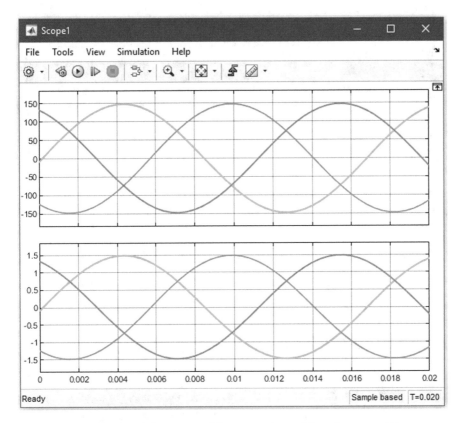

Fig. 16.16 The voltages and the currents of the system observed from the scope window

16.4   Model of Three-Phase Parallel RLC Load with Three-Phase AC Power Source

A Simulink model incorporating a three-phase parallel RLC load with a three-phase AC power source is demonstrated in Fig. 16.17. The model is almost similar to the previous model shown in Fig. 16.13. The only difference is that the Three-Phase Series RLC Load block is replaced by a Three-Phase Parallel RLC Load block. The navigation route of this block is Simulink Library Browser → Simscape → Electrical → Specialized Power Systems → Fundamental Blocks → Elements → Three-Phase Parallel RLC Load. The parameter windows of the Three-Phase Source and the Three-Phase Parallel RLC Load blocks are shown in Fig. 16.18. The real and the reactive power of the system are shown in Fig. 16.19, while the voltages and currents waveforms of the system are illustrated in Fig. 16.20.

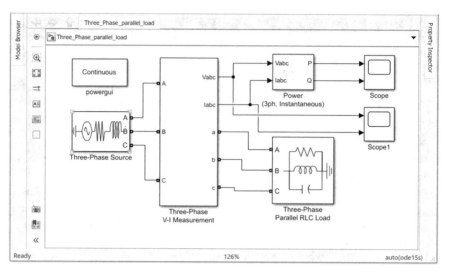

Fig. 16.17 Simulink diagram of a Three-Phase Parallel RLC Load with Three-Phase AC Power Source

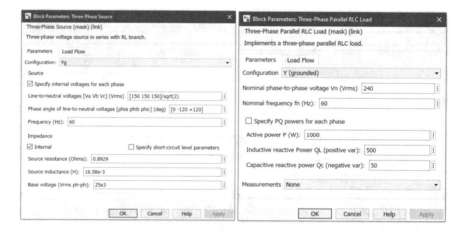

Fig. 16.18 Block parameters of Three-Phase Source and the Three-Phase Parallel RLC Load

16.5   Power Factor Calculation Simulink Model

In the power system, power factor calculation is an important part to analyze different characteristics of a system. In Simulink, a model can be designed to calculate the power factor. An example is given in Fig. 16.21, where the power factor of a system is calculated. The system incorporates a Three-Phase Source with a Three-Phase Parallel RLC Load. To calculate the power factor of the system, the

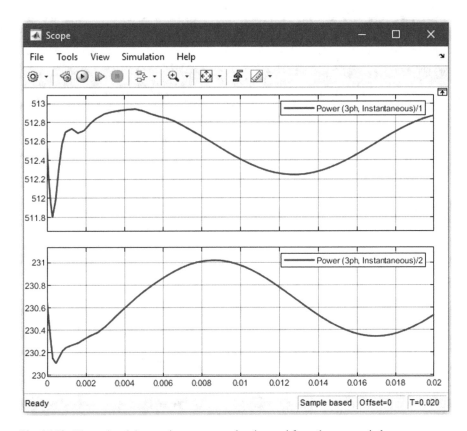

Fig. 16.19 The real and the reactive power graphs observed from the scope window

real (P) and reactive (Q) power of the system are determined by using the Power (3 ph, Instantaneous) block. Later the following formula is utilized to design the rest of the part to calculate the power factor:

$$\text{Power factor} = \frac{\text{Real power}}{\sqrt{\text{Real power}^2 + \text{Reactive power}^2}} = \frac{P}{\sqrt{P^2 + Q^2}} \qquad (16.1)$$

The used blocks for this model can be found from the mentioned navigation routes described in Table 16.5.

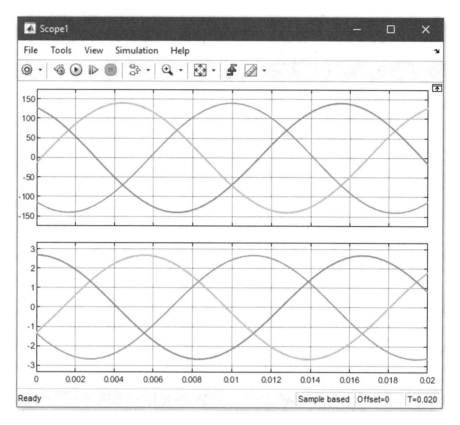

Fig. 16.20 The voltages and currents waveforms observed from the scope window

16.6   Modeling Different Power System Configurations

In the previous sections, wye- and delta-connected three-phase power sources are shown. Similar to the three-phase power source, the load can also be connected in wye and delta forms. Based on the configurations of both three-phase power source and three-phase load, the power system configuration can be categorized in four forms, such as:

1. Y-Y power system configuration.
2. Y-Δ power system configuration.
3. Δ-Δ power system Configuration.
4. Δ-Y power system configuration.

Each of these configurations can be balanced or unbalanced based on their characteristics. In this section, a Y-Y and a $\Delta - \Delta$ configuration of the power system will be demonstrated with the Simulink models for both balanced and unbalanced systems.

Table 16.5 Blocks and navigation routes for modeling the calculation of power factor

Blocks	Navigation path on the Simulink Library Browser
Three-Phase Source	Simscape → Electrical → Specialized Power Systems → Fundamental Blocks → Electrical Sources → Three-Phase Source
Three-Phase V-I Measurement	Simscape → Electrical → Specialized Power Systems → Fundamental Blocks → Measurements → Three-Phase V-I Measurement
Power (3 ph, Instantaneous)	Simscape → Electrical → Specialized Power Systems → Fundamental Blocks → Power Electronics → Pulse & Signal Generators/ Measurements
Three-Phase Parallel RLC Load	Simscape → Electrical → Specialized Power Systems → Fundamental Blocks → Elements → Three-Phase Parallel RLC Load
Product, Product1	Simulink → Math Operations → Product
Add	Simulink → Math Operations → Add
Square Root	Simulink → Quick Insert → Math Operations → Square Root
Divide	Simulink → Math Operations → Divide
Display	Simulink → Sinks → Display
powergui	Simscape → Electrical → Specialized Power Systems → Fundamental Blocks → powergui

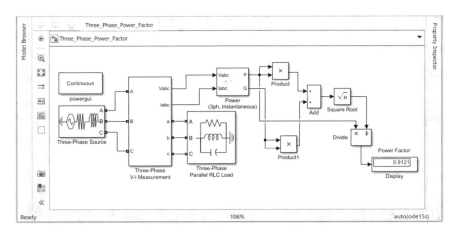

Fig. 16.21 Simulink diagram of a Three-Phase Source with a Three-Phase Parallel RLC Load

16.6.1   Balanced Y-Y Power System Configuration

A balanced Y-Y power system configuration, where the first term "Y" indicates the configuration of the Three-Phase Source, and the second term "Y" refers to the configuration of the three-phase load, is demonstrated in Fig. 16.22. In a balanced system, the amplitudes of the voltages in the Three-Phase Source are similar. Additionally, the impedances of each phase of a three-phase load are identical. The Simulink blocks used in the example shown in Fig. 16.22 are listed in Table 16.6 with their navigation routes.

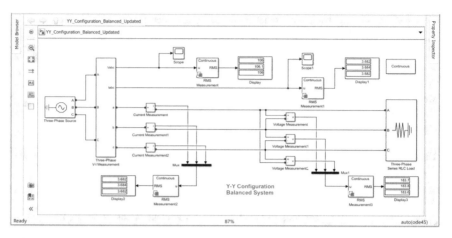

Fig. 16.22 Simulink diagram of a Y-Y configuration balanced system

The customized parameter windows of the Three-Phase Source and the Three-Phase Series RLC Load blocks are shown in Figs. 16.23 and 16.24, respectively. To make the system balanced, the magnitudes of the voltages of each phase are made identical. Moreover, the nominal phase-to-neutral voltages and *PQ* specifications of the three-phase loads are also set to be exactly similar. In Fig. 16.25, the configured parameter windows of the Three-Phase V-I Measurement and the RMS Measurement blocks are demonstrated. For a Y-Y configured balanced power system, the following characteristics should be observed:

Balanced Y-Y power system configuration
Consider the three phases as A, B, and C.
  Line-to-line voltages: V_{AB}, V_{BC}, and V_{CA}
  Phase-to-ground voltages: V_{an}, V_{bn}, and V_{cn}
  For balanced Y-Y configuration:

$$| V_{AB} |=| V_{BC} |=| V_{CA} |$$

$$| V_{an} |=| V_{bn} |=| V_{cn} |$$

$$| V_{AB} |=| V_{BC} |=| V_{CA} |= \sqrt{3} \times | V_{an} |= \sqrt{3} \times | V_{bn} |= \sqrt{3} \times | V_{cn} |$$

Line currents: I_A, I_B, and I_C
Phase currents: I_{an}, I_{bn}, and I_{cn}

$$| I_A |=| I_B |=| I_C |$$

$$| I_{an} |=| I_{bn} |=| I_{cn} |$$

$$| I_A |=| I_B |=| I_C |=| I_{an} |=| I_{bn} |=| I_{cn} |$$

Table 16.6 Blocks and navigation routes for modeling a balanced Y-Y power system

Blocks	Navigation path on the Simulink Library Browser
Three-Phase Source	Simscape → Electrical → Specialized Power Systems → Fundamental Blocks → Electrical Sources → Three-Phase Source
Three-Phase V-I Measurement	Simscape → Electrical → Specialized Power Systems → Fundamental Blocks → Measurements → Three-Phase V-I Measurement
Three-Phase Series RLC Load	Simscape → Electrical → Specialized Power Systems → Fundamental Blocks → Elements → Three-Phase Series RLC Load
Voltage Measurement, Voltage Measurement1, Voltage Measurement2	Simulink → Sinks → Voltage Measurement
Current Measurement, Current Measurement1, Current Measurement2	Simulink → Sinks → Current Measurement
RMS Measurement, RMS Measurement1, RMS Measurement2, RMS Measurement3	Simscape → Electrical → Control → Measurements → RMS Measurement1
Scope, Scope1	Simulink → Sinks → Scope
Mux, Mux1	Simulink → Signal Routing → Mux
Display, Display1, Display2, Display3	Simulink → Sinks → Display
powergui	Simscape → Electrical → Specialized Power Systems → Fundamental Blocks → powergui

To verify whether the above-mentioned characteristics are identical with the simulation results, Table 16.7 is shown to correlate the above-mentioned parameters with the simulation results.

The simulated results verify the characteristics that are defined earlier and can be observed from Fig. 16.22.

The Scope and Scope1 windows demonstrate the phase-to-ground voltages and the phase currents of the system as shown in Fig. 16.26. The line-to-line voltages and the line currents can also be observed similarly by utilizing two other Scopes.

16.6.2 Unbalanced Y-Y Power System Configuration

For an unbalanced Y-Y power system configuration, the impedances of the three-phase load change. If the magnitudes of the Three-Phase Source vary, the system can also become unbalanced. For the same Simulink model shown in Fig. 16.22, the system is made unbalanced by changing the nominal phase-to-neutral voltage of the three-phase load and the magnitudes of the Three-Phase Source from the parameter windows of the blocks. The final Simulink model of the unbalanced Y-Y configured

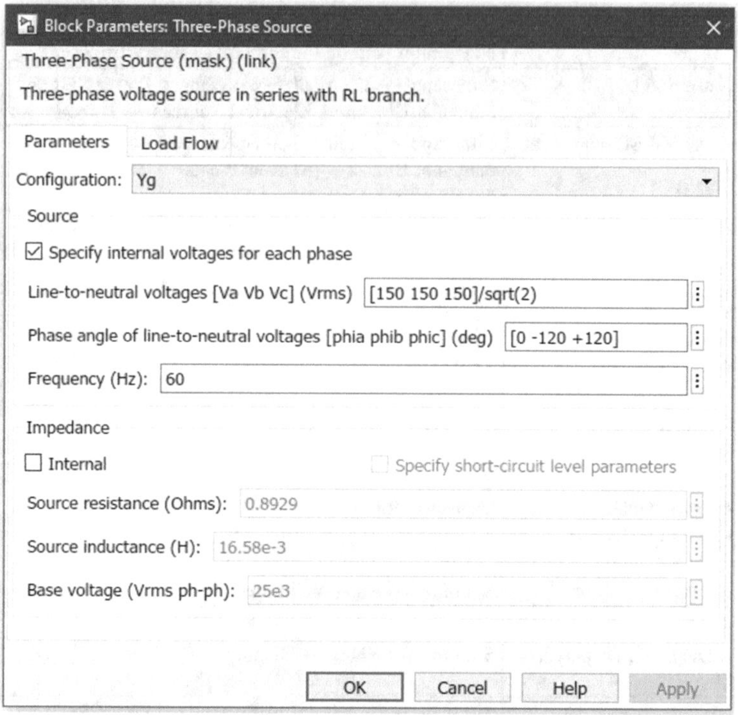

Fig. 16.23 Block parameters of the three-phase source

system is shown in Fig. 16.27. The customized parameter window of the Three-Phase Source block is shown in Fig. 16.28, where the magnitudes of the three phases are made unequal. In Fig. 16.29, the parameter window of the Three-Phase Series RLC Load is shown, where the nominal phase-to-neutral voltages and the *PQ* specifications are also made unequal.

For a Y-Y configured unbalanced power system, the following characteristics should be observed:

Unbalanced Y-Y power system configuration
Consider the three phases as A, B, and C.
   Line-to-line voltages: V_{AB}, V_{BC}, and V_{CA}
   Phase-to-ground voltages: V_{an}, V_{bn}, and V_{cn}
   For unbalanced Y-Y configuration:

(continued)

$$| V_{AB} | {\neq} | V_{BC} | {\neq} | V_{CA} |$$
$$| V_{an} | {\neq} | V_{bn} | {\neq} | V_{cn} |$$
$$| V_{AB} | {\neq} \sqrt{3} {\times} | V_{an} |; \ | V_{BC} | {\neq} \sqrt{3} {\times} | V_{bn} |; | V_{CA} | {\neq} \sqrt{3} {\times} | V_{cn} |$$
$$V_{AB} = V_{an} - V_{bn}; V_{BC} = V_{bn} - V_{cn}; V_{CA} = V_{cn} - V_{an}$$

Line currents: I_{AB}, I_{BC}, and I_{CA}
Phase currents: I_{an}, I_{bn}, and I_{cn}

$$| I_{AB} | {\neq} | I_{BC} | {\neq} | I_{CA} |$$
$$| I_{AB} | {=} | I_{an} |; | I_{BC} | = | I_{bn} |; \ | I_{CA} | {=} | I_{cn} |$$

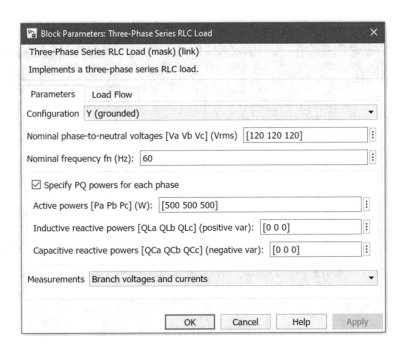

Fig. 16.24 Block parameters of the three-phase series RLC load

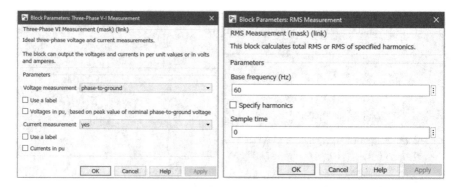

Fig. 16.25 Block parameters of the three-phase V-I measurement and RMS measurement

Table 16.7 Correlation between the Simulink model and the reference parameters of the balanced Y-Y system

Simulink model	Reference
V_{abc} (Three-Phase V-I Measurement Block)	Phase-to-Ground Voltages: V_{an}, V_{bn}, V_{cn}
Display	$\|V_{an}\|$, $\|V_{bn}\|$, $\|V_{cn}\|$ (RMS values)
Display1	$\|V_{AB}\|, \|V_{BC}\|, \|V_{CA}\|$ (RMS values)
I_{abc} (Three-Phase V-I Measurement Block)	Phase currents: I_{an}, I_{bn}, I_{cn}
Display2	$\|I_{an}\|, \|I_{bn}\|, \|I_{cn}\|$ (RMS values)
Display3	$\|I_A\|, \|I_B\|, \|I_C\|$ (RMS values)

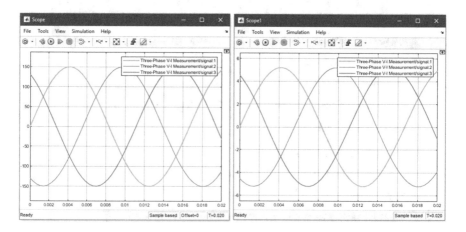

Fig. 16.26 The phase to ground voltages and the phase currents of the system observed from the scope window

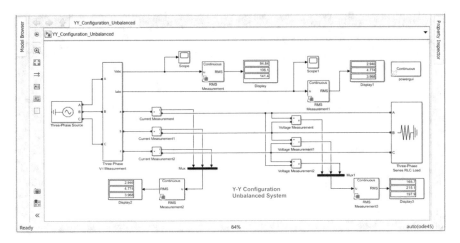

Fig. 16.27 Simulink diagram of a Y-Y configuration unbalanced system

For verifying the above-mentioned characteristics, Table 16.8 can be utilized to make the connection with the Simulink model:

The phase-to-ground voltages and the phase currents of the load can be shown from the Scope windows as shown in Fig. 16.30. From the figure, it is observable that the amplitudes of the three phases vary both for voltages and currents.

16.6.3   Balanced Δ − Δ Power System Configuration

For a balanced Δ − Δ three-phase power system, both the source and the load are configured in the delta configuration. A delta configured Three-Phase Source is already demonstrated previously in Sect. 16.2.2. Here, in this example, an additional delta-connected three-phase load is connected to make the system. The final Simulink model is illustrated in Fig. 16.31. The blocks that are used to create the model are summarized along with their navigation routes in Table 16.9.

The parameter windows of the AC Voltage Source and the Series RLC Branch are shown in Fig. 16.32 to provide a sample. The Series RLC Branch block is used to represent the internal impedance of the AC Voltage Source. The other two AC Voltage Source blocks and the Series RLC Branch blocks are customized in the same manner as well to make the system balanced.

For a balanced Δ − Δ configured power system, the characteristics can be demonstrated as follows:

Balanced Δ − Δ power system configuration
Consider the three phases as A, B, and C.

(continued)

Line-to-line voltages: V_{AB}, V_{BC}, V_{CA}
Phase-to-phase voltages: V_{ab}, V_{bc}, and V_{ca}
For balanced $\Delta - \Delta$ configuration:

$$| V_{AB} | = | V_{BC} | = | V_{CA} |$$

$$| V_{ab} | = | V_{bc} | = | V_{ca} |$$

$$| V_{AB} | = | V_{BC} | = | V_{CA} | = | V_{ab} | = | V_{bc} | = | V_{ca} |$$

Line currents: I_A, I_B, and I_C
Phase-to-phase currents: I_{ab}, I_{bc}, and I_{ca}

$$| I_A | = | I_B | = | I_C |$$

$$| I_{ab} | = | I_{ba} | = | I_{ca} |$$

$$| I_A | = | I_B | = | I_C | = \sqrt{3} \times | I_{ab} | = \sqrt{3} \times | I_{bc} | = \sqrt{3} \times | I_{ca} |$$

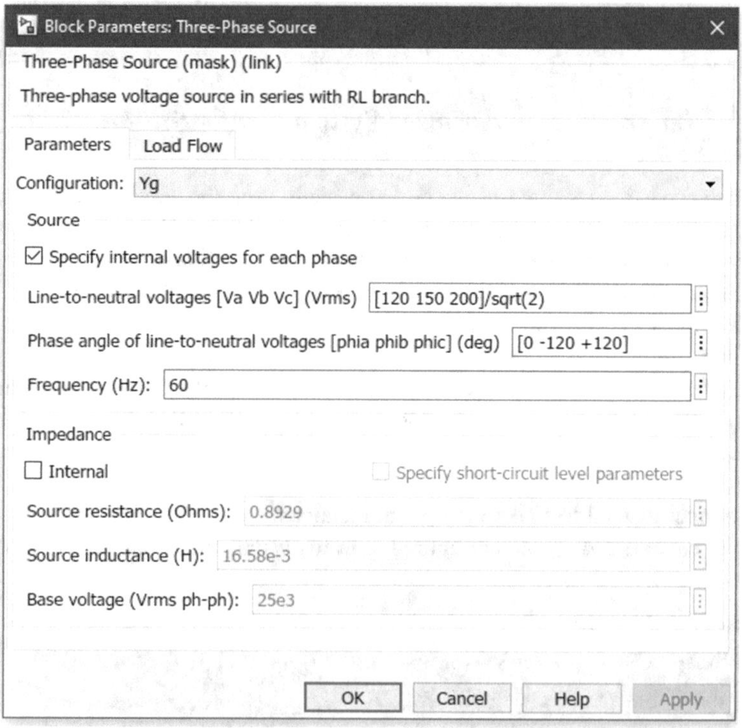

Fig. 16.28 Block parameters of the three-phase source

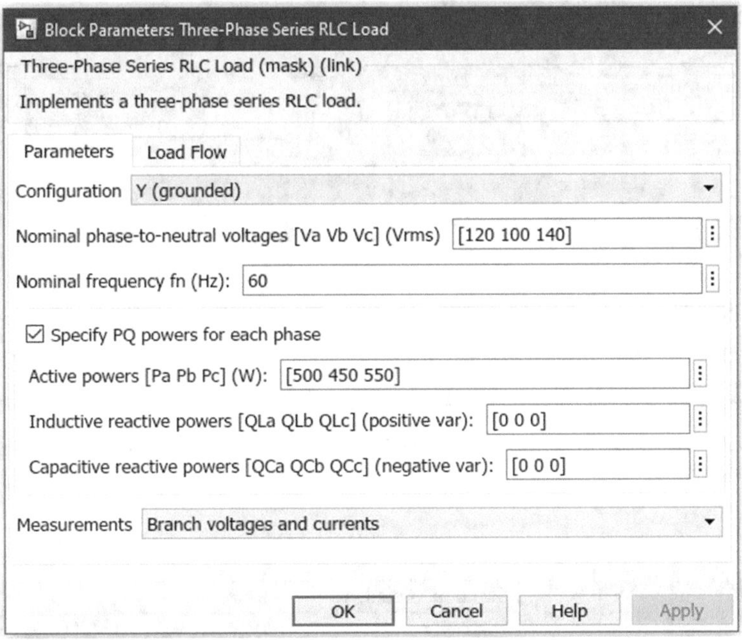

Fig. 16.29 Block parameters of the three-phase series RLC load

In Fig. 16.30, the same characteristics can be observable from the simulation values. To relate the above-mentioned characteristics to the obtained simulation results, Table 16.10 is created:

From the reference of the above table and the simulation results obtained from Fig. 16.31, it can be observed that the phase voltages and the line-to-line voltages are identical. However, the line current is $\sqrt{3}$ times to the phase-to-phase currents per phase. Thus, the characteristics of a balanced Δ-Δ power system configuration can be justified via the Simulink model.

The waveforms of phase-to-phase voltages and phase-to-phase currents of the three-phase load can be seen from the Scope windows as shown in Fig. 16.33.

16.6.4 Unbalanced Δ − Δ Power System Configuration

In an unbalanced Δ-Δ power system configuration, either the magnitudes of the three-phase AC source vary, or the impedances of the three-phase load differ from phase to phase. In Fig. 16.34, an unbalanced Δ-Δ power system is modeled. This model is the same as the previous model showed in Fig. 16.31. The only difference is the customization of the Three-Phase AC Source blocks and the Series RLC Load blocks to make the system unbalanced. The amplitudes of Three-Phase AC Source blocks are selected as 200 V, 150 V, and 100 V to make the system unbalanced (Fig. 16.35). In addition, the Series RLC Load blocks in the three phases are

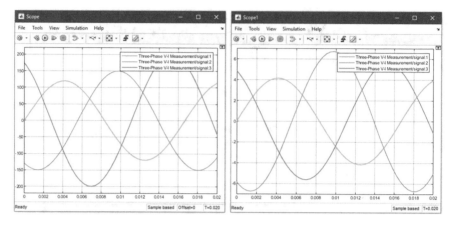

Fig. 16.30 The phase to ground voltages and the phase currents of the load observed from the scope window

Table 16.8 Blocks and navigation routes for modeling an unbalanced Y-Y power system

Simulink model	Reference
Display	$|V_{an}|$, $|V_{bn}|$, $|V_{cn}|$ (RMS values)
Display1	$|V_{AB}|, |V_{BC}|, |V_{CA}|$ (RMS values)
Display2	$|I_{an}|, |I_{bn}|, |I_{cn}|$ (RMS values)
Display3	$|I_{AB}|, |I_{BC}|, |I_{CA}|$ (RMS values)
Scope	V_{an}, V_{bn}, V_{cn}
Scope1	I_{an}, I_{bn}, I_{cn}

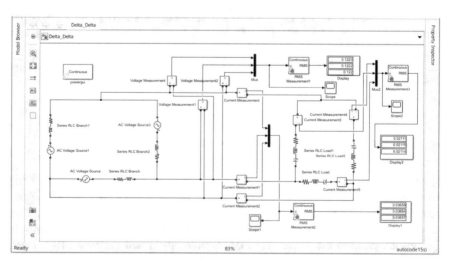

Fig. 16.31 Simulink diagram of a Delta-Delta configuration balanced system

Table 16.9 Blocks and navigation routes for modeling a balanced $\Delta - \Delta$ power system

Blocks	Navigation path on the Simulink Library Browser
AC Voltage Source, AC Voltage Source1, AC Voltage Source2	Simscape $\rightarrow$ Electrical $\rightarrow$ Specialized Power Systems $\rightarrow$ Fundamental Blocks $\rightarrow$ Electrical Sources $\rightarrow$ AC Voltage Source
Series RLC Branch Series RLC Branch1 Series RLC Branch2	Simscape $\rightarrow$ Electrical $\rightarrow$ Specialized Power Systems $\rightarrow$ Fundamental Blocks $\rightarrow$ Elements $\rightarrow$ Series RLC Branch
Voltage Measurement Voltage Measurement1 Voltage Measurement2	Simulink $\rightarrow$ Sinks $\rightarrow$ Voltage Measurement
Current Measurement Current Measurement1 Current Measurement2	Simulink $\rightarrow$ Sinks $\rightarrow$ Current Measurement
RMS Measurement1 RMS Measurement2 RMS Measurement3	Simscape $\rightarrow$ Electrical $\rightarrow$ Control $\rightarrow$ Measurements $\rightarrow$ RMS Measurement
Scope, Scope1	Simulink $\rightarrow$ Sinks $\rightarrow$ Scope
Mux, Mux1	Simulink $\rightarrow$ Signal Routing $\rightarrow$ Mux
Display, Display1, Display2	Simulink $\rightarrow$ Sinks $\rightarrow$ Display
powergui	Simscape $\rightarrow$ Electrical $\rightarrow$ Specialized Power Systems $\rightarrow$ Fundamental Blocks $\rightarrow$ powergui

customized as shown in Fig. 16.36. The nominal phase-to-neutral voltages of each Series RLC Load block are made unequal to ensure that the system is unbalanced.

For an unbalanced $\Delta - \Delta$ configured power system, the characteristics can be demonstrated as follows:

Unbalanced $\Delta - \Delta$ power system configuration
Consider the three phases as A, B, and C.
Line-to-line voltages: V_{AB}, V_{BC}, and V_{CA}
Phase-to-ground voltages: V_{an}, V_{bn}, and V_{cn}
For balanced $\Delta - \Delta$ configuration:

$$| V_{AB} | \neq | V_{BC} | \neq | V_{CA} |$$

$$| V_{ab} | \neq | V_{bc} | \neq | V_{ca} |$$

$$| V_{AB} | = | V_{ab} |; | V_{BC} | = | V_{bc} |; | V_{CA} | = | V_{ca} |$$

Line currents: I_A, I_B, and I_C
Phase-to-phase currents: I_{ab}, I_{bc}, and I_{ca}

(continued)

$$|I_A| \neq |I_B| \neq |I_C|$$
$$|I_{ab}| \neq |I_{bc}| \neq |I_{ca}|$$
$$|I_A| \neq \sqrt{3} \times |I_{ab}|; \ |I_B| \neq \sqrt{3} \times |I_{bc}|; \ |I_C| \neq \sqrt{3} \times |I_{ca}|$$
$$I_A = I_{ab} - I_{bc}; I_B = I_{bc} - I_{ca}; I_C = I_{ca} - I_{ab}$$

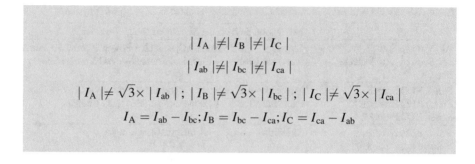

Fig. 16.32 Block parameters of the AC voltage source and the series RLC branch

Table 16.11 helps to relate the characteristics with the simulation results:

From the Scope windows in Fig. 16.37, the phase-to-phase voltages, lines currents, and phase-to-phase currents can be observed as shown in the following figure respectively:

16.7 Electrical Machine

In the power system, the concept of the electrical machine is very essential. An electrical machine can act either as a generator or a motor. In general, the electrical machine can be of two types based on its sources, such as DC Machine and AC Machine. The AC Machine can be further categorized into two types—Asynchronous Machine and Synchronous Machine. In this section, only DC Machine and Asynchronous Machine will be covered by using the Simulink platform to provide the readers a general idea of the applications of the power system in terms of electrical machines.

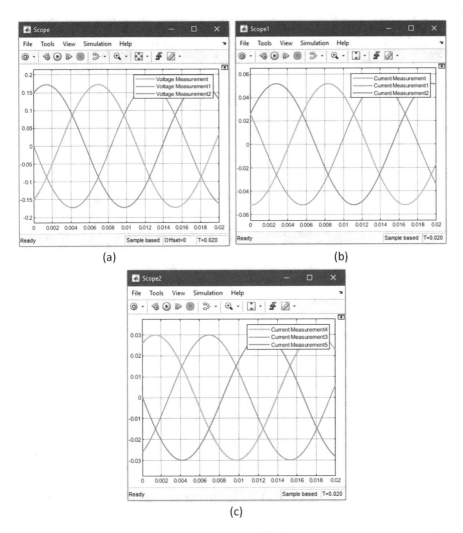

Fig. 16.33 The waveforms of phase-to-phase voltages and phase to phase currents of the three-phase load observed from the scope window

Table 16.10 Correlation between the Simulink model and the reference parameters of the balanced $\Delta - \Delta$ system

Simulink model	Reference
Display	$|V_{AB}|$, $|V_{BC}|$, $|V_{CA}|$ (RMS values)
Display1	$|I_A|, |I_B|$, $|I_C|$ (RMS values)
Display2	$|I_{ab}|, |I_{bc}|$, $|I_{ca}|$ (RMS values)
Scope	Phase-to-phase voltages or line-to-line voltages
Scope1	Line currents (I_A, I_B, I_C)
Scope2	Phase-to-phase currents (I_{ab}, I_{bc}, I_{ca})

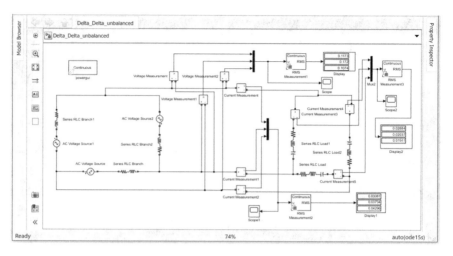

Fig. 16.34 Simulink diagram of a Delta-Delta configuration unbalanced system

16.7.1 DC Machine

In Simulink, a DC Machine block can be utilized both as a generator and a motor. In a DC Machine, two DC voltage sources are required. One DC source is used as a field voltage source and the other acts as an armature voltage source. A Simulink model where the DC Machine block is utilized to act as a motor is illustrated in Fig. 16.38. As a field voltage source, DC Voltage Source1 block is used, the amplitude of which is set as 96 V. For armature voltage source, DC Voltage Source block is utilized set as 480 V. In input port Torque Load (TL) of the DC Machine, a step response is provided via a Step block. The step value is set as +1, which demonstrates that the DC Machine will act as a motor. For a given step input of −1, the DC Machine can be used in the generator mode as well. The mechanical outputs of the motor are displayed both in the Display and the Scope blocks. The required blocks to design the Simulink model shown in Fig. 16.38 are summarized in Table 16.12 with their navigation routes:

The parameter window of the DC Machine is shown in Fig. 16.39. In the Parameter window, the user has the preference to select any Preset model from the dropdown option list. Simulink also provides options to the advanced user to create their own model by specifying the parameters which are shown in the second figure of Fig. 16.39. For getting access to the "Parameters" window, the Preset model should be set as "No." For this particular example, the first Preset model is chosen. All the other parameters are kept in their default mode for this simulation.

The outputs of the DC Machine are the armature current, field current, power, and electrical torque, which are displayed in the Display block. In the Scope window, the output signals can be observed graphically as shown in Fig. 16.40.

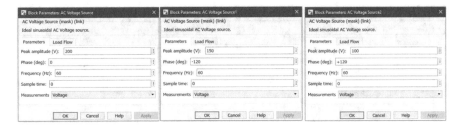

Fig. 16.35 Block parameters of the three AC voltage sources

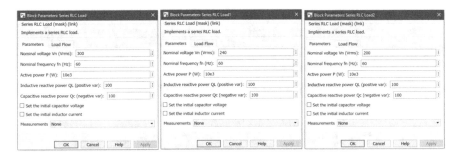

Fig. 16.36 Block parameters of the three series RLC loads

16.7.2 Asynchronous Machine

In an Asynchronous Machine, the AC voltage source is given as a supply. A Simulink model where an Asynchronous Machine is run with the usage of a three-phase wye-connected AC voltage source is shown in Fig. 16.41. For designing the model, the blocks that are used are listed in Table 16.13 with their navigation routes:

In Fig. 16.41, an Asynchronous Machine block is utilized, where the input source is provided by a three-phase wye-connected AC source. The block that is utilized to provide the AC input source is the Three-Phase Source block, which is customized as shown in Fig. 16.42. The RMS voltages for the three phases are set as 480 V.

The Asynchronous Machine block can be customized from its parameter window as shown in Fig. 16.43. The Rotor type of the machine can be selected from three options, such as Squirrel-cage rotor, Wound rotor, or Double squirrel cage rotor. For this example, the Squirrel-cage rotor type is selected. Like DC Machine, a preset model can be selected for the Asynchronous Machine as well. In this example, a preset model of 5 HP, 575 V, 60 Hz, 1750 RPM is selected. If the preset model is selected as "No," the model can be designed manually by the user from the Parameters option. The rest of the parameters are set in their default values.

The outputs of the Asynchronous Machine are shown in two Scope windows for better visibility. In the first Scope window, the rotor speed and the electromagnetic torque can be observed as shown in Fig. 16.44. The stator and rotor currents can be seen in the Scope1 window as demonstrated in Fig. 16.45.

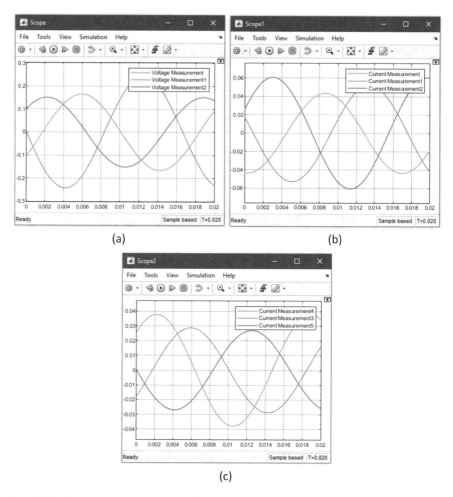

(a)   (b)

(c)

Fig. 16.37 The phase-to-phase voltages, lines currents, and phase-to-phase currents observed from the scope window

Table 16.11 Correlation between the Simulink model and the reference parameters of the unbalanced $\Delta - \Delta$ system

Simulink model	Reference
Display	$\mid V_{AB} \mid$, $\mid V_{BC} \mid$, $\mid V_{CA} \mid$ (RMS values)
Display1	$\mid I_A \mid$, $\mid I_B \mid$, $\mid I_C \mid$ (RMS values)
Display2	$\mid I_{ab} \mid$, $\mid I_{bc} \mid$, $\mid I_{ca} \mid$ (RMS values)
Scope	Phase-to-phase voltages or line-to-line voltages
Scope1	Line currents (I_A, I_B, I_C)
Scope2	Phase-to-phase currents (I_{ab}, I_{bc}, I_{ca})

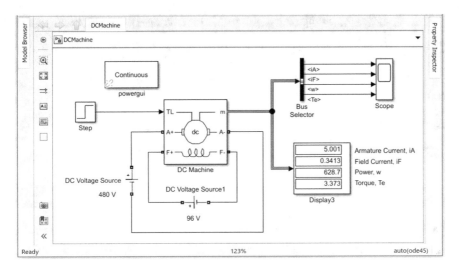

Fig. 16.38 Simulink diagram of a DC machine block

Table 16.12 Blocks and navigation routes for modeling a DC machine

Blocks	Navigation path on the Simulink Library Browser
DC Machine	Simscape → Electrical → Specialized Power Systems → Fundamental Blocks → Machines → DC Machine
DC Voltage Source DC Voltage Source1	Simscape → Electrical → Specialized Power Systems → Fundamental Blocks → Electrical Sources → DC Voltage Source
Step	Simulink → Sources → Step
Bus Selector	Simulink → Signal Routing → Bus Selector
Scope	Simulink → Sinks → Scope
Display	Simulink → Sinks → Display
powergui	Simscape → Electrical → Specialized Power Systems → Fundamental Blocks → powergui

16.8 Conclusion

In this chapter, different types of power sources are modeled using the Simulink platform considering both single-phase and three-phase power sources. Later the chapter focuses mainly on the three-phase AC source due to its wide applicability in the real world. The usage of such three-phase power sources with different series or parallel RLC loads is demonstrated using the Simulink model to understand their characteristics. The power factor calculation of a three-phase system is covered in

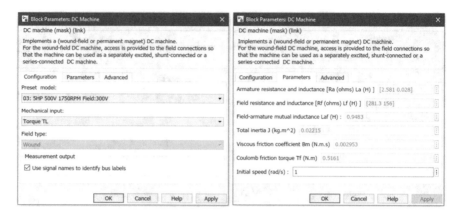

Fig. 16.39 Block parameters of the DC machine

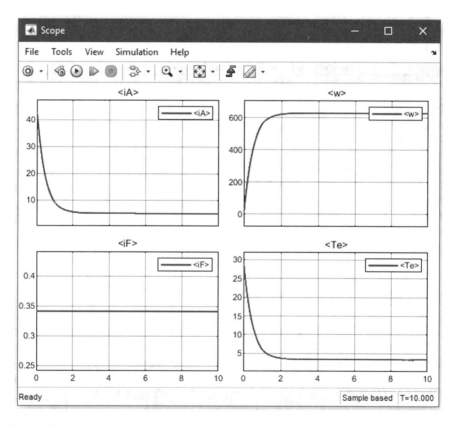

Fig. 16.40 The armature current, field current, power, and electrical torque observed from the scope window

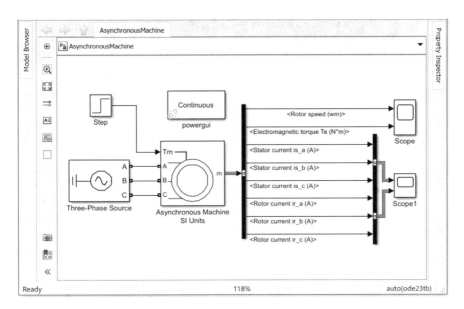

Fig. 16.41 Simulink diagram of an asynchronous machine block

Table 16.13 Blocks and navigation routes for modeling an asynchronous machine

Blocks	Navigation path on the Simulink Library Browser
Asynchronous Machine SI Units	Simscape → Electrical → Specialized Power Systems → Fundamental Blocks → Machines → Asynchronous Machine SI Units
Three-Phase Source	Simscape → Electrical → Specialized Power Systems → Fundamental Blocks → Electrical Sources → DC Voltage Source
Step	Simulink → Sources → Step
Bus Selector	Simulink → Signal Routing → Bus Selector
Scope, Scope1	Simulink → Sinks → Scope
Display	Simulink → Sinks → Display
powergui	Simscape → Electrical → Specialized Power Systems → Fundamental Blocks → powergui

this chapter due to its importance in the power engineering domain. Afterward, different power system configurations, such as Y-Y, $\Delta - \Delta$ configurations, are explained in terms of simulation results obtained via Simulink modeling. To understand the applications of Simulink in the power system domain, the electrical machine section is included in the last part of this chapter. In this section, two different types of machines—DC Machine and Asynchronous Machine—are incorporated with adequate graphical demonstration to grasp the usage of the Simulink in the power system domain. To understand and learn the modeling of different power systems via simulation, this book can be used as a tutorial that incorporates several examples for better comprehensibility.

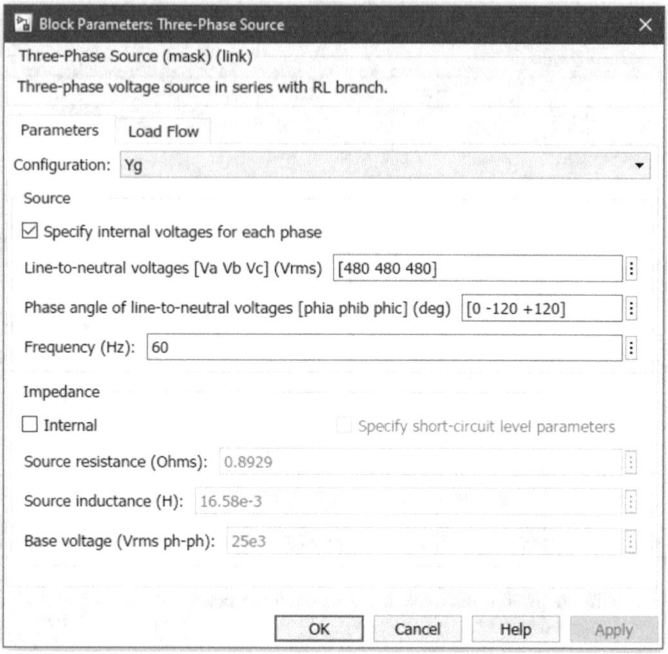

Fig. 16.42 Block parameters of the three-phase source

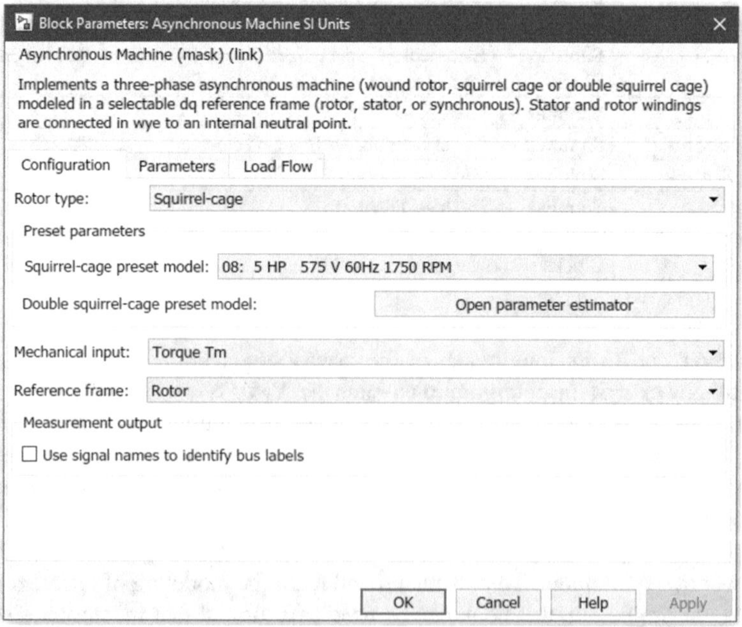

Fig. 16.43 Block parameters of the asynchronous machine SI units

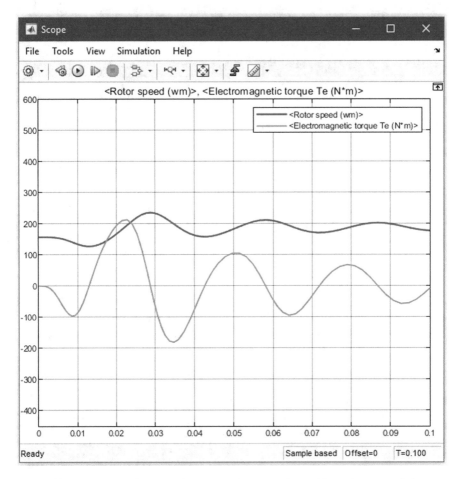

Fig. 16.44 The rotor speed and the electromagnetic torque observed from the scope window

Exercise 16

1. (a) Write down the differences between single-phase and three-phase AC power sources?

(b) Design a Simulink model for a three-phase wye-connected AC power source by utilizing a Three-Phase Source block for the following parameters:

(i) Peak amplitude $= 480$ V per phase; internal impedance, $R = 0.5$ ohms, $L = 0.015$ H.

(ii) RMS voltage $= 480$ V per phase; internal impedance $= 0$.

(c) Show the voltages of the source using a Scope block.

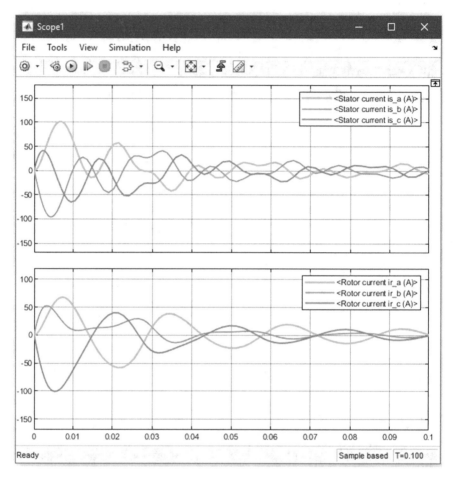

Fig. 16.45 The stator and rotor currents observed from the scope window

(d) With a three-phase AC power source, connect a Three-Phase Series RLC Load. Show the voltages and the currents of the system using Scope.

2. (a) Represent power factor in terms of real and reactive power.

(b) For a delta-connected three-phase AC source with a Three-Phase Parallel RLC Load, simulate a model to calculate the power factor of the system.

(c) Show the real, reactive, and apparent power of the system using a Scope block.

3. (a) What are the characteristics of a balanced Y-Y power system.

(b) Design a balanced Y-Δ power system configuration using Simulink.

(c) Show the different voltages and currents (line and phase) of the entire system.

(d) Make comments on the obtained simulation results to justify the characteristics of the model.

4. (a) What are the differences between a balanced and an unbalanced system?

(b) Design an unbalanced Δ-Y power system using Simulink.

(c) Show the line voltages and currents of the system along with the phase values as well.

(d) Comment on the characteristics of such a model based on the simulation result.

5. (a) What is the categorization of electrical machines?

   (b) Design a Simulink model to run a DC Machine with a preset model of 20HP, 500 V, 1750 RPM, field: 300 V.

   (c) Show the output Power and Torque of the DC Machine using a Scope block.

6. (a) Write down some of the rotor types of Asynchronous Machine.

   (b) Design a Simulink model to run an Asynchronous Machine by using a delta-connected three-phase AC source as the input.

   (c) Show the output parameters of the machine using Scope blocks.

Chapter 17
Application of Simulink in Power Electronics

17.1 Diode

Power electronics, as the name suggests, is the study of the electronics devices that deals with high power, integrating control, energy, and electronics altogether. Electronics devices usually have small voltage and currents (typically millivolt and milliampere), whereas power electronics devices handle high voltage and current of kilovolt and ampere or above. Such devices are at the core of numerous applications, since AC and DC conversions, machine controls, and switching regulations are essential parts of daily electrical activities. A diode is a semiconductor device with two terminals, which can operate as a switch with a specified voltage, called forward voltage. For example, if the forward voltage of a diode is 0.7 V, and if the diode is placed in a series circuit between the switch and load, the circuit will conduct only when the source voltage is above 0.7 V. It is a unidirectional device, which allows current to flow only in a single direction. The major applications of diode include switching, isolation of a system from outside signals, and converting AC parameters into DC parameters.

17.1.1 Diode Characteristics

The diode works in two modes—in forward biased mode when the anode is positive compared to the cathode and in reverse-biased mode when the cathode is positive compared to the anode. The diode conducts when the voltage across the diode is higher than the turn-on voltage of the diode. Typically, the turn-on voltage for the diode is considered to be 0.7 V, which varies depending on the material. In Simulink, diode parameters can be changed to verify the characteristics and make the diode suitable for different power electronics applications. In the Simulink library, diodes for both physical system modeling and Simulink modeling are present. However, the working principle of both devices is the same, and the component for Simulink is used in this chapter.

In order to demonstrate the diode characteristics, a setup has been built using the following components (Table 17.1).

A 5 V AC source with 60 Hz frequency is used in the system with a resistance of 5 Ω. A diode with 0.7 V forward voltage (which can be selected by double-clicking on the diode and changing the forward voltage into 0.7) is chosen for the operation. In this example, the voltage across the source, diode, and resistive load will be demonstrated and compared in understanding the diode behavior.

When the diode is initially placed into the window, it consists of three ports, including an extra measurement port for measuring diode voltages and currents. In this chapter, the measurement ports for the devices are not shown and are disabled by double-clicking on the diode and unchecking the "Show measurement port" checkbox. It is to be noted that the measurement port has been disabled to highlight the actual measurement of the voltage and current (Fig. 17.1).

The discrete powergui block is used and the configuration parameters have been changed to Modeling → Model Settings → Solver → Solver Selection → discrete (no continuous states) (Fig. 17.2).

Connect the components as shown in Fig. 17.3. The voltage measurement devices have been renamed for the ease of demonstration in the scope. Some of the components in this chapter have also been rotated or flipped for better visualization. The components can be rotated or flipped by right-clicking on the block, selecting "Rotate & Flip," and picking the appropriate orientation for the demonstration.

It is seen from the graph in Fig. 17.4 that although the voltage varies sinusoidally from 0 V to 5 V, the output is not the same for the load due to the presence of the diode. As the turn-on voltage for the diode is 0.7 V, the voltage below 0.7 V appears on the second subplot. However, the diode allows the current to flow only when the voltages are above 0.7 V, creating a voltage difference of 4.63 V across the resistive load. A similar characteristic for reverse-biased operation can be obtained by changing the polarity of the diode in the same setup.

Table 17.1 Blocks and navigation routes for demonstrating diode characteristics

Name of the block	Navigation route
powergui	Simscape → Electrical → Specialized Power Systems → Fundamental Blocks
AC Voltage Source	Simscape → Electrical → Specialized Power Systems → Fundamental Blocks → Electrical Sources
Diode	Simscape → Electrical → Specialized Power Systems → Fundamental Blocks → Power Electronics
Voltage Measurement	Simscape → Electrical → Specialized Power Systems → Fundamental Blocks → Measurements
Series RLC Branch	Simscape → Electrical → Specialized Power Systems → Fundamental Blocks → Elements
Scope	Simulink → Sinks

17.1.2   Single-Phase Half-Wave Rectifier

Due to its unidirectional capability, diodes are used for current and voltage rectification—an operation to convert AC into DC. An initial approach for the conversion is to cut the negative half of the voltage through unidirectional switching. The circuit through which this is done for a single phase is called a single-phase half-wave rectifier.

17.1.2.1   Single-Phase Half-Wave Rectifier with R Load

A similar setup used in the previous section is used with a resistive load (R load) for this example, with voltage and current measurement blocks placed as per Fig. 17.5.

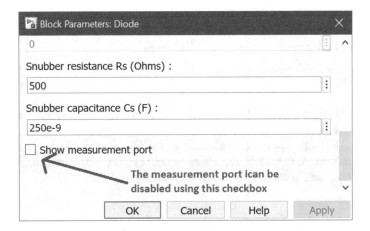

Fig. 17.1 Block parameters of the diode block

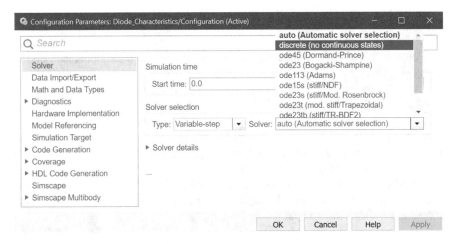

Fig. 17.2 Configuration parameters of the diode

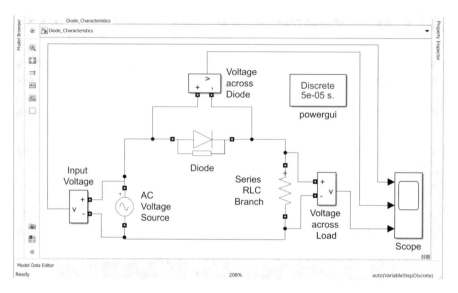

Fig. 17.3 Simulink diagram of the diode in an AC circuit

An AC voltage source of 12 V at 60 Hz, with a resistance of 10 Ω, is used in this setup.

In the circuit, whenever the AC current attempts to flow toward the negative direction in the resistive load, the diode acts as an off-switch and stops the flow. For this reason, the voltage or current value during the negative source voltage appears to be zero, as shown in Fig. 17.6. The circuit is able to work only for the positive half-cycles; hence, the circuit is called a half-wave rectifier.

17.1.2.2 Single-Phase Half-Wave Rectifier with RL Load

The value of the inductance as a load changes the output waveform. To demonstrate the output with inductance load, the "Series RLC Branch" is modified to be an "RL" branch by double-clicking on it. Two values of inductances, i.e., 1 mH and 500 mH, have been considered for comparing the impact of low and high inductance on the setup (Fig. 17.7).

It is seen that at a low inductance, the branch behaves like the R branch. As inductance works as an energy storage device, at a low value, the inductor does not store much energy, and thus, the whole branch behaves somewhat similar to the R branch (Fig. 17.8). But at a high inductance value, the inductor stores more energy to allow a constant flow of current. This extra energy is dissipated through high voltage across the load. Hence, the output waveform deviates from that of the R load (Fig. 17.9). Sometimes, the excess energy trapped in the load may damage the diode, for which a bypass path through another diode called a freewheeling diode is used in parallel with the load.

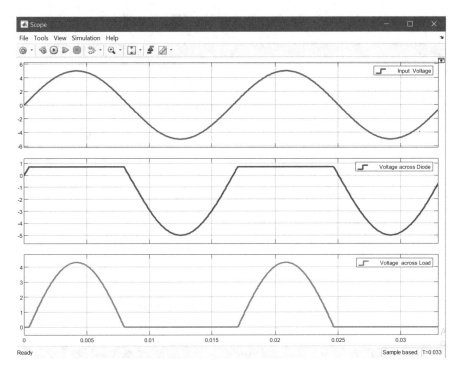

Fig. 17.4 The input voltage and the voltages across the diode and the load observed from the scope window

17.1.2.3 Single-Phase Half-Wave Rectifier with RC Load

Like the inductor, the presence of a capacitor in the diode circuit also impacts the output waveforms. To demonstrate the output with capacitance load, the "Series RLC Branch" is modified to be an "RC" branch by double-clicking on it. Two values of capacitors, i.e., 1 μF and 100 μF, have been considered for comparing the impact of low and high capacitance on the setup (Fig. 17.10).

Unlike inductors, capacitors attempt to maintain a constant voltage. For a lower capacitance value, the output voltage still fluctuates, as shown in Fig. 17.11, but for a higher value, the output voltage almost smoothens, as visible from the waveform of the output voltage in Fig. 17.12.

17.1.3 Single-Phase Full-Wave Rectifier

Unlike half-wave circuits, this rectifier can convert both the cycles of the AC sine wave into a unidirectional signal (pulsating DC). A two-diode with center-tap transformer or a four-diode with center-tap or linear transformer is used for this circuit. In this system, one set of the diodes are forward biased and two of the other

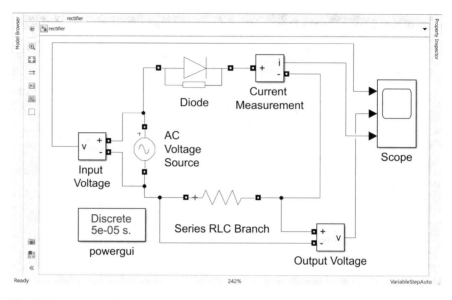

Fig. 17.5 Simulink diagram of a Single-Phase Half-wave Rectifier with resistive load

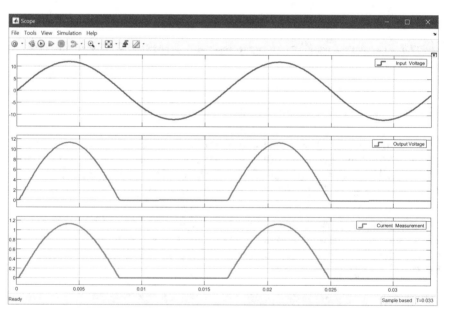

Fig. 17.6 The input and output voltages and the current measurement observed from the scope window

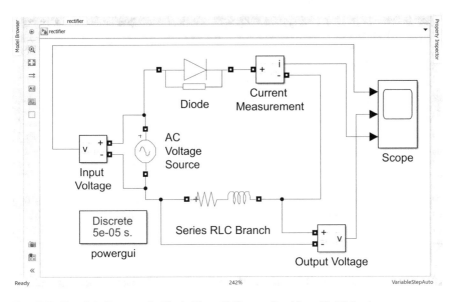

Fig. 17.7 Simulink diagram of a Single-Phase Half-wave Rectifier with RL load

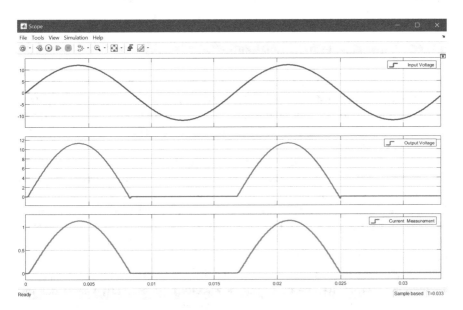

Fig. 17.8 The input and output voltages and the current measurement observed from the scope window

diodes remain in the reverse-biased mode to let the first half-cycle flow in the positive direction. During the flow of the negative half-cycle, the diodes alter their modes, providing the circuital path for another flow in the positive direction—converting the complete input sinusoidal into a full-wave rectified output.

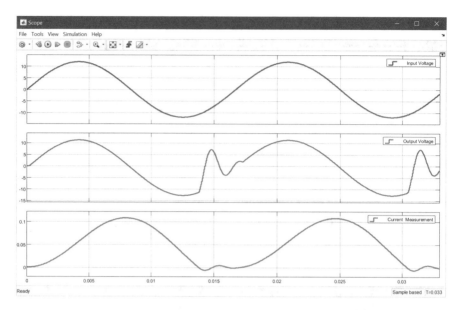

Fig. 17.9 The input and output voltages and the current measurement for high inductance value observed from the scope window

17.1.3.1   Two-Diode Full-Wave Rectifier

In a center-tap transformer, one of the coils in a winding is used to take the output for voltage or current rectification or transformation. Such transformers are widely used as dual or split supplies, or for rectification. This transformer also helps to perform full-wave rectification using only two diodes; hence, this setup is utilized in laboratories for experimentations (Fig. 17.13).

The setup is simulated in Fig. 17.14 with an 110 V 60 Hz AC voltage source and with a resistance of 10 Ω. The linear transformer is taken from the path Simscape → Electrical → Specialized Power Systems → Fundamental Blocks → Elements → Linear Transformer. The transformer parameters are changed as follows for a 40 VA 60 Hz 110/12 V transformer. In order to make it center-tapped, winding two and three are shortened with a ground (from the same library path as the linear transformer).

Notice that with the R load, another ground has also been connected to make sure the full-wave rectification. From the output waveforms, it is clearly visible that the input voltage of 110 V is rectified into a full-wave pulsating DC of 12 V (Fig. 17.15).

17.1.3.2   Four-Diode Full-Wave Rectifier

Four-diode full-wave rectifiers are used in relatively high-power applications. In this configuration, the loads are not needed to be connected to the ground (Fig. 17.16).

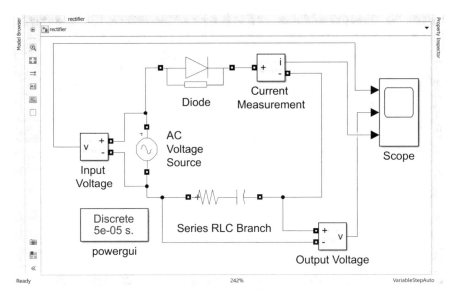

Fig. 17.10 Simulink diagram of Single-Phase Half-wave Rectifier with RC load

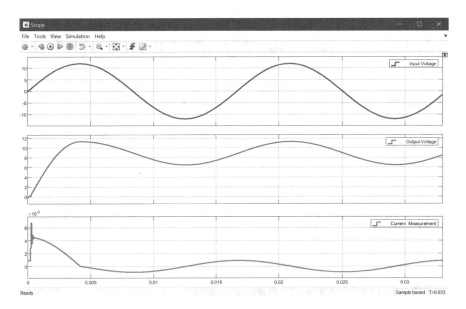

Fig. 17.11 The input and output voltages and the current measurement observed from the scope window

Notice the connection of the AC voltage source, the first terminal of which is connected with the first set of diodes, while the other terminal is connected with the

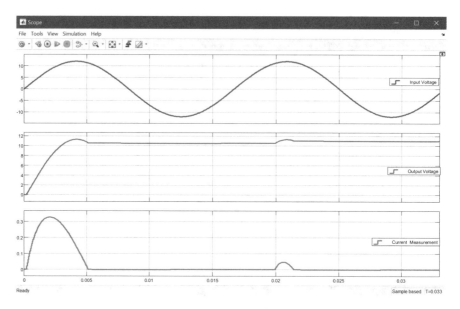

Fig. 17.12 The input and output voltages and the current measurement for high capacitance value observed from the scope window

second set of diodes as shown in the diagram. The same AC voltage source of 110 V 60 Hz is used with 10 Ω resistive load. The transformer with the previously mentioned ratings is used, but the third winding has been disabled by double-clicking on the "Linear Transformer" and unchecking the box for the "Three windings transformer." Note that in order to work with the two-diode full-wave rectifier, a center-tapped transformer must be used, but in the case of the four-diode counterpart, either a center-tapped or double winding transformer can be utilized. The output waveforms are similar to the two-diode configuration, as shown in Fig. 17.17.

In both these examples, it is seen that the output voltage does not fully achieve 12 V. It is because of some voltage drop across other electrical elements in the circuit, especially across the diode, which has a forward voltage of 0.7 V for this example. By decreasing the value, a more accurate output voltage can be acquired.

17.1.4   Three-Phase Full-Wave Rectifier

A three-phase full-wave rectifier performs full-wave rectification but simultaneously for all three phases. A six-diode configuration with a Three-Phase Source is used for this purpose. As shown in Fig. 17.18, the phase angles for each of the AC voltage source are chosen at a phase difference of 120°. A resistive load of 100 Ω has been considered for this setup. The voltage and frequency of the AC source are chosen to

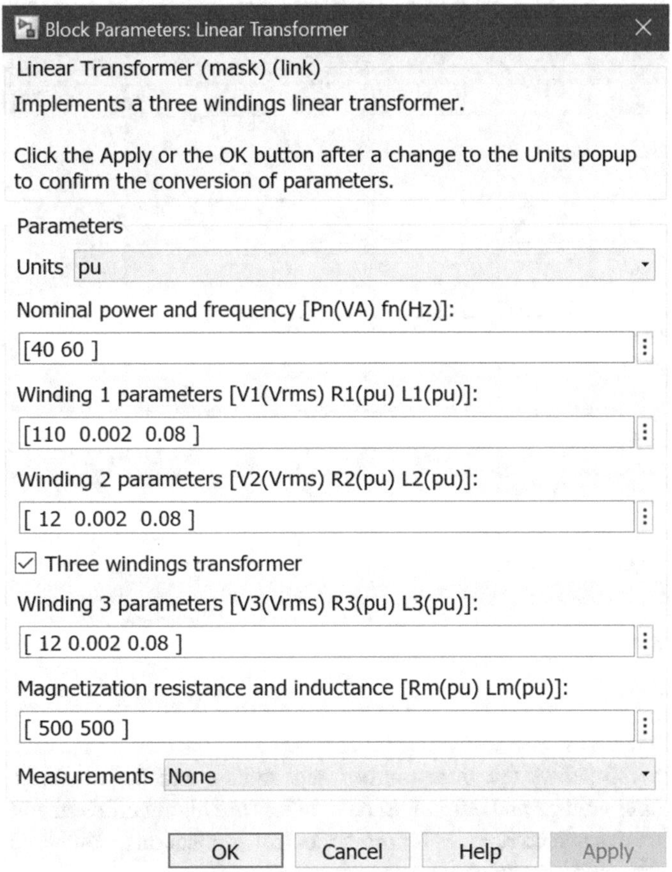

Block Parameters: Linear Transformer ✕

Linear Transformer (mask) (link)

Implements a three windings linear transformer.

Click the Apply or the OK button after a change to the Units popup to confirm the conversion of parameters.

Parameters

Units pu ▾

Nominal power and frequency [Pn(VA) fn(Hz)]:

[40 60] ⋮

Winding 1 parameters [V1(Vrms) R1(pu) L1(pu)]:

[110 0.002 0.08] ⋮

Winding 2 parameters [V2(Vrms) R2(pu) L2(pu)]:

[12 0.002 0.08] ⋮

☑ Three windings transformer

Winding 3 parameters [V3(Vrms) R3(pu) L3(pu)]:

[12 0.002 0.08] ⋮

Magnetization resistance and inductance [Rm(pu) Lm(pu)]:

[500 500] ⋮

Measurements None ▾

OK   Cancel   Help   Apply

Fig. 17.13 Block parameters of the linear transformer

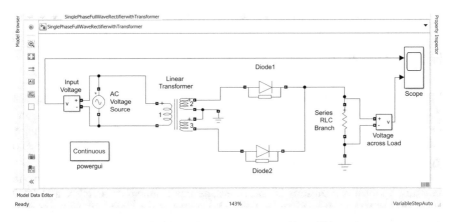

Fig. 17.14 Simulink diagram of Single-Phase Full-Wave Rectifier with transformer

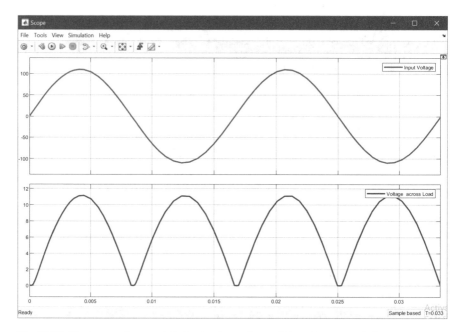

Fig. 17.15 The input voltage and the load voltage observed from the scope window

be 110 V and 60 Hz, respectively. The complete setup in Simulink for the rectifier is shown in Fig. 17.19.

Figure 17.20 shows the three-phase input voltage and the resulting full-wave rectified output voltage and current. Such rectifiers are highly efficient and are vastly used in industries and companies for motor control applications, voltage stabilization and protection, and energy storage charging.

From the graph, it is visible that the voltage across the load is rectified and the current waveform shares its resemblance with the voltage waveform.

17.2   Transistor

Transistors are three-terminal devices that can transfer signals from lower resistance to higher resistance. The device is used for regulation, amplification, generating, and controlling electrical signals. Such a wide range of applications has resulted in many variations of transistors. Based on the constructions, transistors are mainly of two types, called bipolar junction transistors (BJTs) and field-effect transistors (FETs). Both of these transistors can be designed in Simulink, and some of the built-in transistor models are present in the Simscape environment. The NPN and PNP variations of BJTs, along with metal-oxide-semiconductor field-effect transistors (MOSFETs) and insulated gate bipolar transistors (IGBTs), are discussed in detail in the following subsections.

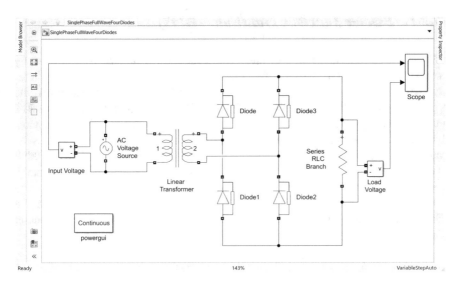

Fig. 17.16 Simulink diagram of Single-Phase Full-Wave rectifier with Four Diodes

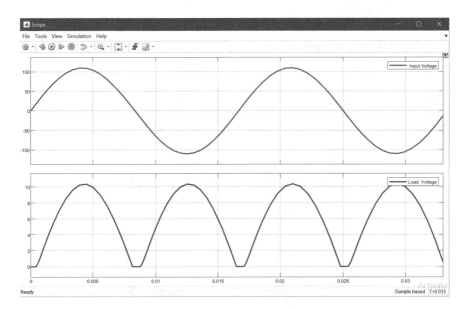

Fig. 17.17 The input voltage and the load voltage observed from the scope window

17.2.1 Bipolar Junction Transistors (BJTs)

BJTs are transistors consisting of three terminals and two junctions controlled by both electron and hole carriers. The transistor goes through two semiconductor layers, called the p-layer and the n-layer. Depending on the constructions, the transistor is called either an NPN or a PNP transistor. The three terminals of a BJT are named emitter, base, and collector. BJT is known as a current-controlled device; the small amount of current provided from the base to the emitter (in NPN) paves the path for a higher current to flow from collector to emitter (in NPN). In PNP transistors, the current flows are in the opposite direction than NPN transistors. The I-V characteristics for NPN and PNP transistors are shown using Simulink to understand their behaviors better. The following elements are taken from the "Simulink Library Browser" for constructing the circuits in this subsection (Table 17.2).

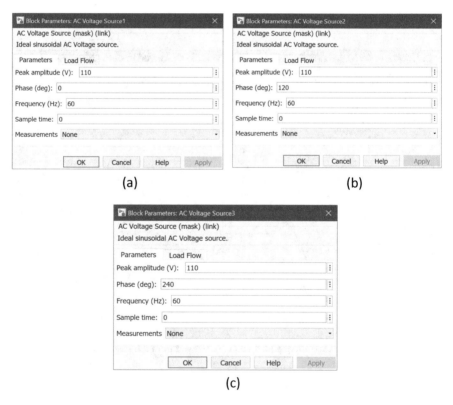

Fig. 17.18 Block parameters of the three AC voltage sources

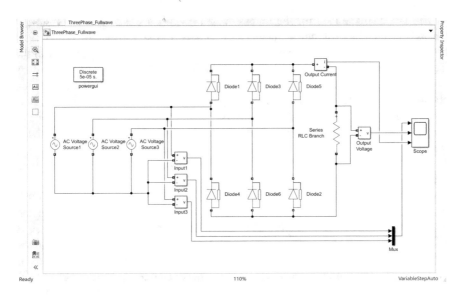

Fig. 17.19 Simulink diagram of a Three-Phase Full-Wave Rectifier

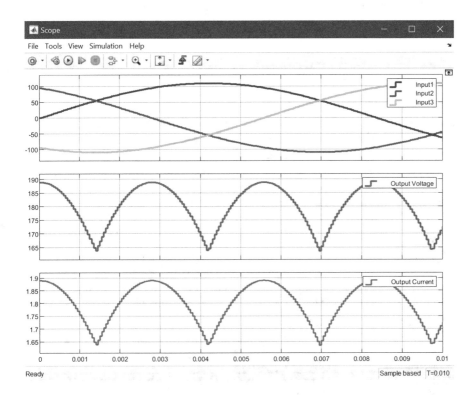

Fig. 17.20 The three input voltages, the output voltage, and the output current observed from the scope window

The I-V characteristic for the transistor is a graph for the collector current (I_c) versus the collector-emitter voltage (V_{ce}). A small base current of 0.003 A is provided in the "DC Current Source" block (named as Ib), and a slope of 4 in the ramp (named as V_{ce}) at the input of the "Controlled Voltage Source" block is given, to observe the collector current (I_c) for a period of 5 s. As the bipolar transistors are physical components (i.e., from the Simscape environment), conversion blocks such as PS-Simulink and Simulink-PS converters are used for signal conversion. Since an I-V characteristic is to be demonstrated, an "XY Graph" with a limit of 0–5 in the x-axis and 0–0.3 in the y-axis is used. The circuit is connected as per Fig. 17.21.

The I-V characteristics for the NPN transistor for 0.003 A base current, therefore, appear to be as follows (Fig. 17.22):

The PNP transistor demonstrates a similar characteristic to that of the NPN transistor, but since the current flow is the opposite, the values also appear negative to that of the NPN transistor. A small base current of −0.003 A is provided in the "DC Current Source" block (named as Ib), and a slope of 5 in the ramp (named as V_{ce}) at the input of the "Controlled Voltage Source" block is given, to see the collector current (I_c) for a period of 5 s. Additionally, in the Ramp block, the start time is 0, and the initial output is selected to be −5. The "XY Graph" with a limit of −4 to 0 on the x-axis and −0.3 to 0.05 in the y-axis are used. The circuit is connected as per Fig. 17.23 with the components as follows (Fig. 17.24 and Table 17.3):

17.2.2 MOSFET

The metal-oxide-semiconductor field-effect transistor (MOSFET) is a FET that has three terminals, which are named drain, source, and gate. Unlike typical BJT, FET is controlled by a gate signal. Whenever the gate signal is greater than zero, the MOSFET turns on irrespective of the drain-source voltage. However, if the current

Table 17.2 Blocks and navigation routes for modeling NPN bipolar transistor

Name of the block	Navigation route
Solver Configuration	Simscape → Utilities
NPN Bipolar Transistor	Simscape → Electrical → Semiconductors & Converters
DC Current Source	Simscape → Foundation Library → Electrical → Electrical Sources
Current Sensor	Simscape → Foundation Library → Electrical → Electrical Sensors
Controlled Voltage Source	Simscape → Foundation Library → Electrical → Electrical Sources
PS-Simulink Converter	Simscape → Utilities
Simulink-PS Converter	Simscape → Utilities
Ramp	Simulink → Sources
XY Graph	Simulink → Sinks
Electrical Reference (ERef)	Simscape → Foundation Library → Electrical → Electrical Elements

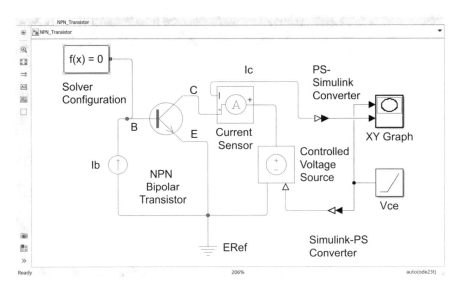

Fig. 17.21 Simulink diagram of a circuit containing an NPN transistor

flowing through it is negative with no gate signal, MOSFET turns off. MOSFETs are also divided into two types, N-Channel MOSFET and P-Channel MOSFET. There is more than one instance of MOSFET block in the Simscape library. The N-Channel MOSFET is utilized in this example.

In order to build a MOSFET circuit for characterization, the following components were used (Table 17.4).

A gate voltage of 4.5 V has been set in the DC Voltage Source block named V_g. A Slope of 1 for the Ramp block (named as V_{ds}), with other parameters ("start time" and "initial output") as 0, has been set. For showing the V-I graph clearly, the parameters in the XY Graph block have been changed as X-min 0, X-max 4, Y-min 0, and Y-max 20, with the default sample time. In the setup outlined in Fig. 17.25 and the graph in Fig. 17.26, the V-I characteristics for a MOSFET are shown for a stop time of 5 s. It is seen that the current across the MOSFET increases as the voltage increases.

17.2.3 IGBT

Similar to MOSFET, Insulated Gate Bipolar Transistor (IGBT) is a transistor that has three terminals. These are made with four layers of alternating P- and N-type semiconductors. Their terminals are termed as collector, emitter, and gate. An IGBT merges the features of both BJT and MOSFET. An IGBT turns on when a positive gate signal is applied, and its collector-emitter voltage is positive and greater than its forward voltage. The device turns off during a zero gate signal and stays in

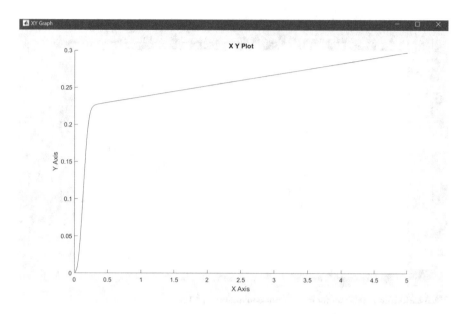

Fig. 17.22 The I-V characteristic curve of the NPN transistor

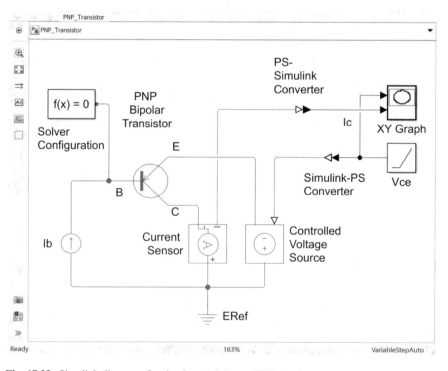

Fig. 17.23 Simulink diagram of a circuit containing an PNP transistor

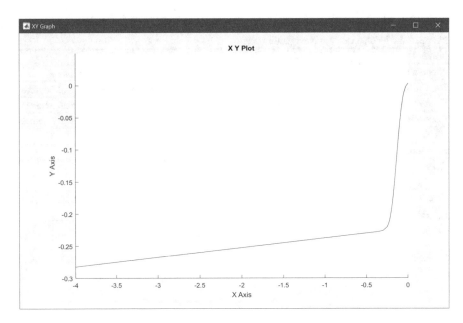

Fig. 17.24 The I-V characteristic curve of the PNP transistor

Table 17.3 Blocks and navigation routes for modeling PNP bipolar transistor

Name of the block	Navigation route
Solver Configuration	Simscape → Utilities
PNP Bipolar Transistor	Simscape → Electrical → Semiconductors & Converters
DC Current Source	Simscape → Foundation Library → Electrical → Electrical Sources
Current Sensor	Simscape → Foundation Library → Electrical → Electrical Sensors
Controlled Voltage Source	Simscape → Foundation Library → Electrical → Electrical Sources
PS-Simulink Converter	Simscape → Utilities
Simulink-PS Converter	Simscape → Utilities
Ramp	Simulink → Sources
XY Graph	Simulink → Sinks
Electrical Reference (ERef)	Simscape → Foundation Library → Electrical → Electrical Elements

off-state during a negative collector-emitter voltage. In Simulink, there are two IGBTs as physical components (named as "IGBT (Ideal, Switching)" and "N-Channel IGBT"), and as Simulink components (named as "IGBT" and "IGBT/Diode"). The IGBT/Diode block consists of a diode in parallel with the IGBT called antiparallel diode to block current flow in the reverse direction. Like MOSFET blocks, the IGBT blocks contain snubber circuits for enhanced protection and performance.

In order to build an IGBT circuit for characterization, the following components were used (Table 17.5).

Table 17.4 Blocks and navigation routes for demonstrating MOSFET characteristics

Name of the block	Navigation route
Solver Configuration	Simscape → Utilities
N-channel MOSFET	Simscape → Electrical → Semiconductors & Converters
DC Current Source	Simscape → Foundation Library → Electrical → Electrical Sources
Current Sensor	Simscape → Foundation Library → Electrical → Electrical Sensors
Controlled Voltage Source	Simscape → Foundation Library → Electrical → Electrical Sources
PS-Simulink Converter	Simscape → Utilities
Simulink-PS Converter	Simscape → Utilities
Ramp	Simulink → Sources
XY Graph	Simulink → Sinks
Electrical Reference (ERef)	Simscape → Foundation Library → Electrical → Electrical Elements

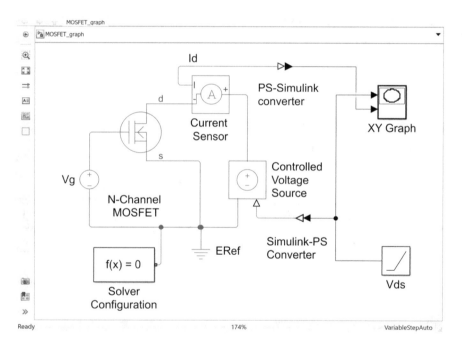

Fig. 17.25 Simulink diagram of a circuit containing an N-channel MOSFET

In the setup outlined in Fig. 17.27 and the graph in Fig. 17.28, the ideal V-I characteristics for an IGBT are shown. The DC Voltage Source block named as "V_{ge}" is selected to be 4.5 V. The Ramp block annotated as "V_{ce} ramp" is used to update the Controlled Voltage Source block with a "Slope" of 1, "Start time" of 0, and "Initial output" of 0. For showing the V-I graph clearly, the parameters in the XY Graph block have been changed as X-min 0, X-max 5, Y-min 0, and Y-max

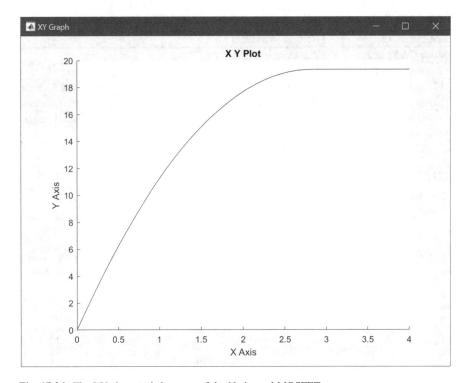

Fig. 17.26 The I-V characteristic curve of the N-channel MOSFET

400, with the default sample time. The simulation has been run for 5 s for a better representation.

17.3   Operational Amplifier

As mentioned in Chap. 8, an operational amplifier or Op-amp is an active device that can amplify any input signals, perform mathematical operations, and filter signals. A standard Op-amp has five important ports, which are used to invert input signal as necessary, connect with power signals, and deliver the output signal. The calculations for the inverting amplifier, non-amplifier, differentiator, and integrator were performed using MATLAB in Chap. 8. In this section, Simulink models for those Op-amp circuits will be prepared, and the formulas will be verified to understand Op-amp applications better. Op-amp is a physical device in Simulink, which is available from the following library:

Simscape → Foundation Library → Electrical → Electrical Elements

Table 17.5 Blocks and navigation routes for demonstrating IGBT characteristics

Name of the block	Navigation route
Solver Configuration	Simscape → Utilities
N-channel IGBT	Simscape → Electrical → Semiconductors & Converters
DC Current Source	Simscape → Foundation Library → Electrical → Electrical Sources
Current Sensor	Simscape → Foundation Library → Electrical → Electrical Sensors
Controlled Voltage Source	Simscape → Foundation Library → Electrical → Electrical Sources
PS-Simulink Converter	Simscape → Utilities
Simulink-PS Converter	Simscape → Utilities
Ramp	Simulink → Sources
XY Graph	Simulink → Sinks
Electrical Reference (ERef)	Simscape → Foundation Library → Electrical → Electrical Elements

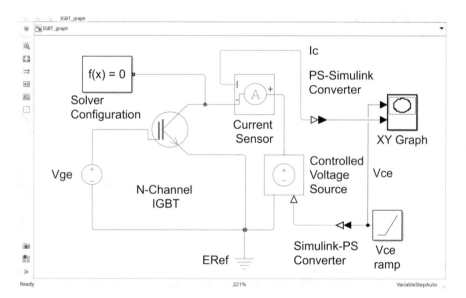

Fig. 17.27 Simulink diagram of a circuit containing an N-channel IGBT

17.3.1 *Inverting Amplifier*

As the name suggests, the circuit amplifies the input signal through the Op-amp but inverts the gain at the same time. The formula for determining the output voltage for the inverting amplifier is:

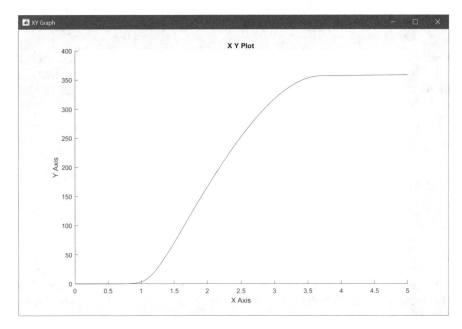

Fig. 17.28   The I-V characteristic curve of the N-channel IGBT

$$V_{\text{out}} = -\frac{R_2}{R_1} V_{\text{in}}$$

For an AC source with 12 V 60 Hz; two resistors of values 10 Ω and 50 Ω (R_1 and R_2), respectively; and a load resistance of 1 Ω (named as Resistor3 in the circuit), the output voltage for the inverter amplifier can be calculated as:

$$V_{\text{out}} = -\frac{50}{10} * 12 \text{ V} = -60 \text{ V}$$

The minus sign indicates the negative polarity of the waveform. The circuit is constructed with the mentioned specifications according to Fig. 17.29. The Op-amp is flipped for the convenience of connection. The Op-amp can be flipped by right-clicking on it, selecting "Rotate & Flip," selecting "Flip Block," and then "Up-Down." Upon running the simulation and observing the graph from the Scope, it will be seen that the output waveform resembles the mathematical calculation performed above, with a peak value of 60 V and in the opposite direction with respect to the input signal (Fig. 17.30).

17.3.2 Non-inverting Amplifier

Unlike an inverter amplifier, a non-inverting amplifier amplifies the input signal without inverting the signal. The formula for the output voltage can be written as follows:

$$V_{out} = \left(1 + \frac{R_2}{R_1}\right) * V_{in}$$

For the same setup as in the previous circuit (12 V circuit with 10 Ω R_1 resistor, 50 Ω R_2 resistor, and load resistance of 1 Ω), the output voltage can be calculated as:

$$V_{out} = \left(1 + \frac{50}{10}\right) * 12\ V = 72\ V$$

That means the voltage across the 1 Ω load resistor will be 72 V. The positive sign indicates that the output voltage is in a similar direction with the input voltage. The circuit diagram and the voltage waveforms are shown in Fig. 17.31 and Fig. 17.32, respectively.

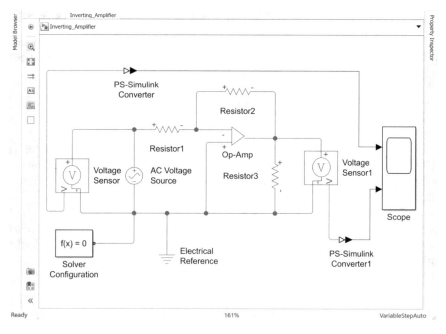

Fig. 17.29 Simulink diagram of a circuit containing an operational amplifier

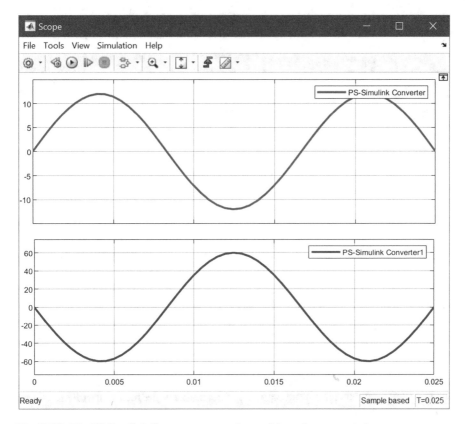

Fig. 17.30 The PS Simulink Converter outputs observed from the scope window

17.3.3 Differentiator Circuit

In a differentiator circuit, the Op-amp performs differentiation on the system and provides the resultant signal as the output signal. The formula for the resultant voltage can be represented as follows:

$$V_{out} = -RC\left(\frac{dV_{in}}{dt}\right)$$

In the designed circuit in Fig. 17.33, the input AC voltage source is set to be 12 V 60 Hz, with the value of capacitor and resistance being 0.001 F and 10 Ω, respectively, and with a load resistance of 1 Ω only. The input and output waveform is shown in Fig. 17.34.

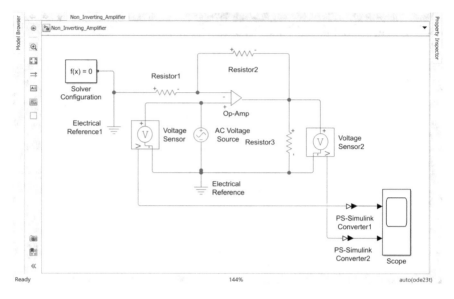

Fig. 17.31 Simulink diagram of a circuit containing a non-inverting amplifier

17.3.4   Integrator Circuit

In an integrator circuit, the Op-amp performs integration on the system and provides the resultant signal as the output signal. The formula for the resultant voltage can be represented as follows:

$$V_{\text{out}} = -\frac{1}{RC}\left(\int V_{\text{in}}dt\right)$$

In the designed circuit in Fig. 17.35, the input AC voltage source is set to be 12 V 60 Hz, with the value of capacitor and resistance being 0.001 F and 10 Ω, respectively, and with a load resistance of 1 Ω only. The input and output waveform is shown in Fig. 17.36.

17.4   Control Devices

Power electronics devices can be broadly classified into controllable and uncontrollable devices. The two-terminal devices, such as diodes, are considered uncontrolled because there is no way to control the current flow or voltage across such devices. On the other hand, three-terminal devices, such as MOSFET, IGBT, and thyristor, are controlled devices, as their third terminal controls the current flow between the other two terminals. Since the characteristics of MOSFET and IGBT have been discussed

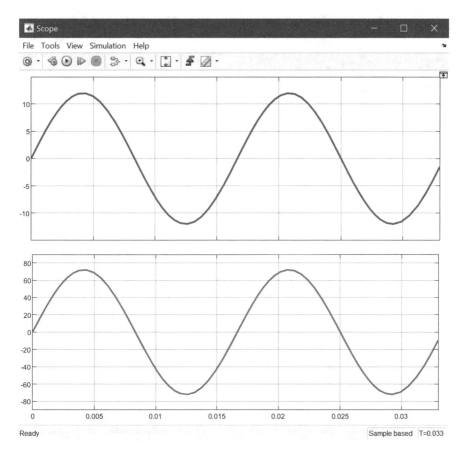

Fig. 17.32 The voltage outputs observed from the scope window

in the previous section, this chapter will build circuits with thyristors and gate turn-off (GTO) thyristor and show the characteristics through controlling the sinusoidal inputs. The model for thyristor and GTO is readily available in Simulink and will be used in other sections of this chapter.

17.4.1   Pulse Generation

Pulse generation is important for control devices in power electronics, as these signals guide how the current flow between the devices will flow. The pulses are usually square waves with a unit amplitude, varying from 1 (on-state) and 0 (off-state) only, similar to the digital systems. This signal is provided at the gate terminal of the control devices. There are two types of pulse in Simulink that can be

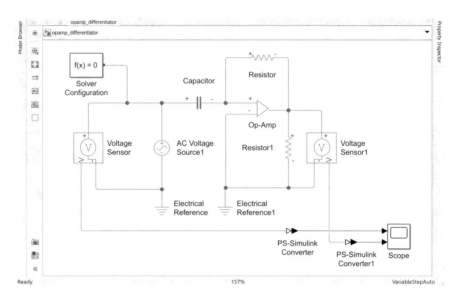

Fig. 17.33 Simulink diagram of a differentiator circuit

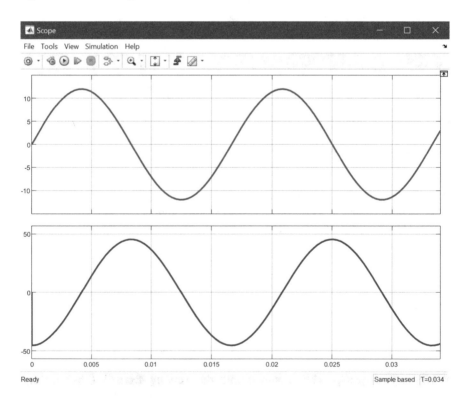

Fig. 17.34 The differentiator circuit input and output observed from the scope window

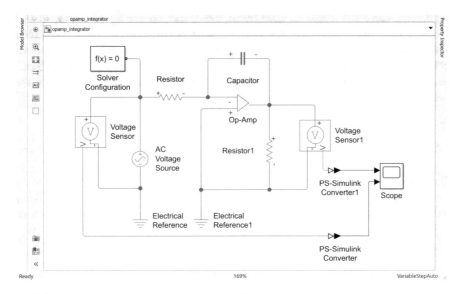

Fig. 17.35 Simulink diagram of an integrator circuit

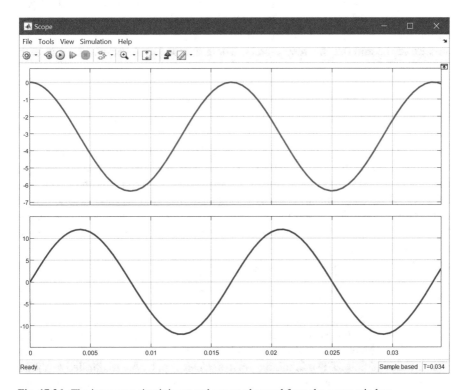

Fig. 17.36 The integrator circuit input and output observed from the scope window

used with these devices, i.e., time-based pulse and sample-based pulse. In this chapter, time-based pulse using simulation time is utilized. The most commonly used pulse generator is available in Simulink → Source. There are options for selecting the amplitude, pulse period, pulse width, and phase delay for time-based pulses. For every setup, the pulse characteristics are determined and provided to the control device for the desired performance.

17.4.1.1 Duty Cycle

Duty cycle is the percentage time a pulse stays on-state, with respect to the total cycle time. If a pulse has a total cycle time of T, and the pulse stays on-state and off-state for a time period of T_1 and T_2, respectively, then the duty cycle can be represented as:

$$\text{Duty cycle} = \frac{T_1}{T_1 + T_2} = \frac{T_1}{T}$$

For the following Simulink example, 20% and 80% duty cycles are shown. A PWM Generator block is used to show that a DC input of constant value can be considered the on-time for generating a pulse. Constant values of 0.2 and 0.8 are used to generate a duty cycle of 20% and 80%, respectively, in these examples. After setting the components as shown in Fig. 17.37, the output waveform shows an on-time of 0.2 and 0.8 ms, respectively, where the total cycle time is 1 ms in each case (Fig. 17.38).

17.4.1.2 Pulse Modulation

Pulse modulation is the transmission of pulsating signals in different forms. The variations in different forms of pulse help to produce pulses of different nature (e.g., different amplitude, width, or position). These pulses trigger (turn-on) the controlled devices to generate output waveforms as desired. A modulating wave is provided as a reference to modify a given pulse (career pulse) to obtain modulated pulses. By using modulating wave and career pulse, pulses can be modulated with three techniques mentioned as follows:

1. Pulse-Amplitude Modulation (PAM)

    In PAM, the amplitude of the career pulse is changed, keeping the width and frequency the same. The amplitude of the modulated signal is dependent on the modulating waveform. In Fig. 17.39, a pulse of constant amplitude and width is generated using the Pulse Generator block. Sinusoidal wave generated from the Sine Wave block is used as the modulated signal to vary the amplitude of the pulse sinusoidally. The modulating wave and career pulse are multiplied to obtain such variations of amplitude. Figure 17.40 shows the PAM technique and corresponding output for the given signal. It is to be noted that, for this example,

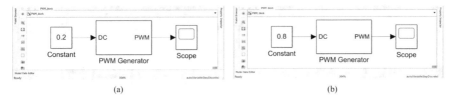

Fig. 17.37 Use of the PWM generator blocks in Simulink

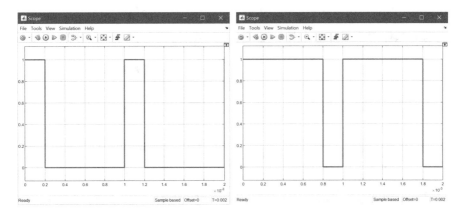

Fig. 17.38 The output of the PWM generator blocks observed from the scope window

the amplitude and frequency (rad/s) for the setup are chosen to be 1 and 5, respectively. The rest of the parameters are 0 for the Sine Wave block. For the Pulse Generator block, the amplitude, period, pulse width, and phase delay are chosen to be 1, 0.1, 50, and 0, respectively.

2. Pulse-Width Modulation (PWM)

In PWM, the width of the career pulse is changed, keeping the amplitude the same. In order to produce a pulse of the same amplitude but with the varying wave, the sawtooth wave and sine wave of different amplitudes have been superimposed on each other. At the points where the amplitude of the sawtooth wave is higher than the sine wave is considered to be the on-time for the pulse. The varying nature of both the waves generates a pulse with different widths.

In order to demonstrate the modulation, a Sine Wave block (Simulink → Sources) and a Sawtooth Generator block (Simscape → Electrical → Specialized Power Systems → Fundamental Blocks → Power Electronics → Pulse & Signal Generators) are used, the output waveforms of which are subtracted using an Add block. The output is then compared to a "Compare to Zero" block (Simulink → Logic and Bit Operations), which compares the subtracted result and returns the pulse when the resultant is less than zero. For this example, the amplitude and frequency of the Sine Wave block are chosen to

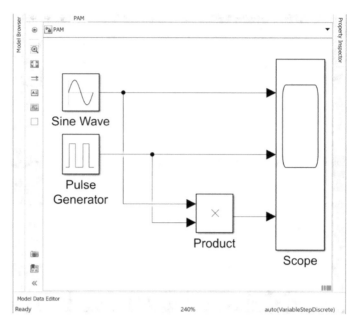

Fig. 17.39 Pulse Amplitude Modulation (PAM) in Simulink

be 0.75 and 60 rad/s, the rest of the parameters (bias, phase, and sample time) being zero. For the Add block, the sign is changed to "+−" to perform subtraction. The operator in the Compare to Zero block is changed to "<=." After setting the model up as in Fig. 17.41, the waveforms are visible as shown in Fig. 17.42.

3. Pulse-Position Modulation (PPM)

In PPM, the position of the career pulse is changed with respect to a reference waveform, keeping the amplitude and width the same. In order to demonstrate PPM, the PWM output obtained from the previous example is used as a reference waveform, which considers the rising edge of the PWM signal as the starting position for the PPM output pulse (Fig. 17.43).

The setup is similar to the PWM, but a new block Monostable (available at Simscape → Electrical → Specialized Power Systems → Control & Measurements → Logic) is used to provide a rising pulse whenever a change in the PWM logic is detected. The edge detection parameter for the Monostable block is selected to be "Rising," with a pulse duration of 0.002 s, and the initial condition of previous input and sample time of zero. The output waveform for PPM is shown in Fig. 17.44, which shows the change in position with respect to the rising edge of the PWM signal.

17.4.1.3   Determining the Firing Angle

For a particular frequency of f for a system, the time period can be calculated as $T = \frac{1}{f}$. Since one cycle corresponds to 360° if the phase of the signal is to be delayed

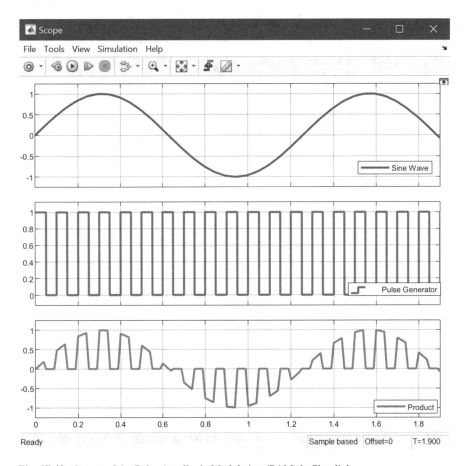

Fig. 17.40 Output of the Pulse Amplitude Modulation (PAM) in Simulink

for x°. Then the phase angle is calculated to be $\frac{x}{360}^\circ$, making the phase delay to be $x * \frac{1}{T} * \frac{1}{360}$. The turn-on time for a control device from the starting of the waveform can be calculated through the delay angle, also called the firing angle. Therefore, for a 60 Hz system, if a control device has a firing angle of 45°, the phase delay for the system can be calculated as $45 * \frac{1}{60} * \frac{1}{360} = 0.002083$ s. Using this calculation, the "phase delay" parameter for the pulse generator in Simulink is determined based on the required firing angle.

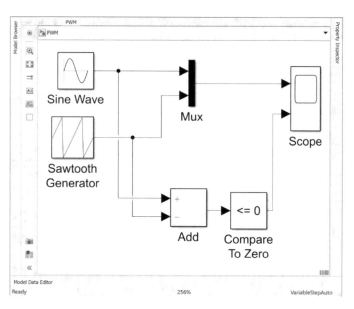

Fig. 17.41 Pulse Width Modulation (PWM) in Simulink

17.4.2 Controlled Rectification with Thyristor

A thyristor is a three-terminal device with four layers of alternating p- and n-type semiconductors. A thyristor is also known as a silicon-controlled rectifier (SCR). These devices are different than transistors because of their usage in more high-power applications for having high current and voltage ratings. Thyristors have three terminals called anode, cathode, and gate. Through the gate terminal, the current across the anode and cathode is controlled. This subsection will control a sinusoidal input with a firing angle of $60°$, provided by a 240 V 60 Hz AC voltage source, connected to a 0.5 Ω resistor in series, as shown in Fig. 17.45. The following components are taken from the Simulink library to arrange the circuit (Table 17.6).

The parameters for the Pulse Generator block to produce an angle of $60°$ are planned as follows (Fig. 17.46).

The output voltage waveform shows firing at $60°$, after which the wave follows the positive half-cycle of the input and turns off for the negative half-cycle of the input at the zero crossings of the current (Fig. 17.47).

17.4.3 Controlled Rectification with GTO

The gate turn-off (GTO) thyristor is a controlled device similar to a thyristor, which is controlled upon applying a gate signal at the gate of the device. Similar to conventional thyristors, the three terminals of GTO are named anode, cathode, and

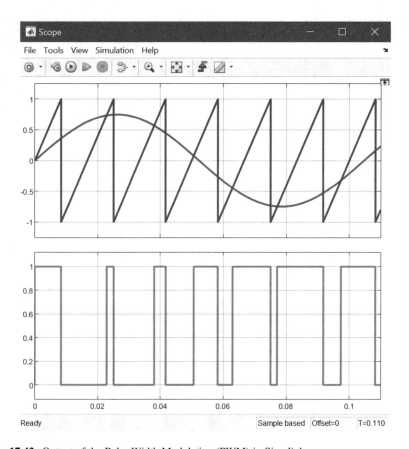

Fig. 17.42 Output of the Pulse Width Modulation (PWM) in Simulink

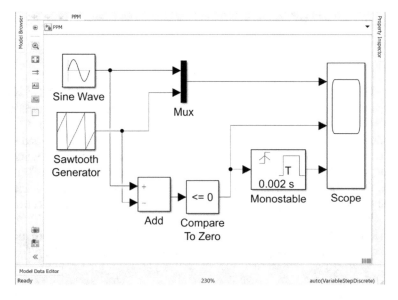

Fig. 17.43 Pulse Position Modulation (PPM) in Simulink

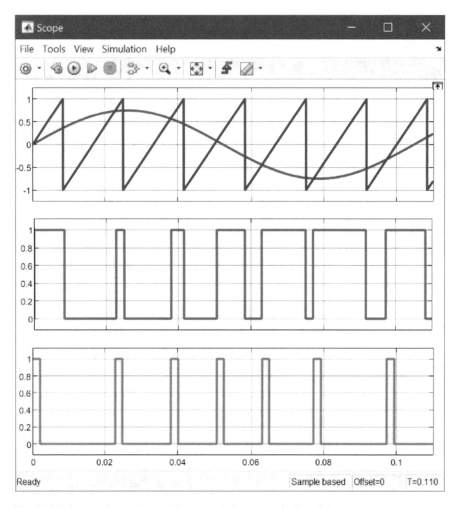

Fig. 17.44 Output of the Pulse Position Modulation (PPM) in Simulink

gate. The GTO thyristor turns on when a positive gate signal is provided. However, GTO thyristor is different from a conventional thyristor because it can be turned off anytime upon applying a zero gate signal.

The characteristics of GTO thyristor are studied with the same circuit and same parameter to that of the thyristor. The thyristor is replaced with the GTO, and the input and output voltages are visualized through the Scope block. The GTO is available from the same navigation path as the thyristors (Simscape → Electrical → Specialized Power Systems → Fundamental Blocks → Power Electronics) (Fig. 17.48).

Since a pulse width of 5% is selected for this example, the gate pulse will turn on at the mentioned phase delay (for this example, it is at 60°), and stay on for 5% of the

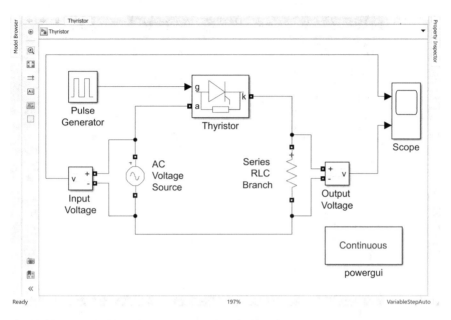

Fig. 17.45 Controlled Rectification with Thyristor in Simulink

Table 17.6 Blocks and navigation routes for modeling thyristor-based controlled rectifier

Name of the block	Navigation route
Powergui	Simscape → Electrical → Specialized Power Systems → Fundamental Blocks
AC Voltage Source	Simscape → Electrical → Specialized Power Systems → Fundamental Blocks → Electrical Sources
Series RLC Branch	Simscape → Electrical → Specialized Power Systems → Fundamental Blocks → Elements
Thyristor	Simscape → Electrical → Specialized Power Systems → Fundamental Blocks → Power Electronics
Pulse Generator	Simulink → Sources
Voltage Measurement	Simscape → Electrical → Specialized Power Systems → Fundamental Blocks → Measurements
Scope	Simulink → Commonly Used Blocks

period. It is seen from the graph that although the GTO turns on at the mentioned firing angle, the GTO turns off as soon as the gate pulse falls to 0, at 5% pulse width. If the pulse width is increased, the gate will stay on for more period, and turn off as soon as the gate pulse falls to 0 (Fig. 17.49).

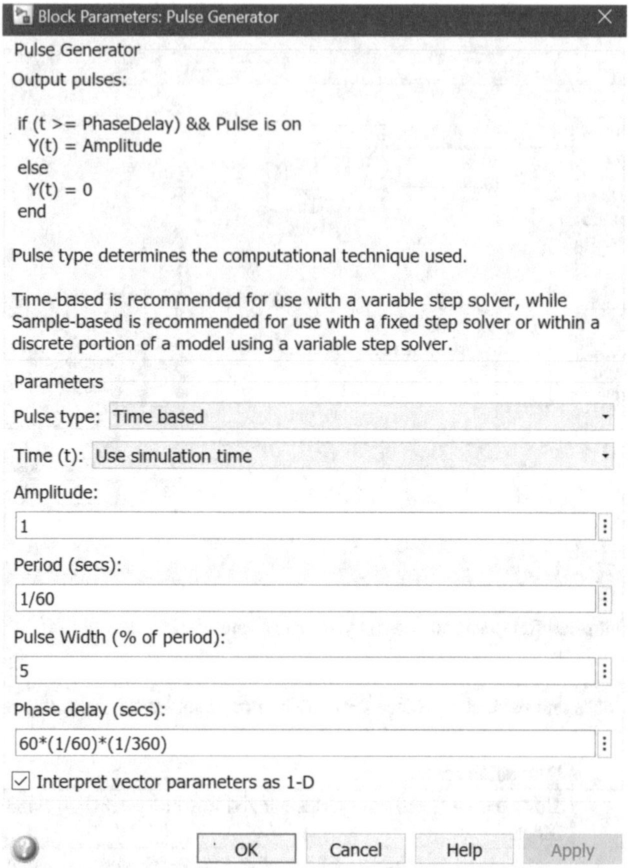

Fig. 17.46 Block parameters of the pulse generator

17.5 Facts

Flexible AC transmission systems (FACTS) are a growing technology in the domain of power engineering, where static power electronics components are used to enhance the quality of AC transmission. FACTS devices enhance the controlling capacity during AC transmission, thus improving power transfer capability and quality. In addition, these devices have the specifications to control megawatt-range power using control devices. Among numerous applications of FACTS devices, voltage regulation, transmission parameter control, and line loading capacity are notable.

In Simulink, there are four power electronics-based FACTS devices dedicated to experimenting with the power compensation technique through phasor-type FACTS devices. Static synchronous compensator (STATCOM), static synchronous series compensator (SSSC), static var compensator (SVC), and unified power flow

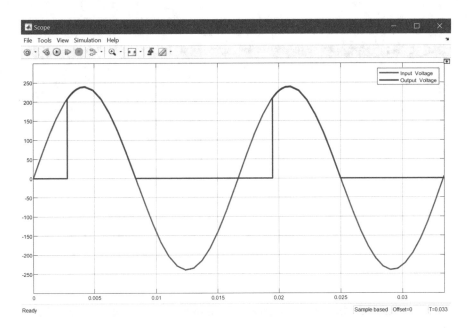

Fig. 17.47 Output of the Controlled Rectification with Thyristor

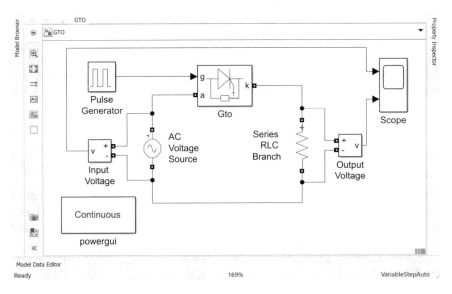

Fig. 17.48 Controlled Rectification with GTO in Simulink

controller (UPFC) are the blocks for the purpose, which are available in the navigation path Simscape → Electrical → Specialized Power Systems → FACTS → Power-Electronics Based Facts.

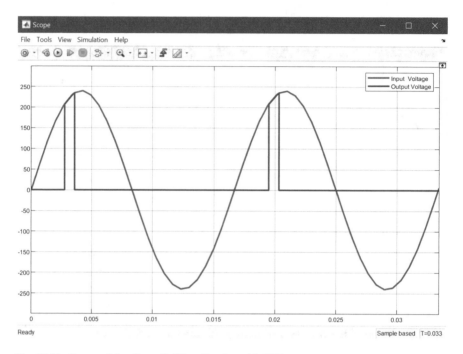

Fig. 17.49 Output of the Controlled Rectification with GTO

The construction of these devices is relatively complex and requires some prior knowledge of reference frame transformation and phase-locked loop for designing phasor models with these blocks. Therefore, these concepts are briefly explored, and only the operation of SVC is demonstrated in this section for the sake of simplicity.

17.5.1 Reference Frame Transformation

The three-axis frame where three-phase quantities of the AC circuit are represented is called abc frame. For the ease of mathematical calculation, the three-phase quantities of the AC circuits are sometimes required to be represented with respect to two stationary axes (e.g., for calculating the reference signals for space vector modulation control of three-phase inverter circuits). The two-axis stationary frame is called the alpha-beta-zero frame. The transformation from abc frame to alpha-beta-zero frame is known as Clarke transformation.

There is another reference frame with a rotating reference frame, unlike the alpha-beta-zero frame, which is stationary. The frame is called direct-quadrature-zero (also known as dq0 and dqz) frame. The purpose of the frame is to rotate the reference frames for the AC waveforms so that the AC quantities appear as DC quantities, which helps simplify complex calculations. The transformation from abc frame to dq0 frame is called Park transformation. This transformation is used in performing induction motor control.

Simulink has built-in blocks for straightforward transformation without diving deeper into complicated mathematical calculations. Figure 17.50 shows the Clarke and Park transformation using the components and their navigation path as shown below. There are four other blocks to perform transformation between alpha-beta-0 to dq0 and inverse transformations for each conversion (Table 17.7).

A Three-Phase Source is used to generate three-phase voltage. The parameters for the Three-Phase Source are mentioned as follows (Fig. 17.51):

The voltage measurement parameter of the Three-Phase V-I Measurement block is selected to be "phase-to-ground" for measuring the phase voltage, by omitting the current measurement option (Fig. 17.52).

The Ramp block used in the setup is provided with a slope of 2*pi*60, with starting time and initial output as 0. Figure 17.53 shows the output waveforms for the reference frame. In the first subplot, the abc quantities are shown; in the second subplot, the alpha (red) and beta (blue) components are shown (the green wave being 0). Finally, in the third subplot, the direct (red) and quadrature (blue) components of the dq0 frame are shown, representing the DC equivalent components of abc in a rotating dq0 frame.

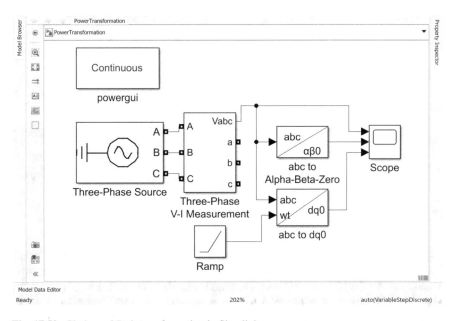

Fig. 17.50   Clarke and Park transformation in Simulink

Table 17.7 Blocks and navigation routes for demonstrating reference frame transformation

Name of the Bbock	Navigation route
Powergui	Simscape → Electrical → Specialized Power Systems → Fundamental Blocks
Three-Phase Source	Simscape → Electrical → Specialized Power Systems → Fundamental Blocks → Electrical Sources
Three-Phase V-I Measurement	Simscape → Electrical → Specialized Power Systems → Fundamental Blocks → Measurements
abc to Alpha-Beta-Zero	Simscape → Electrical → Specialized Power Systems → Control & Measurements → Transformations
abc to dq0	Simscape → Electrical → Specialized Power Systems → Control & Measurements → Transformations
Ramp	Simscape → Sources
Scope	Simulink → Commonly Used Blocks

Fig. 17.51 Block parameters of the three-phase source

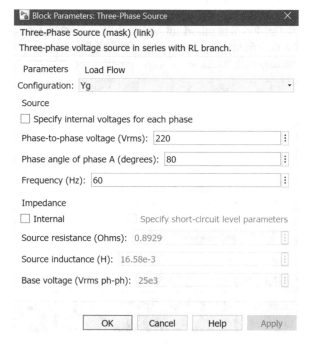

17.5.2 Phase-Locked Loop (PLL)

A phase-locked loop (PLL) is a control algorithm that determines the frequency and phase angle of a sinusoidal input. PLL is used for frequency matching between two systems, after which there remains a constant phase difference, hence "locking" the phase. PLL consists of a phase detection mechanism, a PID controller, and an oscillator to generate the phase angle information. A low-pass filter is also present

Block Parameters: Three-Phase V-I Measurement ✕

Three-Phase VI Measurement (mask) (link)

Ideal three-phase voltage and current measurements.

The block can output the voltages and currents in per unit values or in volts and amperes.

Parameters

Voltage measurement | phase-to-ground | ▾

☐ Use a label

☐ Voltages in pu, based on peak value of nominal phase-to-ground volt...

Current measurement | no | ▾

| OK | Cancel | Help | Apply |

Fig. 17.52 Block parameters of the three-phase V-I measurement

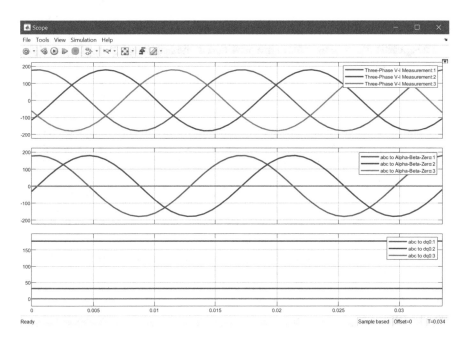

Fig. 17.53 The output waveforms of the Clarke and Park transformation

in the system to obtain the frequency information of the sinusoidal input. In FACTS devices, PLL gain plays a vital role in stabilizing system performance.

In order to visualize how PLL reflects the change in frequency and phase, a setup is created in the Simulink environment. The purpose is to change the frequency from 60 Hz to 61 Hz and observe how PLL responds to the change in frequency. For

Table 17.8 Blocks and navigation routes for designing phase-locked loop

Name of the block	Navigation route
Three-Phase Programmable Generator	Simscape → Electrical → Specialized Power Systems → Fundamental Blocks → Power Electronics → Pulse & Signal Generators
Selector	Simulink → Signal Routing
Bus Selector	Simulink → Signal Routing
PLL	Simscape → Electrical → Specialized Power Systems → Control & Measurements → PLL
PLL (3 ph)	Simscape → Electrical → Specialized Power Systems → Control & Measurements → PLL
Scope	Simulink → Commonly Used Blocks

establishing the setup, the following components are utilized from the Simulink library (Table 17.8):

For a three-phase signal with specific amplitude, phase, and frequency, generated by the Three-Phase Programmable Generator, the behavior of the single-phase PLL, the three-phase PLL, and the ideal system through bus selector is shown in Fig. 17.54.

The specifications of the blocks are provided in the following figures (Figs. 17.55, 17.56, and 17.57):

In the first area (green), only the first phase information from the Three-Phase Programmable Generator is obtained through a selector. The Selector is modified so that the input port size becomes 3 and the index becomes 1. After double-clicking on the Selector block, the index and input port size are, therefore, changed into 1 and 3, respectively, to obtain the symbol as shown in Fig. 17.54. The single-phase information is fed to the single-phase PLL through the Selector to see the frequency and phase angle information using Scope1.

In the second area (cyan), all the three-phase information is directly fed at the input of the three-phase PLL to see the output results in Scope2. Finally, in the third area (purple), the frequency and phase angle information is extracted from four other signals using a Bus Selector block, where only the "Freq (Hz)" and "wt (rad)" signals are selected from the left portion for visualization through Scope3 (Fig. 17.58).

The first graph shows the frequency response for the single-phase PLL. The response is relatively slower than that of the three-phase PLL because of the lack of phase information. However, both the responses are close to the ideal behavior (Fig. 17.59).

17.5.3 Static Var Compensator

The Static Var Compensator (SVC) is a power electronics device for FACTS applications. The SVC is used in an AC transmission system to provide or absorb reactive power to regulate voltage. During low voltages, SVC becomes capacitive

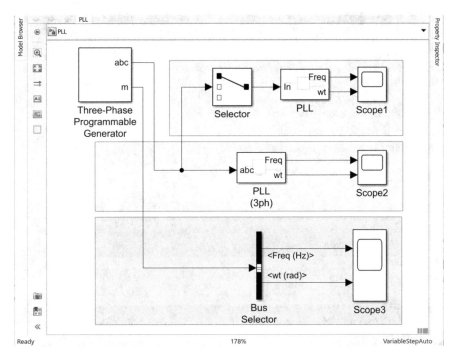

Fig. 17.54 Phase Locked Loop (PLL) in Simulink

Fig. 17.55 Block parameters of the three-phase programmable generator

Block Parameters: Three-Phase Programmable Generator ✕

Three-Phase Programmable Generator (mask) (link)

Generate a set of three-phase sinusoidal signals. Time variation for amplitude, phase, and frequency of the fundamental component can be programmed. In addition, two harmonics can be superimposed on the fundamental.

Parameters

Positive-sequence [Amplitude, Phase (degrees), Freq. (Hz)]:

[100, 0, 60]

Time variation of: Frequency

Type of variation: Step

Step magnitude:

0.5

Variation timing (s) [Start, End]:

[0.2 1.2]

☐ Harmonic generation

Sample time:

0

OK   Cancel   Help   Apply

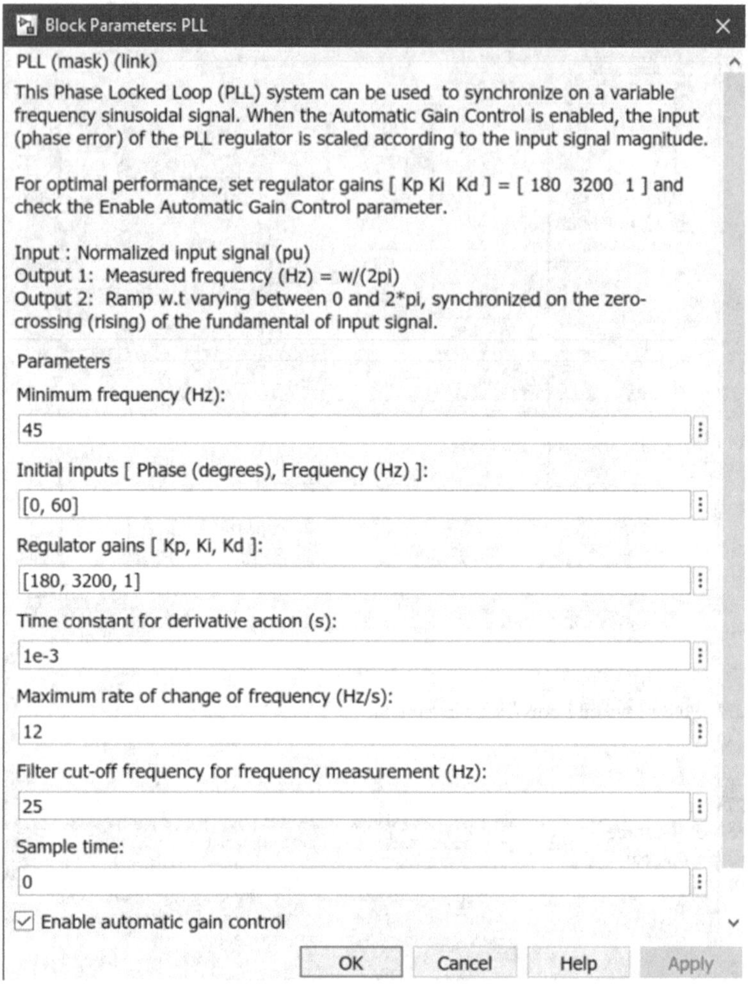

Fig. 17.56 Block parameters of the PLL

and provides reactive power to compensate for the reactive power, hence the name "Var Compensator." For high voltages, it absorbs reactive power to maintain the voltage at the terminal. Switching for delivering or absorbing reactive power in this manner is performed by thyristor switches, for which a PLL-supported synchronizing mechanism provides the pulse.

For demonstrating how an SVC compensates reactive power, the setup shown in Fig. 17.60 is built using the following components from the Simulink Library Browser (Table 17.9):

The objective of the setup is to compare the performance of the system with and without SVC. A Three-Phase Programmable Voltage Source block with specific parameters mentioned in Fig. 17.61 is used as the main source. Only the positive

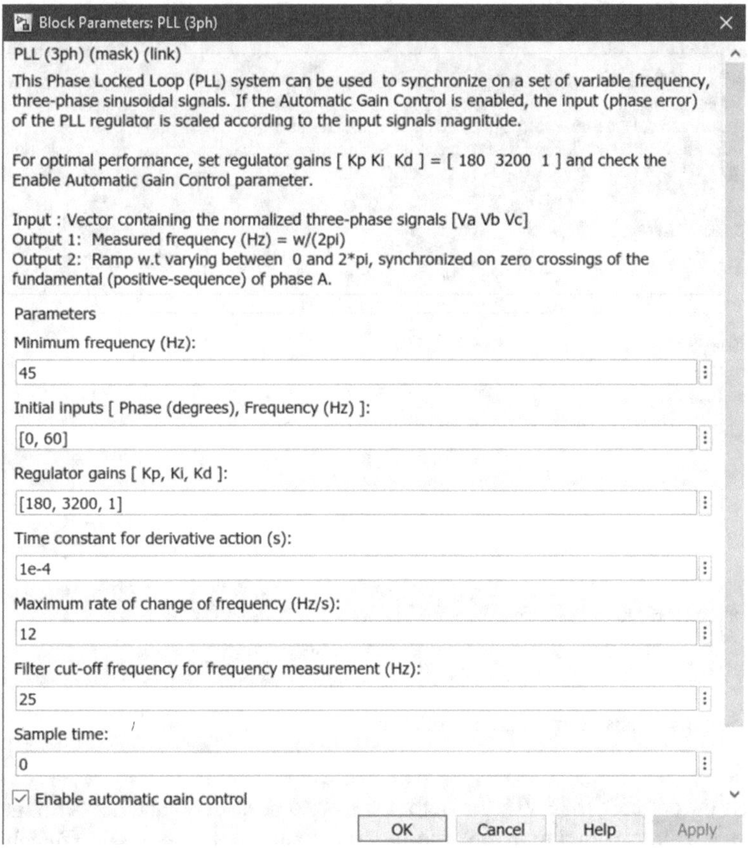

Fig. 17.57 Block parameters of the three-phase PLL

sequence input voltage is measured through the blocks "Three-Phase V-I Measurement1" block (Fig. 17.62) and "Sequence Analyzer (Phasor)1" (Fig. 17.63)at the first input of the Scope (blue color in Fig. 17.68). A Terminator block at the end of the second output in the Sequence Analyzer block is used to cap the output and block any error message from Simulink for keeping the terminal disconnected.

Through a Three-Phase Series RLC Branch, a Three-Phase Series RLC Load is connected with the specifications provided in Figs. 17.64 and 17.65. The positive sequence voltage taken from the "Three-Phase V-I Measurement2" block is actually the voltage without the SVC (as shown by the red line in the graph). This indicates that the voltage drops significantly compared to the input voltage. SVC will be used at this instance to improve the voltage level and make the output voltage as close as possible to the input voltage.

The following parameters are used in the Static Var Compensator block (Fig. 17.66). Only the control voltage "Control VM (pu)" is chosen through a Bus Selector block, as shown in Fig. 17.67.

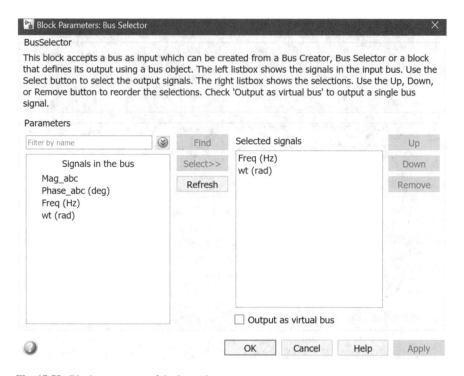

Fig. 17.58   Block parameters of the bus selector

After introducing the SVC, the output is seen to be closely matched with the input voltage. A gain is added with the output of the control voltage signal to match better with the input voltage (Fig. 17.68).

17.6   Modeling of Converters

In power electronics, it is necessary to work with various waveforms and convert those to any suitable level of amplitude, frequency, or phase as per the requirement. Converters help to serve the purpose by taking the input and transforming specific features according to the behavior of the circuital elements. In this section, a brief overview of the significant DC-DC, DC-AC, AC-DC, and AC-AC converter modeling will be discussed with the help of Simulink simulations so that the readers can obtain a sound knowledge of the working principles of each of the converters and can design a converter with a proper understanding of how the circuital elements impact the output.

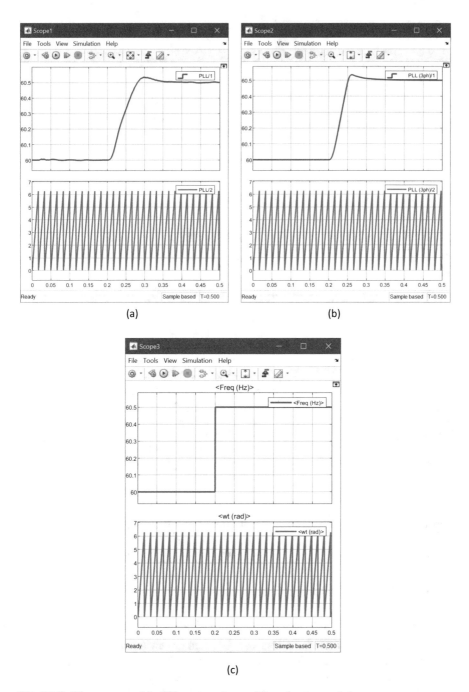

Fig. 17.59 The response of the PLL system observed from the scope window

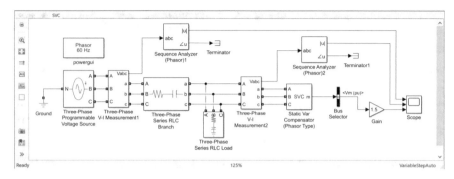

Fig. 17.60 Static Var Compensator (SVC) in Simulink

Table 17.9 Blocks and navigation routes for modeling static var compensator

Name of the block	Navigation route
powergui	Simscape → Electrical → Specialized Power Systems → Fundamental Blocks
Three-Phase Programmable Voltage Source	Simscape → Electrical → Specialized Power Systems → Fundamental Blocks → Electrical Sources
Three-Phase V-I Measurement1, Three-Phase V-I Measurement2	Simscape → Electrical → Specialized Power Systems → Fundamental Blocks → Measurements
Three-Phase Series RLC Branch	Simscape → Electrical → Specialized Power Systems → Fundamental Blocks → Elements
Three-Phase Series RLC Load	Simscape → Electrical → Specialized Power Systems → Fundamental Blocks → Elements
Static Var Compensator (Phasor Type)	Simscape → Electrical → Specialized Power Systems → FACTS → Power-Electronics Based FACTS
Sequence Analyzer (Phasor)1, Sequence Analyzer (Phasor)2	Simscape → Electrical → Specialized Power Systems → Control & Measurements → Measurements
Terminator	Simulink → Sinks
Bus Selector	Simulink → Commonly Used Blocks
Gain	Simulink → Commonly Used Blocks
Scope	Simulink → Commonly Used Blocks
Ground	Simscape → Electrical → Specialized Power Systems → Fundamental Blocks → Elements

17.6.1   Model of DC-DC Converters

DC-DC converters convert a DC input to a DC output of a different voltage, as required for the application. These can be used as switching-mode regulators, converting an unregulated DC voltage into a regulated DC output voltage. The voltages can either be lower or higher than the input voltages. The switching for such conversion is performed using transistors with a pulse generated at a particular

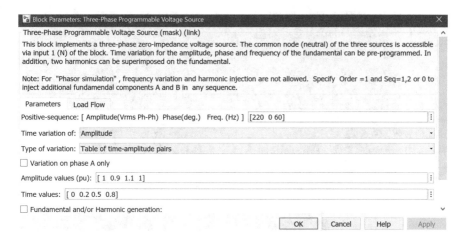

Fig. 17.61 Block parameters of the three-phase programmable voltage source

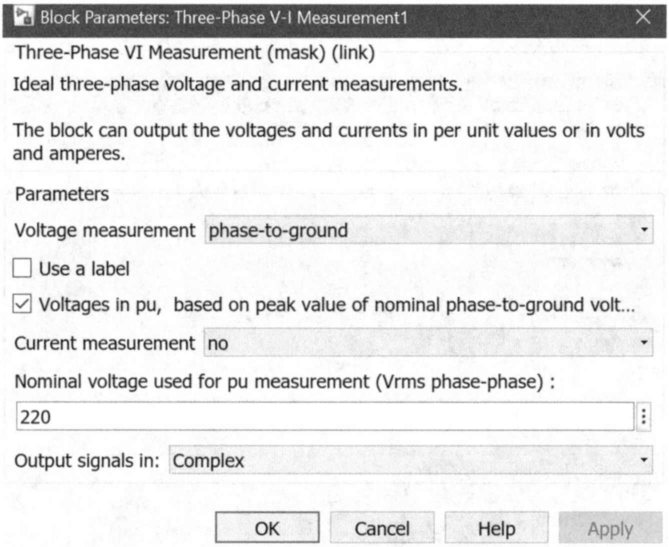

Fig. 17.62 Block parameters of the three-phase V-I measurement 1

frequency. Buck converter and boost converter are the two most known switching-mode regulators. The former decreases an input DC voltage, whereas the latter increases the voltage. The simulation model for these converters is shown in the following subsections.

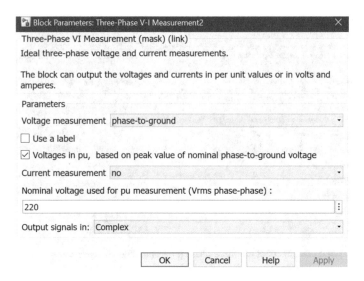

Fig. 17.63 Block parameters of the three-phase V-I measurement 2

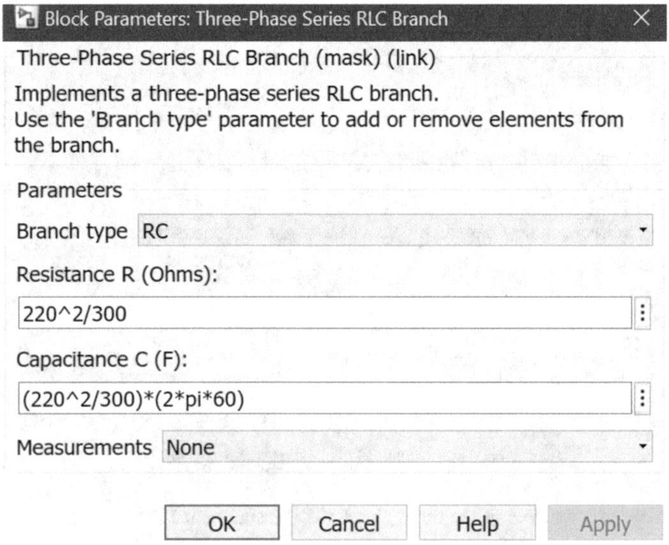

Fig. 17.64 Block parameters of the three-phase series RLC branch

17.6.1.1   Buck Converter

Buck converter provides a lower DC voltage than the source DC voltage. If a voltage of V_a is desired from a source voltage of V_s, a pulse with duty cycle k is is to be

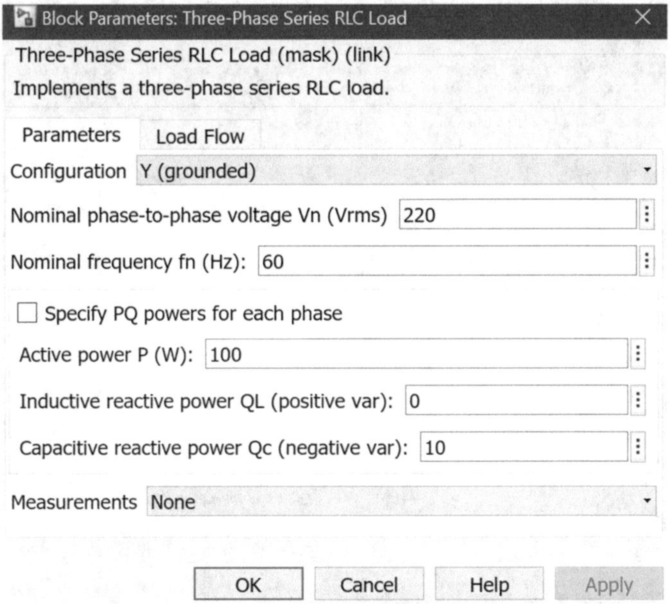

Fig. 17.65 Block parameters of the three-phase series RLC load

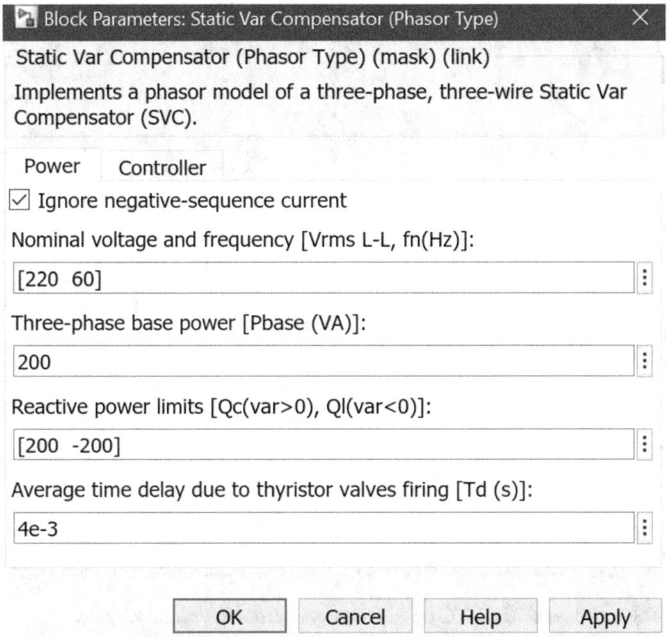

Fig. 17.66 Block parameters of the SVC (phasor type)

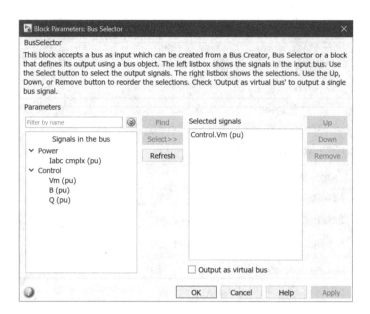

Fig. 17.67 Block parameters of the bus selector

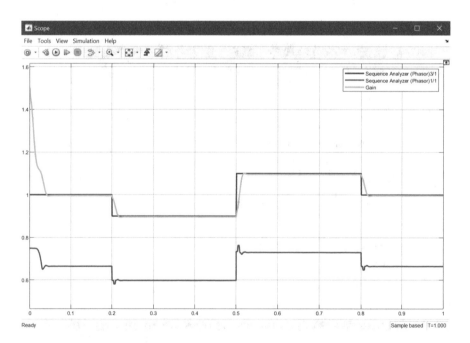

Fig. 17.68 The response of the SVC system observed from the scope window

provided at the gate of the transistor. For buck converter, the relation between the parameters can be written as follows:

$$V_a = k\,V_s$$

According to the formula above, for the desired voltage of 8 V from an input voltage of 12 V, the duty cycle for the transistor should be:

$$k = \frac{8}{12} = 0.667 \text{ s}$$

The following components are used to demonstrate the modeling of the buck converter (Table 17.10). The circuit is then built as per Fig. 17.69.

Here, the DC voltage is 12 V, the inductor is 10 mH, the capacitor is 1 μF, and the resistor is 50 Ω. In the pulse generator, the amplitude is set to be 1, with a pulse period of $\frac{1}{10000}$ s, as the switching frequency is considered to be 10,000 Hz. The pulse width in percentage is the duty cycle, which is set as 66.67%, as calculated from the value of k from the equation. After running the simulation for 0.05 s, the output voltage is found out to be nearing 8 V, as shown in the graph below (Fig. 17.70).

As it is seen from the graph, the output is not exactly 8 V. It is because there are some voltage drops in other switching components such as the MOSFET or diode. For example, the output voltage drops due to a default forward voltage of 0.8 V in the diode. To verify this, change the forward voltage of the diode from 0.8 V to a minimum value (e.g., 0.1 V). It will be seen that the output voltage becomes closer to 8 V as there is a very little voltage drop across the diode.

Table 17.10 Blocks and navigation routes for designing buck converter

Name of the block	Navigation route
powergui	Simscape → Electrical → Specialized Power Systems → Fundamental Blocks
DC Voltage Source	Simscape → Electrical → Specialized Power Systems → Fundamental Blocks → Electrical Sources
MOSFET	Simscape → Electrical → Specialized Power Systems → Fundamental Blocks → Power Electronics
Diode	Simscape → Electrical → Specialized Power Systems → Fundamental Blocks → Power Electronics
Pulse Generator	Simulink → Sources
Series RLC Branch	Simscape → Electrical → Specialized Power Systems → Fundamental Blocks → Elements
Voltage Measurement	Simscape → Electrical → Specialized Power Systems → Control & Measurements → Measurements
Scope	Simulink → Commonly Used Blocks

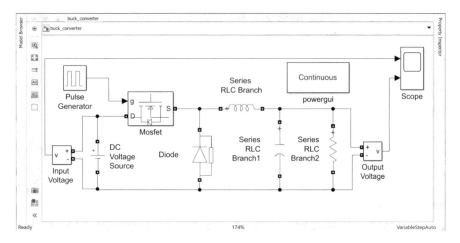

Fig. 17.69   Buck converter in Simulink

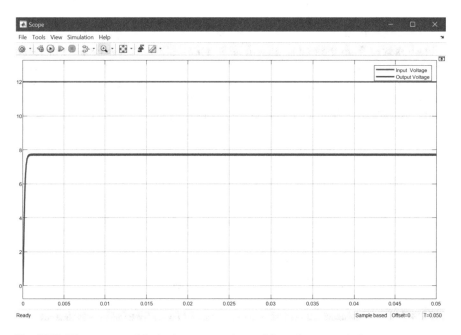

Fig. 17.70   The response of the buck converter observed from the scope window

17.6.1.2   Boost Converter

As the name suggests, a boost converter provides a higher DC voltage than the source DC voltage. For an input voltage of V_s, which is to be boosted to an output voltage of V_a, with a pulse with duty cycle k at the gate of the transistor, the relation is shown as:

$$V_a = (1 - k) V_s$$

According to the formula above, for the desired voltage of 5 V from an input voltage of 15 V, the duty cycle for the transistor should be:

$$k = 1 - \frac{5}{15} = 0.667 \text{ s}$$

The same components that were used to design the buck converter are used in this simulation as well. With these components, the circuit can be built as follows (Fig. 17.71).

Here, the input DC voltage of 5 V is expected to be boosted into 15 V with an amplitude of 1, pulse period of $\frac{1}{10000}$ s, and pulse width of 66.67% at the pulse generator. The value of the inductor is chosen to be 333 mH, the capacitor to be 10 μF, and the resistor to be 500 Ω. After running the simulation for 0.2 s, the output voltage is found out to be nearing 15 V, as shown in the graph below. Similar to the previous case, a small voltage drops across the diode, if minimized by reducing the forward voltage, will help to increase the output voltage to 15 V (Fig. 17.72).

17.6.1.3 Buck-Boost Converter

Buck-boost converter changes the input DC voltage depending on the pulse width (duty cycle) of the pulse. If the pulse is within 1–50%, the converter works as a buck converter. If the pulse is within 51–100%, the converter works as a boost converter.

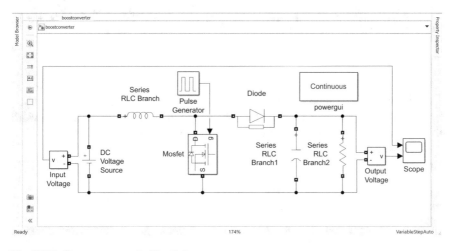

Fig. 17.71 Boost converter in Simulink

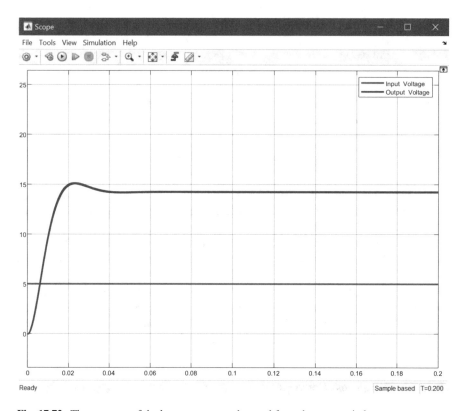

Fig. 17.72 The response of the boost converter observed from the scope window

A special characteristic of the buck-boost converter is that it reverses the polarity of the output voltage. For a source voltage of V_s, the output voltage V_a can be represented with respect to the duty cycle k as follows:

$$V_a = -V_s * \frac{k}{1 - k}$$

In this example, a source voltage of 10 V is considered. For demonstrating buck operation, a pulse width of 25% and for demonstrating boost operation, a pulse width of 75% are to be used. As per the formula, the output voltages will be:

For $k = 0.25$,

$$V_a = -10V * \frac{0.25}{1 - 0.25} = -3.33V$$

For $k = 0.75$,

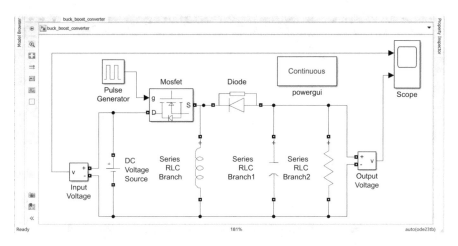

Fig. 17.73 Buck-boost converter in Simulink

$$V_\mathrm{a} = -10\mathrm{V} * \frac{0.75}{1 - 0.75} = -30\mathrm{V}$$

The minus sign indicates that the output voltage will be in opposite direction with respect to the input voltage.

The same components that were used to design the buck converter are used in this simulation as well. With these components, the circuit is then built as per Fig. 17.73.

Here, an input DC voltage of 10 V is used as the input for the converter. A 10 mH inductor, a 500 μF capacitor, and a 10 Ω resistor are used. Pulse with amplitude of 1, pulse period of $\frac{1}{10,000}$ s, is provided at the gate of the MOSFET, at first with 25% pulse width for an output voltage of −3.33 V (buck operation), and then with 75% pulse width for an output voltage of −30 V (boost operation). After running the simulation for 0.1 s, the output for both the pulse widths can be demonstrated as follows (Figs. 17.74 and 17.75):

17.6.2 Model of DC-AC Converter

DC-AC converters are called inverters, as they change the incoming DC signal to a fixed or variable AC output signal. Inverters can either be voltage source or current source, inverting either voltage or current, respectively. PWM signals are usually utilized to vary the gain of the inverter to obtain an output signal of desired magnitude or frequency. The ideal output waveform should be sinusoidal, but due to harmonics (lower frequency components of the fundamental signal) or noise from external or internal sources, it is not easy to obtain the pure sinusoidal output. For this reason, there are different inverter schemes to minimize harmonics as much as possible for nearing the sinusoidal waveform. In this subsection, the basic layout of

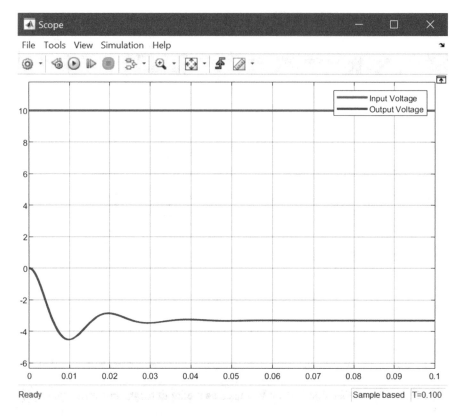

Fig. 17.74 The response of the buck operation observed from the scope window

single-phase and three-phase inverters will be shown without diving deeper into advanced architectures for simplicity.

17.6.2.1 Single-Phase Half-Wave Bridge Inverter

Single-phase half-wave inverter is a voltage source bridge inverter made with two control devices, where one of the devices remains in on-state for half of the time period and the other remains in on-state for another half of the time period. In the setup shown in Fig. 17.76, two "IGBT/Diode" blocks have been used since the block consists of both IGBT and an antiparallel diode. Two DC sources or three-wire DC sources, or two capacitors, are additionally used in this setup. The other blocks that are used in this setup are shown in the following Table 17.11.

A DC voltage source of 48 V is used, with two capacitors of value 100 mF each. A small resistor of 1 Ω is connected in series with the DC voltage source, as connecting a capacitor with a DC voltage source would initially short-circuit the source. Load resistance of 5 Ω is considered for the circuit. Two pulses are provided

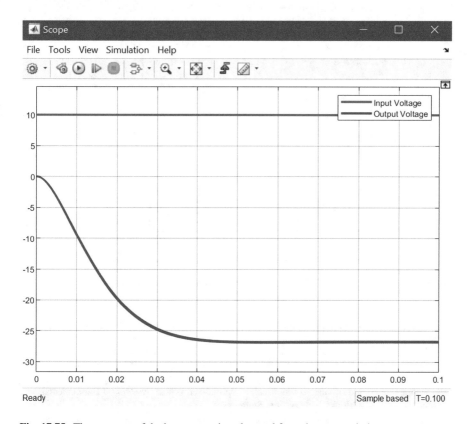

Fig. 17.75 The response of the boost operation observed from the scope window

at the gates of two IGBTs, one directly connected and the other with a "NOT" logical operator in the middle. The amplitude of the pulse generator is 12, with a period of 0.0167 s, a pulse width of 50%, and with no phase delay. While connecting the circuit, the polarity of the measurement devices is to be matched because changing their polarity may show the waveforms with the same magnitude and frequency but in the opposite direction. The input DC voltage and the output waveforms of voltage and current are shown in Fig. 17.77.

17.6.2.2 Single-Phase Full-Wave Inverter

Unlike a single-phase half-wave inverter, the full-wave inverter can be used with two-wire DC voltage sources without the need for any capacitors. However, the setup requires the use of four control devices with antiparallel diodes, with the gate pulses in the particular order shown in Fig. 17.78. Using the same components as for

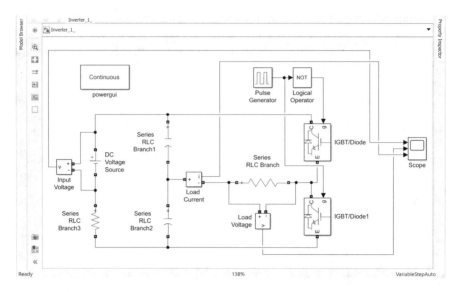

Fig. 17.76 Simulink diagram of a Single-Phase Half-Wave Bridge Inverter

Table 17.11 Blocks and navigation routes for designing single-phase half-wave bridge inverter

Name of the block	Navigation route
powergui	Simscape → Electrical → Specialized Power Systems → Fundamental Blocks
DC Voltage Source	Simscape → Electrical → Specialized Power Systems → Fundamental Blocks → Electrical Sources
IGBT/Diode	Simscape → Electrical → Specialized Power Systems → Fundamental Blocks → Power Electronics
Pulse Generator	Simulink → Sources
Logical Operator	Simulink → Logic and Bit Operations
Series RLC Branch	Simscape → Electrical → Specialized Power Systems → Fundamental Blocks → Elements
Voltage Measurement	Simscape → Electrical → Specialized Power Systems → Control & Measurements → Measurements
Current Measurement	Simscape → Electrical → Specialized Power Systems → Control & Measurements → Measurements
Scope	Simulink → Commonly Used Blocks

the half-wave inverter (except the capacitors), the configuration is connected as follows.

The voltage source is 48 V and the load resistance is 5 Ω for this setup. The parameters for the pulse generator are kept similar to that of the half-wave inverter. The input voltage, load voltage, and current are visible from the scope (Fig. 17.79).

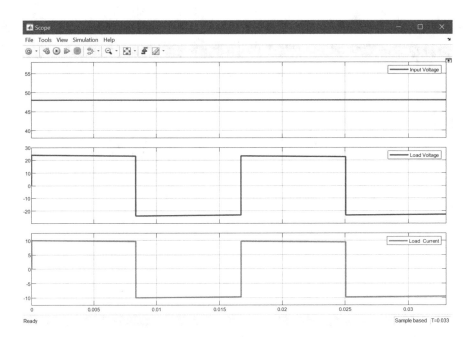

Fig. 17.77 Input voltage, load voltage, and load current of Single-Phase Half-Wave Bridge Inverter observed from the scope window

17.6.2.3   Three-Phase Inverter

A three-phase inverter converts a DC voltage into a three-phase AC supply using six control devices with six different pulses. Each arm of the inverter can be delayed either for 120° or 180° to obtain the three-phase output. In the setup shown in Fig. 17.80, 180° conduction mode is used where each of the control devices is activated at an interval of 60°. The list of components used in this simulation are as follows (Table 17.12):

An input DC voltage of 48 V is provided, with a "Three-Phase Series RLC Load." The default parameters for the load are used in this example. The "IGBT/Diode" and "Pulse Generator" blocks should be placed in the order as mentioned in the image. Since each pulse will be at a 60° delay, a varying pulse width is provided in each signal. For all six Pulse Generator blocks, the amplitude is set as 1, the period as $\frac{1}{60}$ s (as the frequency is considered to be 60 Hz), and the pulse wide as 50%. The phase delay for each pulse is varied as follows:

For "Pulse Generator 1," the phase delay is $0 * \left(\frac{1}{60}\right) * \left(\frac{1}{360}\right)$.
For "Pulse Generator 2," the phase delay is $60 * \left(\frac{1}{60}\right) * \left(\frac{1}{360}\right)$.
For "Pulse Generator 3," the phase delay is $120 * \left(\frac{1}{60}\right) * \left(\frac{1}{360}\right)$.
For "Pulse Generator 4," the phase delay is $180 * \left(\frac{1}{60}\right) * \left(\frac{1}{360}\right)$.

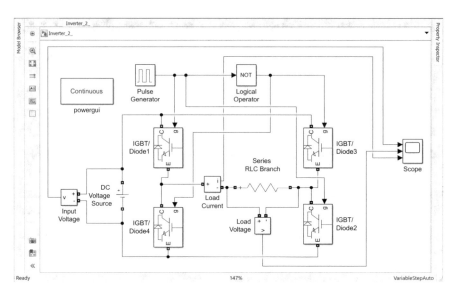

Fig. 17.78 Simulink diagram of a Single-Phase Full-Wave Inverter

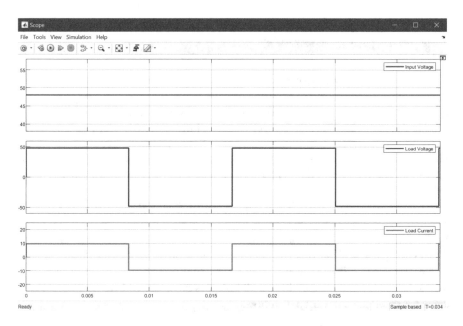

Fig. 17.79 Simulink diagram of a Single-Phase Full-Wave Inverter

For "Pulse Generator 5," the phase delay is $240 * \left(\frac{1}{60}\right) * \left(\frac{1}{360}\right)$.

For "Pulse Generator 6," the phase delay is $300 * \left(\frac{1}{60}\right) * \left(\frac{1}{360}\right)$.

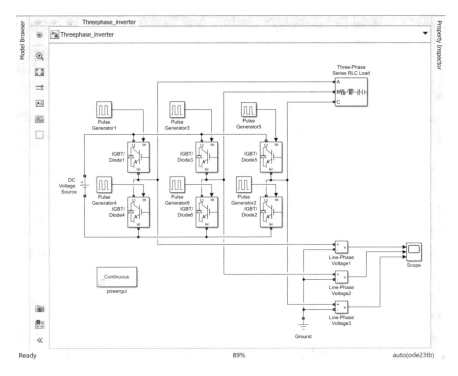

Fig. 17.80   Simulink diagram of a Three-Phase Inverter

Table 17.12   Blocks and navigation routes for designing three phase inverter

Name of the block	Navigation route
powergui	Simscape → Electrical → Specialized Power Systems → Fundamental Blocks
DC Voltage Source	Simscape → Electrical → Specialized Power Systems → Fundamental Blocks → Electrical Sources
IGBT/Diode	Simscape → Electrical → Specialized Power Systems → Fundamental Blocks → Power Electronics
Pulse Generator	Simulink → Sources
Three-Phase Series RLC Load	Simscape → Electrical → Specialized Power Systems → Fundamental Blocks → Elements
Voltage Measurement	Simscape → Electrical → Specialized Power Systems → Control & Measurements → Measurements
Scope	Simulink → Commonly Used Blocks

The line-to-phase voltage is measured for each arm using "Voltage Measurement" block and "Scope." It is seen from Fig. 17.81 that the voltages appear in three levels as per the equivalent resistance of the Y (grounded) load. There are spikes visible in the first portion of the waves due to the internal parameters of diodes,

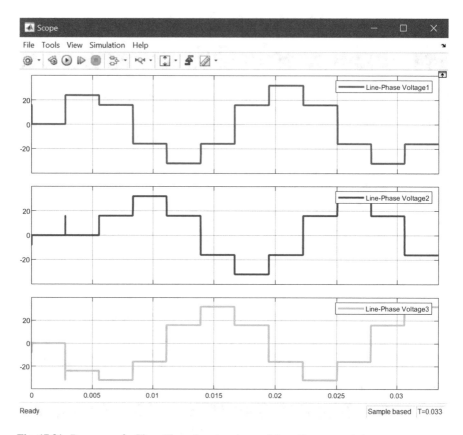

Fig. 17.81 Response of a Three-Phase Inverter observed from the scope window

which can be considered to be negligible for this example. Such inverters are vastly used in high voltage applications and electrical machines in industries.

17.6.3 Model of AC-DC Converter

In previous sections, AC-DC conversion with uncontrolled rectifiers was demonstrated. Control devices, such as thyristors and GTOs, were used to show the characteristics of the devices by changing the firing angle in the latter sections. With the help of a similar setup from the previous subsection, the AC-DC converters can be modeled for single-phase and three-phase full-wave AC signals to be converted into DC voltage of desired nature. These AC-DC converters are used in motor speed controls, designing uninterruptible power supplies, and charging energy storage systems.

17.6.3.1 Single-Phase Full-Wave Converter

For designing a single-phase full-wave converter, four thyristors are required. The Table 17.13 enlists the components used for designing the converter.

For this example, one pair of the thyristor is triggered at an angle of 45°, whereas the other pair of the thyristor is triggered at an angle of 225°. An AC voltage source with 110 V at 60 Hz is used, and an R load of 5 Ω resistor is selected as the load. The model is constructed as shown in Fig. 17.82.

In each pulse generator, an amplitude of 1, a period of $\frac{1}{60}$ s, and a pulse width of 5% are mentioned. The phase delay for each pulse generator is mentioned as follows:

For "Pulse Generator 1," the phase delay is $45 * \left(\frac{1}{60}\right) * \left(\frac{1}{360}\right)$.
For "Pulse Generator 2," the phase delay is $225 * \left(\frac{1}{60}\right) * \left(\frac{1}{360}\right)$.

It is seen from the output waveform that a controlled DC output voltage with the same magnitude is obtained. Upon changing the triggering angle for the thyristors, it is possible to vary the waveform as desired for the application (Fig. 17.83).

17.6.3.2 Three-Phase Full-Wave Converter

Similar to a single-phase converter, the three-phase converter converts a three-phase AC input into a controlled DC output voltage, controlled by three pairs of thyristors. For demonstrating such a converter, the same components as mentioned in the single-phase full converter are utilized: three AC voltage sources with 220 V 60 Hz, with a phase of 0, 120, and 240, respectively. A 1 Ω R load is used to see the output voltage across. In each "Pulse Generator," the amplitude is selected to be 1, the period is selected to be $\frac{1}{60}$ s as the system frequency is 60 Hz, and the pulse width is selected to be 5%. The phase delay for each of the pulse generators is mentioned below.

Table 17.13 Blocks and navigation routes for designing single-phase full-wave converter

Name of the block	Navigation route
powergui	Simscape → Electrical → Specialized Power Systems → Fundamental Blocks
AC Voltage Source	Simscape → Electrical → Specialized Power Systems → Fundamental Blocks → Electrical Sources
Thyristor	Simscape → Electrical → Specialized Power Systems → Fundamental Blocks → Power Electronics
Pulse Generator	Simulink → Sources
Series RLC Branch	Simscape → Electrical → Specialized Power Systems → Fundamental Blocks → Elements
Voltage Measurement	Simscape → Electrical → Specialized Power Systems → Control & Measurements → Measurements
Scope	Simulink → Commonly Used Blocks

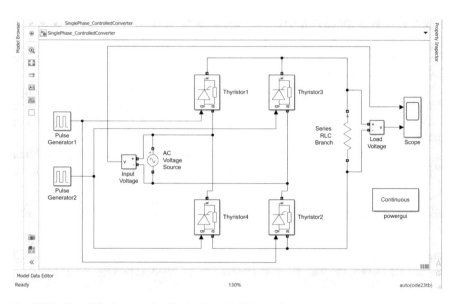

Fig. 17.82 Simulink diagram of a Single-Phase Full-Wave Converter

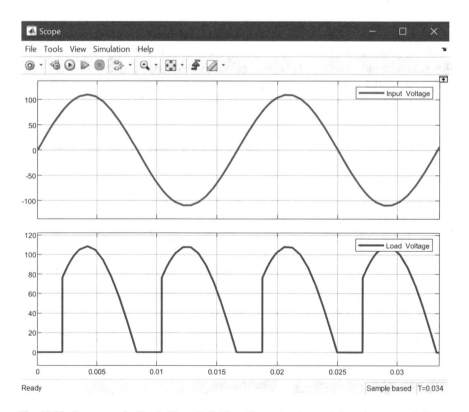

Fig. 17.83 Response of a Single-Phase Full-Wave Converter observed from the scope window

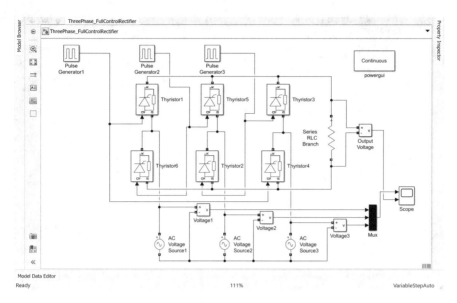

Fig. 17.84 Simulink diagram of a Three-Phase Full-Wave Converter

For "Pulse Generator 1," the phase delay is $45 * \left(\frac{1}{60}\right) * \left(\frac{1}{360}\right)$.

For "Pulse Generator 2," the phase delay is $165 * \left(\frac{1}{60}\right) * \left(\frac{1}{360}\right)$.

For "Pulse Generator 3," the phase delay is $285 * \left(\frac{1}{60}\right) * \left(\frac{1}{360}\right)$.

The circuit is set up as shown in Fig. 17.84.

After running the simulation for 0.0335 s, the output waveform is found out to be as follows. The input three-phase voltage is summarized into a controlled DC output, which can further be controlled by changing the phase delay in each pulse generator (Fig. 17.85).

17.6.4 Model of AC-AC Converter

AC-AC converters convert the magnitude and frequency of an AC input voltage as desired. Electrical transformers also serve the same purpose, which either "steps up" or "steps down" an input AC voltage with the same frequency. The major difference between a converter and a transformer lies in the method of AC conversion. In converters, an input signal is disintegrated into the shape of the desired output, but in the case of a transformer, the length of the input wave is modified without chopping the input waveform. Moreover, converters can change the input frequency, whereas transformers cannot change the frequency. Since transformer blocks can easily change the voltage level depending on the turns ratio, this section discusses the converter that changes the frequency of the input AC voltage. There are typically

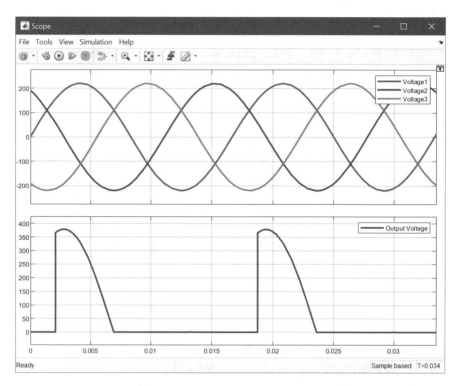

Fig. 17.85 Response of a Three-Phase Full-Wave Converter observed from the scope window

three types of cycloconverters, single-phase cycloconverter, three-phase to single-phase cycloconverter, and three-phase to three-phase cycloconverter. Since the structure of the three-phase cycloconverters can easily be understood using three single-phase cycloconverters, only the formation of a single-phase cycloconverter is discussed in this section.

17.6.4.1   Single-Phase Cycloconverter

Cycloconverters are AC-AC converters that convert AC signals of particular frequency and voltage into a different frequency and voltage as required by the system. Figure 17.86 shows a single-phase cycloconverter. The components that are used for modeling the converter are tabulated as follows (Table 17.14):

In this system, the voltage of the AC voltage source is chosen to be 220 V 60 Hz. Two sets of converters, called P-converter and N-converter, are combined in this single-phase cycloconverter with four ideal switches. The left portion of the cycloconverter consisting of thyristors 1–4 is called the P-converter, which provides the positive half-cycles of the input waveform. The right portion of the cycloconverter, consisting of thyristors 5–8, is called the N-converter, which

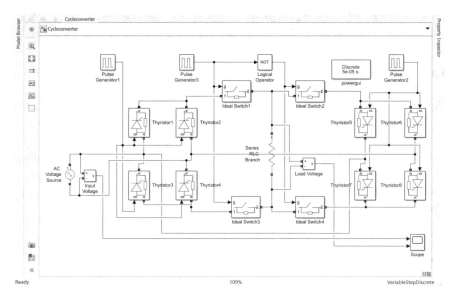

Fig. 17.86 Simulink diagram of a Single-Phase Cycloconverter

Table 17.14 Blocks and navigation routes for modeling single-phase cycloconverter

Name of the block	Navigation route
powergui	Simscape → Electrical → Specialized Power Systems → Fundamental Blocks
AC Voltage Source	Simscape → Electrical → Specialized Power Systems → Fundamental Blocks → Electrical Sources
Thyristor	Simscape → Electrical → Specialized Power Systems → Fundamental Blocks → Power Electronics
Ideal Switch	Simscape → Electrical → Specialized Power Systems → Fundamental Blocks → Power Electronics
Pulse Generator	Simulink → Sources
Logical Operator	Simulink → Commonly Used Blocks
Series RLC Branch	Simscape → Electrical → Specialized Power Systems → Fundamental Blocks → Elements
Voltage Measurement	Simscape → Electrical → Specialized Power Systems → Control & Measurements → Measurements
Scope	Simulink → Commonly Used Blocks

provides the negative half-cycles of the input waveform. In this circuit, three pulse generators are used: "Pulse Generator 1" provides a pulse to the P-converter, "Pulse Generator 2" provides a pulse to the N-converter, and "Pulse Generator 3" and its opposite pulse through the logical operator "NOT" are provided in the "Ideal Switch 1" to "Ideal Switch 4" as shown in the figure below.

A resistor of 1 Ω is used as the load for this circuit. The powergui block has been changed to "Discrete" with the default sample time. The parameters for the pulse generators are given as follows:

For "Pulse Generator 1" and "Pulse Generator 2," the amplitude is 1, the period is
   0.00835 s, the pulse width is 50%, and the phase delay is 0.
For "Pulse Generator 3," the amplitude is 1, the period is 0.0334, the pulse width is
   50%, and the phase delay is 0.

The period in "Pulse Generator 3" is a multiple of 0.0167 since the source voltage is 60 Hz. If the period is twice that of 0.0167 (0.0334, which is used in this example), two positive half-cycles and two negative half-cycles will be generated as the output waveform (as shown in Fig. 17.87). If the period is increased, the output waveform would also have an increased number of positive and negative cycles.

17.7   Conclusion

Power electronics is one of the crucial aspects in the domain of electrical engineering, as it bridges the gap between low-power and high-power applications by working as an interface. In this chapter, basic components of power electronics such as diodes, bipolar junction transistors, MOSFETs, IGBTs, GTOs, and Op-amps have been introduced, and their characteristics have been demonstrated using Simulink. Intermediate concepts of power frame transformation and phase-locked loop have been briefly explained with examples so that the readers may work with the advanced concepts of flexible AC transmission systems. Modeling of DC-DC converter and switching regulators, inverters, AC-DC converters, and AC-AC cycloconverters with guided simulations will help the readers distinguish their performances and choose the appropriate converter well-suited to the applications. The chapter will work as a compendium of power electronics concepts and mathematical examples to work on advanced topics in the domain.

Exercise 17

1. (a) Name some of the applications of diode.
   (b) State the major differences between single-phase half- and single-phase full-wave rectifiers.
   (c) Design a single-phase full-wave rectifier with (i) two diodes and (ii) four diodes which will take a 220 V input and provide an output of 24 V.
   (d) Replicate the three-phase full-wave rectifier with (i) RL load and (ii) RC load. Change the values of the inductor and capacitor from low to high. What are the impacts of these variations on the output voltage and current?
2. (a) Define transistor. What is the difference between a diode and a transistor?

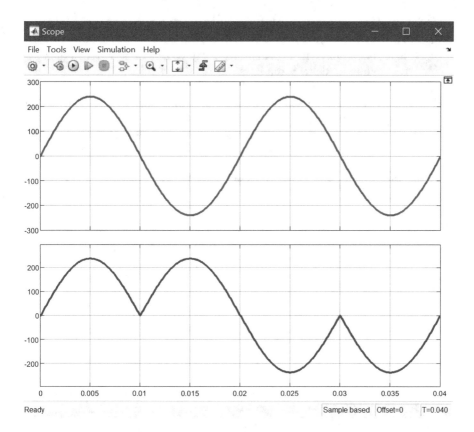

Fig. 17.87 Response of a Single-Phase Cycloconverter observed from the scope window

    (b) Demonstrate the BJT characteristics for (i) NPN transistor for a base voltage of 0.004 V and (ii) PNP transistor for a base voltage of −0.004 V.
3. (a) Write down the formula for determining the output voltage of Op-amp differentiator and integrator circuit from the circuital elements.
    (b) Design an Op-amp-based inverter circuit with appropriate resistors to obtain a voltage of 50 V from an input of 5 V.
    (c) Design an Op-amp-based non-inverter circuit with appropriate resistors to obtain a voltage of 70 V from an input of 7 V.
4. (a) What is the difference between uncontrolled and controlled rectification?
    (b) Perform controlled rectification with (i) MOSFET with a firing angle of 45° and 90°, and (ii) GTO with firing angle at 45° at pulse width 10%, and with a firing angle of 90° at pulse width 25%. Compare the waveforms for each case and explain the device characteristics.
5. (a) Replicate the example of reference frame transformation shown in Sect. 17.5.1. Perform alpha-beta-zero to dq0 transformation using Simulink block and verify the output from the result provided in the example.

(b) Design a buck converter to obtain a regulated DC voltage of 5 V from a source of 24 V.

(c) Design a boost converter to obtain a regulated voltage of 24 V from a source of 5 V. Use arbitrary circuital parameters.

6. (a) What are the major applications of AC-DC converters?

(b) Replicate the three-phase inverter model shown in Sect. 17.6.2.3. Change the DC voltage to 24 V. Determine the phase-to-phase voltage of the system.

(c) Design a single-phase cycloconverter as created in Sect. 17.6.4, which produces the output voltage with the same waveform as follows.

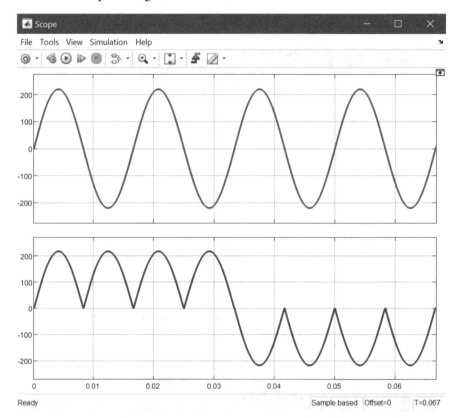

Chapter 18
Application of Simulink in Renewable Energy Technology

18.1 Solar Photovoltaics

Solar energy is one of the most used renewable energy in power system applications. The term "photovoltaics" originated from the combination of two terms—photon and voltage. Solar photovoltaics converts solar lights, i.e., photons into electricity, or in other terms—voltage. This phenomenon is named as "photovoltaic effect" and is considered as the fundamental concept behind the recent progress of solar photovoltaics in the power domain. A single photovoltaic cell can generate a certain DC voltage and current with limited power wattage. By combining multiple PV cells, a PV array can be formed for usage in small- or medium-scale deployment. For contributing to a large-scale power system, PV panels are used that can be formed by combining multiple PV arrays.

18.1.1 Mathematical Model of PV Cell

To understand the mathematical model of a PV cell, first consider the following representation of a single solar cell circuit consisting of a current source, two photodiodes, and resistors:

In the circuit of Fig. 18.1, the current of the current source is represented by I_{ph}, and the two diodes D1 and D2 are connected in parallel with the current source. The parallel resistor is named as R_p, and series resistance is represented by R_s. The final output current from this solar cell is I, which can be represented by the following mathematical equation:

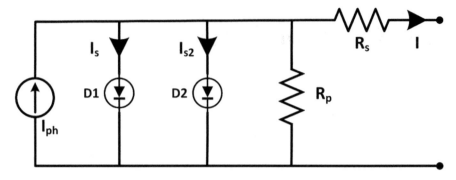

Fig. 18.1 Equivalent circuit of a PV cell

$$I = I_{\text{ph}} - I_s \cdot (I_d - 1) - I_{s2} \cdot (I_{d2} - 1) \frac{V + I \cdot R_s}{R_P}$$

Here, $I_d = e^{\frac{V+I \cdot R_s}{N \cdot V_t}}$; $I_{d2} = e^{\frac{V+I \cdot R_s}{N_2 \cdot V_t}}$ and, $V_t = \frac{kT}{q}$

The final output power of the cell can be defined by the following equation:

Power, $P = V \times I$

$$= V \times \left[I_{\text{ph}} - I_s \cdot \left(e^{\frac{V+I \cdot R_s}{N \cdot V_t}} - 1 \right) - I_{s2} \cdot \left(e^{\frac{V+I \cdot R_s}{N_2 \cdot V_t}} - 1 \right) - \frac{V + I \cdot R_s}{R_P} \right]$$

Here, I_s represents the current flowing through the first parallel diode, D1; I_{s2} indicates the current through diode D2; V_t represents the thermal voltage; k is the Boltzmann constant; T is the temperature; q represents the charge of an electron; N is the quality factor of D1; N_2 is the quality factor of D2; and V refers to the output voltage of the cell.

The mathematical model of the solar PV cell is modeled in the Simulink as illustrated in Fig. 18.2. The PV and IV characteristic curves of the PV cell are determined and plotted using the XY Graph blocks in the Simulink.

To understand the designed mathematical model, all the blocks that are utilized in the design are summarized with their navigation routes in the Table 18.1. The mathematical representation of each block based on their output is also enlisted in the block for better comprehensibility.

In the simulation, multiple variables are used, which are provided in the simulation using different blocks via parameter customization. The values of the parameters that are considered for the example given in Fig. 18.2 are listed in the Table 18.2:

At the end, the characteristic curves are plotted by utilizing two XY Graph blocks, which are customized from their parameter window for fitting the curves within the window with better visibility. The customization of these two blocks is shown in Fig. 18.3. The output of the simulation, i.e., the two characteristic curves, is illustrated in Fig. 18.4. In Fig. 18.4, the first figure represents the PV characteristic curve, while the second figure indicated the VI characteristic curve of the solar cell.

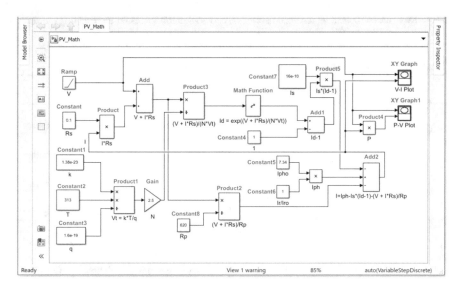

Fig. 18.2 Simulink mathematical model of the solar PV cell

18.1.2 PV Panel Design from Solar Cell

Before building a PV panel, it is important to learn about the solar cell. In the previous section, the mathematical model of a single solar cell is demonstrated. The same modeling can be done by utilizing the Solar Cell block, which can be customized from its parameter window. By implementing the Solar Cell block, the PV and VI characteristic curves are shown in the demonstration (Fig. 18.5). The blocks that are utilized to design the model are summarized in the following table with their navigation routes (Table 18.3).

The parameter window of the Solar Cell block is shown in Fig. 18.6, from where the parameters of the solar cell can be customized. The defining mathematical equation of the Solar Cell block can be seen from the description of the parameter window. In the parameter window, there are three settings—Cell Characteristics, Configuration, and Temperature Dependence. In each of these settings, multiple parameters are available that can be customized. Under the Cell Characteristics options, parameters such as short-circuit current, open-circuit voltage, irradiance, quality factor, and series resistance can be configured manually to define the characteristics of a particular solar cell. In the Configuration setting, the number of cells in series can be defined. The last option of settings of the Solar Cell block called Temperature Dependence has several customizable parameters, which is shown in Fig. 18.7.

A PS Constant block is connected to the input port of the Solar Cell block. This PS Constant block signifies the irradiance value of the solar cell. From its parameter window, the constant value is set as 1000 (Fig. 18.8), which refers that the irradiance of the solar cell is 1000 W/m^2.

Table 18.1 Mathematical representation, blocks, and navigation routes for representing the mathematical model for the PV cell

Mathematical representation	Block name	Navigation route
V	Ramp	Simulink Library Browser → Simulink → Sources → Ramp
R_s	Constant	Simulink Library Browser → Simulink → Sources → Constant
k	Constant1	Simulink Library Browser → Simulink → Sources → Constant
T	Constant2	Simulink Library Browser → Simulink → Sources → Constant
q	Constant3	Simulink Library Browser → Simulink → Sources → Constant
1	Constant4	Simulink Library Browser → Simulink → Sources → Constant
I_{pho}	Constant5	Simulink Library Browser → Simulink → Sources → Constant
I_r / I_{ro}	Constant6	Simulink Library Browser → Simulink → Sources → Constant
I_s	Constant7	Simulink Library Browser → Simulink → Sources → Constant
R_p	Constant8	Simulink Library Browser → Simulink → Sources → Constant
$I * R_s$	Product	Simulink Library Browser → Simulink → Math Operations → Product
$V_t = \frac{kT}{q}$	Product1	Simulink Library Browser → Simulink → Math Operations → Product
$\frac{V + I * R_s}{R_p}$	Product2	Simulink Library Browser → Simulink → Math Operations → Product
$\frac{V + I * R_s}{N * V_t}$	Product3	Simulink Library Browser → Simulink → Math Operations → Product
P	Product4	Simulink Library Browser → Simulink → Math Operations → Product
$I_s * (I_d - 1)$	Product5	Simulink Library Browser → Simulink → Math Operations → Product
$I_d = e^{\left(\frac{V + I * R_s}{N * V_t}\right)}$	Math Function	Simulink Library Browser → Simulink → Math Operations → Math Function
$V + I * R_s$	Add	Simulink Library Browser → Simulink → Math Operations → Add
$I_d - 1$	Add1	Simulink Library Browser → Simulink → Math Operations → Add
$I = I_{ph} - I_s * (I_d - 1) - \left(\frac{V + I \cdot R_s}{R_p}\right)$	Add2	Simulink Library Browser → Simulink → Math Operations → Add
N	Gain	Simulink Library Browser → Simulink → Math Operations → Gain
V-I Plot	XY Graph	Simulink Library Browser → Simulink → Sinks → XY Graph
P-V Plot	XY Graph1	Simulink Library Browser → Simulink → Sinks → XY Graph

Table 18.2 Parameter values considered for the example

Parameters	Considered value
R_s	0.1 ohms
k	1.38e-23 J/K
T	313 K
q	1.6e-19 C
I_{pho}	7.34 A
I_r/I_{ro}	1
I_s	16e-10 A
R_p	620 ohms
N	2.5

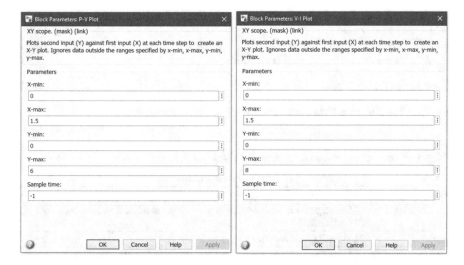

Fig. 18.3 Block parameters of the P-V plot and the V-I plot

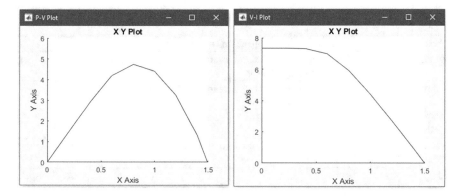

Fig. 18.4 The P-V plot and the V-I plot

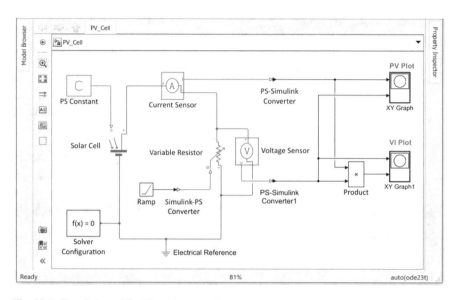

Fig. 18.5 Simulink model of the solar PV cell

Table 18.3 Blocks and navigation path for PV panel design from solar cell

Blocks	Navigation path on the Simulink Library Browser
Solar Cell	Simscape → Electrical → Sources → Solar Cell
PS Constant	Simscape → Foundation Library → Physical Signals → Sources → PS Constant
Ramp	Simulink → Sources → Ramp
Variable Resistor	Simscape → Foundation Library → Electrical → Electrical Elements → Variable Resistor
PS-Simulink Converter	Simscape → Utilities → PS-Simulink Converter
Simulink-PS Converter	Simscape → Utilitie → Simulink-PS Converter
Product	Simulink → Math Operations → Product
Voltage Sensor	Simscape → Foundation Library → Electrical → Electrical Sensors → Voltage Sensor
Current Sensor	Simscape → Foundation Library → Electrical → Electrical Sensors → Current Sensor
XY Graph, XY Graph1	Simulink → Sinks → XY Graph
Solver Configuration	Simscape → Electrical → Specialized Power Systems → Fundamental Blocks → powergui

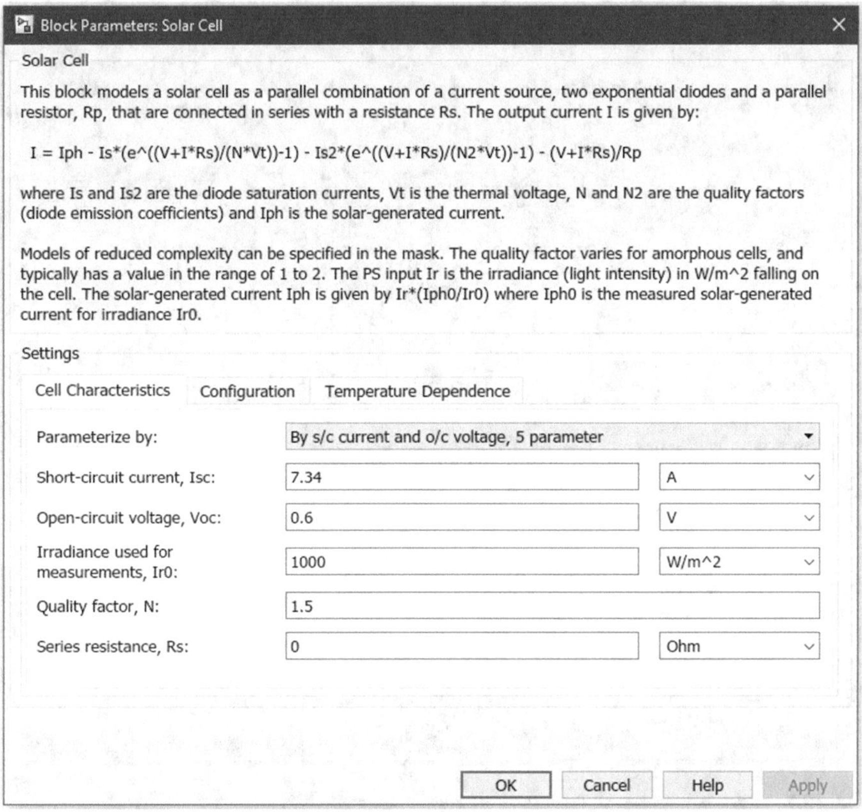

Fig. 18.6 Block parameters of the solar cell: cell characteristics

To complete the circuit, a Variable Resistor block is connected in series with the Solar Cell block. The Variable Resistor block responds with the ramp signal provided in the input port via the Ramp block. Both of these blocks are kept in their default parameter mode for this example. A Current Sensor is connected in series and a Voltage Sensor is connected in parallel to the variable resistor to measure the current and voltage of the circuit. It is to be noted that Simulink-PS Converter is used between the Ramp block and the Variable Resistor block, as these two blocks belong to separate libraries. Both Current and Voltage Sensors are physical systems; therefore, to display their results in a XY Graph block, PS-Simulink Converters are utilized. The Product block is used to multiply the voltage and current to determine the overall power of the solar cell. In the XY Graph and XY Graph1 blocks, VI and PV characteristic plots can be seen after simulating the model. The scaling of these XY Graph and XY Graph1 block is altered from their respective parameter windows as shown in Fig. 18.9.

The output curves are illustrated in Fig. 18.10, where the first curve represents the PV characteristic curve and the second curve refers to the VI characteristic curve.

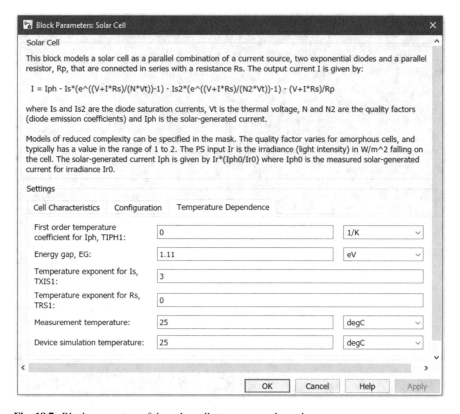

Fig. 18.7 Block parameters of the solar cell: temperature dependence

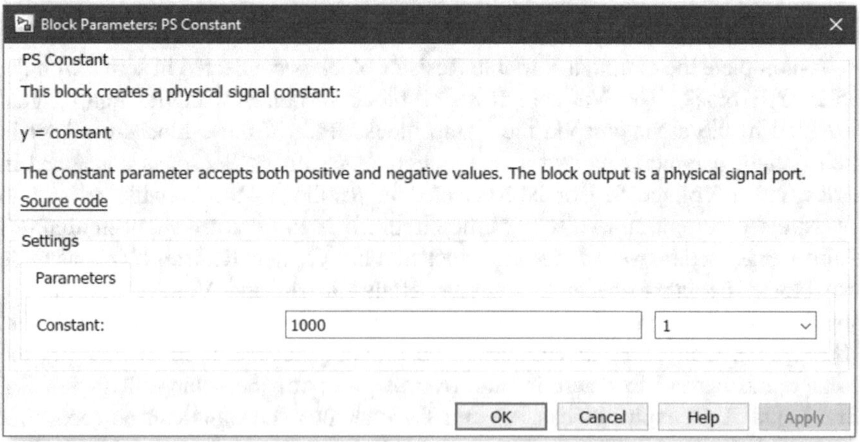

Fig. 18.8 Block parameters of the PS constant

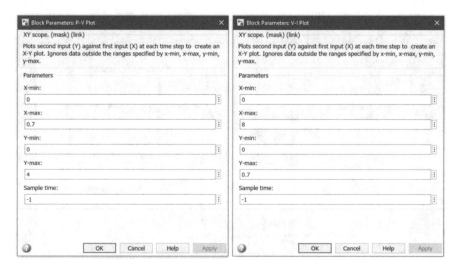

Fig. 18.9 Block parameters of the P-V plot and the V-I plot

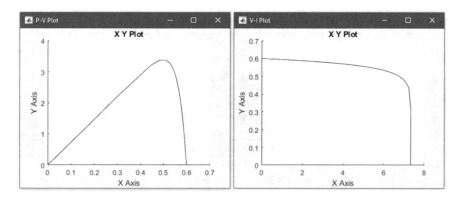

Fig. 18.10 The P-V plot and the V-I plot

18.1.3   PV Panel Design with PV Array

Instead of creating PV panels from the solar cells, it is more convenient to utilize PV arrays, where different solar cells are connected in series and parallel configurations to create a PV array. By utilizing multiple PV arrays, a PV panel can be designed. A Simulink model is constructed in Fig. 18.11 to show the performance of a PV array via its PV and VI characteristic curves. The navigation routes of the utilized blocks in this example are listed in the Table 18.4:

The parameter window of the PV Array block is shown in Fig. 18.12, where multiple parameters can be adjusted for specific PV Array module. In this block, several module options are available under the Module dropdown box option. For

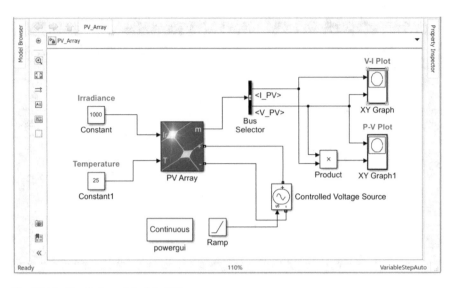

Fig. 18.11 Simulink model of the PV array

Table 18.4 Blocks and navigation path for PV panel design with PV array

Blocks	Navigation path on the Simulink Library Browser
PV Array	Simscape → Electrical → Specialized Power Systems → Renewables → Solar → PV Array
Constant, Constant1	Simulink → Sources → Constant
Controlled Voltage Source	Simscape → Electrical → Specialized Power Systems → Fundamental Blocks → Electrical Sources → Controlled Voltage Source
Ramp	Simulink → Sources → Ramp
Bus Selector	Simulink → Signal Routing → Bus Selector
Product	Simulink → Math Operations → Product
XY Graph, XY Graph1	Simulink → Sinks → XY Graph
powergui	Simscape → Electrical → Specialized Power Systems → Fundamental Blocks → powergui

this example, "SunPower SER-220P" module is selected. The number of parallel strings and the series-connected modules per string are both set as 1. If the user does not want to use any specific module from the dropdown list, and want to create a completely new user-defined module by customizing its parameters, it can be done by selecting "User-defined" option from the dropdown option list of the Module. It will provide the users an opportunity to customize user-defined modules by defining the parameters.

The PV Array block has two input parameters—Irradiance and Temperature. These two values are provided to the PV Array block by using two Constant blocks. The Constant block that is referring to the Irradiance value is customized by setting

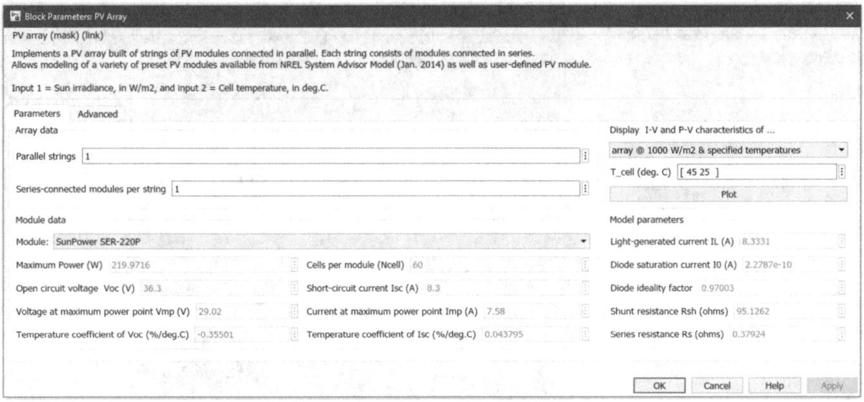

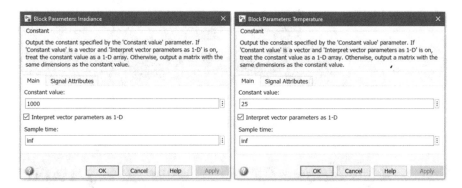

Fig. 18.12 Block parameters of the PV array

Fig. 18.13 Block parameters of the irradiance and the temperature

the constant value 1000. In the same manner, the temperature is provided by the Constant1 block by assigning the constant value 25. Both of these customizations are shown in the Fig. 18.13:

With the PV Array block, a Controlled Voltage Source is connected in series, which is controlled by a source block Ramp. The Source type of the Controlled Voltage Source is selected as AC, and all other values of different parameters are assigned to zero (Fig. 18.14).

The output port of the PV Array is connected to the Bus Selector block. After connecting the block, double-click on the Bus Selector block, which will create the appearance of the window shown in Fig. 18.15. In this parameter window, the right box named Signals in the bus corresponds to the available output signals of the PV Array. The left box named Selected Signals represents the signals that are selected to appear from the output ports of the Bus Selector block. To shift any signal from the right box to the left box, first select a specific signal from the right box, and click on the "Select" option, which will shift the selected signal to the left box. In this same

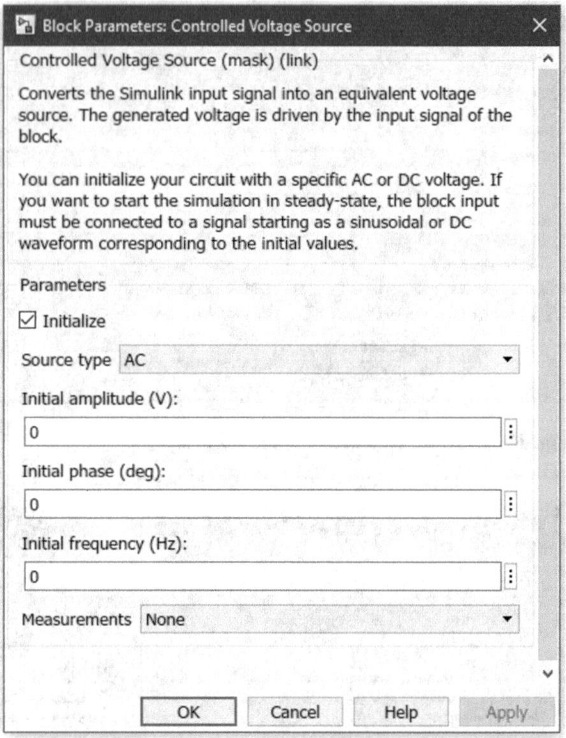

Fig. 18.14 Block
parameters of the controlled
voltage source

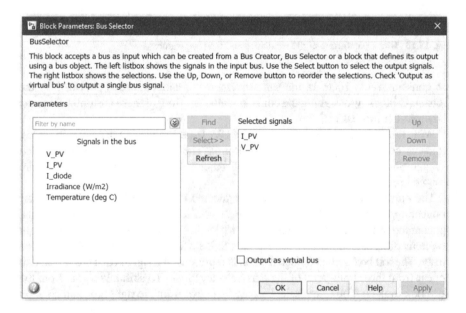

Fig. 18.15 Block parameters of the bus selector

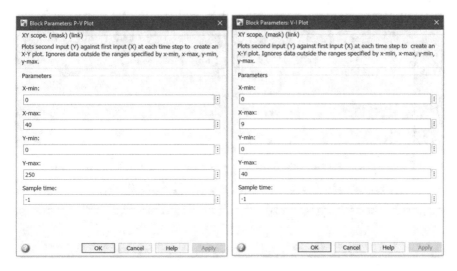

Fig. 18.16 Block parameters of the P-V plot and the V-I plot

manner, V_PV and I_PV signals are selected to shift to the left box. After clicking the Apply button, it can be observed that the Bus Selector has two output ports, as two signals are selected and shifted into the left box. One of the signals represents the voltage and the other represents the current of the PV Array.

A Product block is used to calculate the power from the voltage and current values of the PV Array circuit. Two XY Graph blocks are utilized to produce the characteristic curves of both PV and VI. The scaling of these two curves in the XY Graphs is customized from their respective parameter window as shown in Fig. 18.16. The output curves can be observed from Fig. 18.17, which are similar to the standard PV and VI characteristic curves of a solar system.

18.1.4   Case Study: Grid-Connected PV Array

An example is provided in Fig. 18.18, where a PV panel is utilized to generate electricity and connected with a utility grid system. The output voltage of a PV array is DC; hence, before connecting it to a grid system, it is mandatory to convert the DC voltage into three-phase AC voltage. For doing this, the first basic step is to boost the voltage level of the PV array using a DC-DC Boost Converter. In this example for simplicity, the Boost circuit is avoided. Instead, the number of PV array in series and parallel strings is increased to get comparatively higher voltage from the PV panel. A three-phase inverter is also necessary to convert the DC voltage into a three-phase AC voltage. Finally, the obtained three-phase AC voltage is connected with the grid to complete the design. Therefore, the entire design can be classified into three

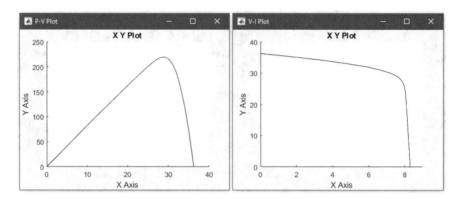

Fig. 18.17 The P-V plot and the V-I plot

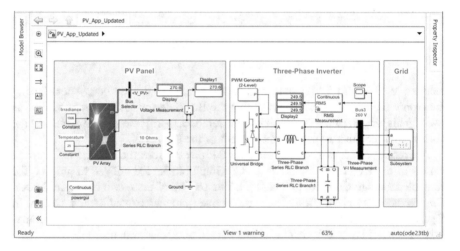

Fig. 18.18 Simulink model of a Grid-Connected PV Array

parts—PV panel, three-phase inverter, and grid design—which are separated by using three different Area boxes in the design shown in Fig. 18.18 for the sake of better comprehensibility.

Before explaining the customization of each block, a list of all the blocks with their navigation routes for this design is shown in the Table 18.5:

The PV panel design of this case study can be found in the first area block, which is named accordingly. The core component of this area block is the PV Array. The parameter window of this block is shown in Fig. 18.19. The "SunPower SER-220P" PV Array module is chosen for this simulation. The number of parallel strings is set as 5 and the number of series-connected modules per string is assigned to 30. The PV Array has two input ports—Irradiance and Temperature. By using two Constant blocks, these two inputs are provided numerically. The value of the irradiance is

Table 18.5 Blocks and navigation path for the designed grid-connected PV array

Blocks	Navigation path on the Simulink Library Browser
PV Array	Simscape → Electrical → Specialized Power Systems → Renewables → Solar → PV Array
Constant, Constant1	Simulink → Sources → Constant
Bus Selector	Simulink → Signal Routing → Bus Selector
Universal Bridge	Simscape → Electrical → Specialized Power Systems → Fundamental Blocks → Power Electronics → Universal Bridge
PWM Generator (Two Level)	Simscape → Electrical → Specialized Power Systems → Fundamental Blocks → Power Electronics → Pulse & Signal Generators
Three-Phase Series RLC Branch, Three-Phase Series RLC Branch1	Simscape → Electrical → Specialized Power Systems → Fundamental Blocks → Elements → Three-Phase Series RLC Branch
Three-Phase V-I Measurements, Three-Phase V-I Measurements1, Three-Phase V-I Measurements2	Simscape → Electrical → Specialized Power Systems → Fundamental Blocks → Measurements → Three-Phase V-I Measurements
Three-Phase PI Section Line, Three-Phase PI Section Line1	Simscape → Electrical → Specialized Power Systems → Fundamental Blocks → Elements → Three-Phase PI Section Line
Three-Phase Transformer (Two Windings), Three-Phase Transformer (Two Windings)1, Three-Phase Transformer (Two Windings)1	Simscape → Electrical → Specialized Power Systems → Fundamental Blocks → Elements → Three-Phase Transformer (Two Windings)
Three-Phase Series RLC Load, Three-Phase Series RLC Load1	Simscape → Electrical → Specialized Power Systems → Fundamental Blocks → Elements → Three-Phase Series RLC Load
Three-Phase Source	Simscape → Electrical → Specialized Power Systems → Fundamental Blocks → Electrical Sources → Three-Phase Source
RMS Measurement	Simscape → Electrical → Control → Measurement
Ground	Simulink → Sources → Ground
Scope, Scope1, Scope2	Simulink → Sinks → Scope
Display, Display1, Display2	Simulink → Sinks → Display
powergui	Simscape → Electrical → Specialized Power Systems → Fundamental Blocks → powergui

provided to be 1000 W/m^2, while the temperature value is set as 25°C. With the positive and negative output port of the PV array, a Series RLC Branch configured for 10 ohm resistance only is connected. With the mechanical output port of the PV Array, a Bus Selector is connected. The parameter window of the Bus Selector is given in Fig. 18.20, where the left box indicated the available output signals and the right-side box incorporates the selected signal. For this example, only the voltage is the point of interest; hence, the V_PV signal is selected to be shifted on the right side. From the Display block, the generated DC output voltage from the PV panel can be observed, which is 270.6 V.

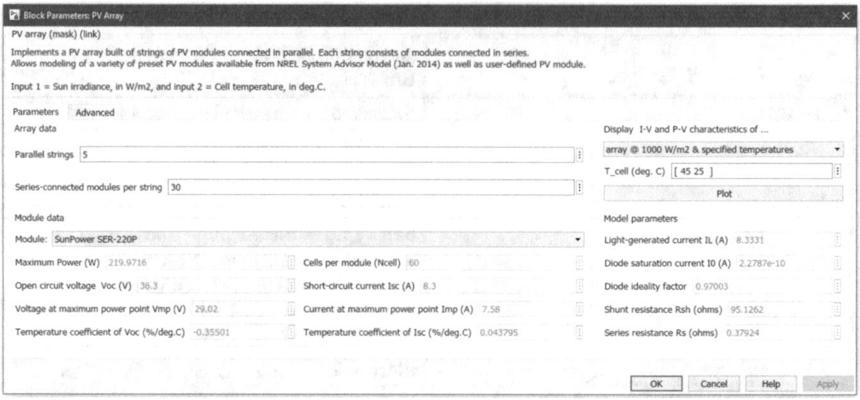

Fig. 18.19 Block parameters of the PV array

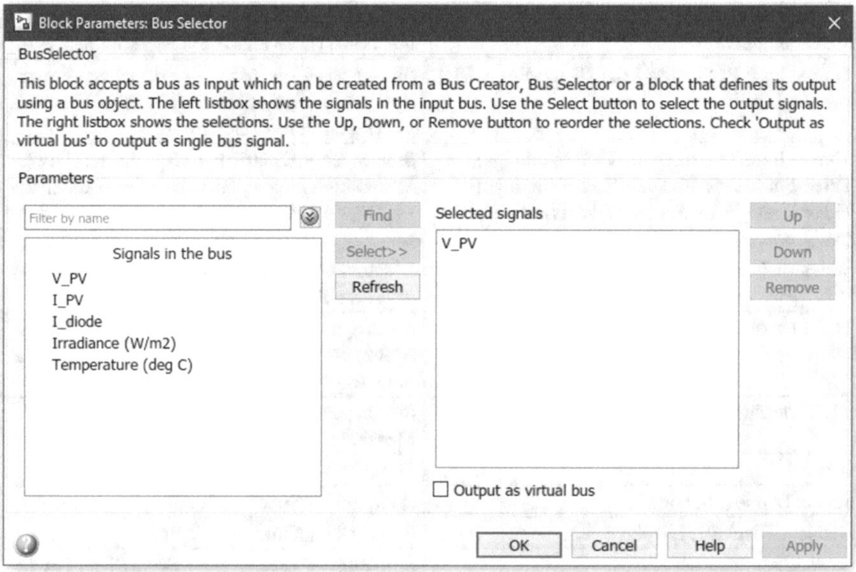

Fig. 18.20 Block parameters of the bus selector

The next step of the design is the three-phase inverter, which is marked by the second area block. To design a three-phase inverter, a Universal Bridge block is utilized. The parameter window of this block is given in Fig. 18.21. The first step of customizing this block is to assign the number of bridge arms to be 3 and to select the type of power electronic device as IGBT/Diodes. Later the other parameters such as the Snubber resistance and capacitance can be configured as shown in the figure below:

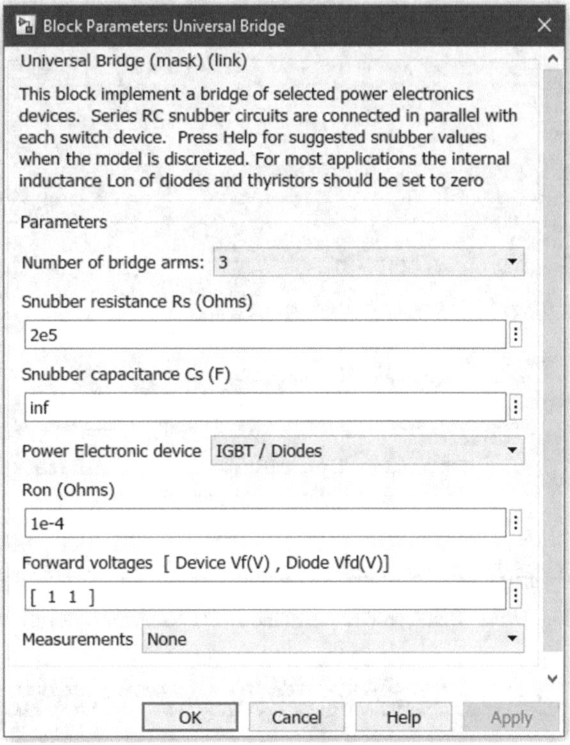

Fig. 18.21 Block parameters of the universal bridge

The two output DC voltage terminals of the PV panel are connected with the positive and negative terminals of the Universal Bridge. The output of this block is the three terminals that represent the three phases. The Universal block requires a gate signal, which is provided by a PWM Generator (two-Level) block. The customization of the PWM Generator (two-Level) block is shown in Fig. 18.22.

To filter the harmonics of the output of the inverter, an LC filter requires to be used. For doing so, two Three-Phase Series RLC Branch blocks are utilized. The first RLC Branch is configured for inductance only, with a value of 30 mH, while the second RLC Branch is configured for capacitance only with a value of 100 uF. The parameter window of these two blocks is shown below (Fig. 18.23):

After this LC filter, a Three-Phase V-I Measurement block is utilized to represent as a Bus3 of RMS voltage 240 V. The phase-to-neutral voltages of the three phases can be observed from the Scope window connected with this Bus. The parameter window of the Three-Phase V-I Measurement block is shown in Fig. 18.24. An RMS Measurement block is connected with the Three-Phase V-I Measurement block to display the RMS voltage value of the phase-to-ground voltages of three phases, which appear to be 249.5 V for each phase.

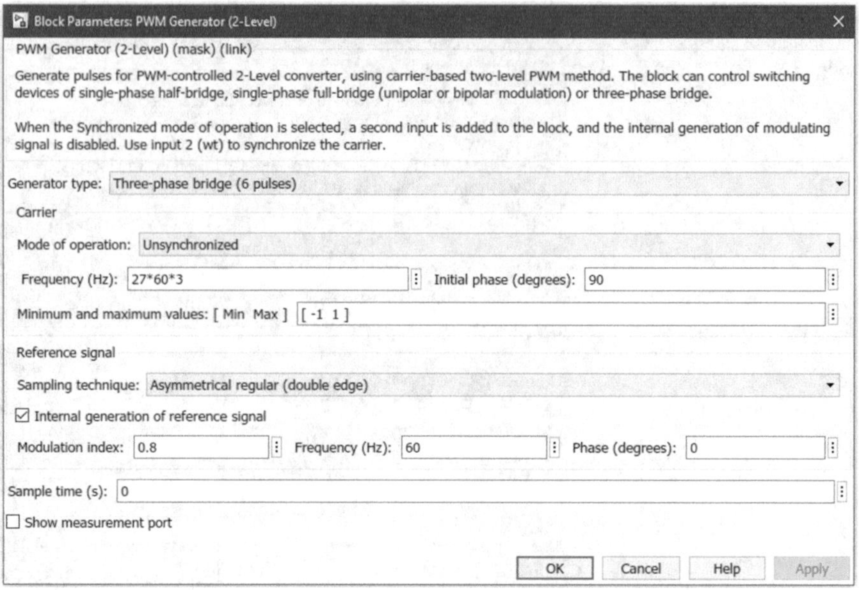

Fig. 18.22 Block parameters of the PWM generator (2-level)

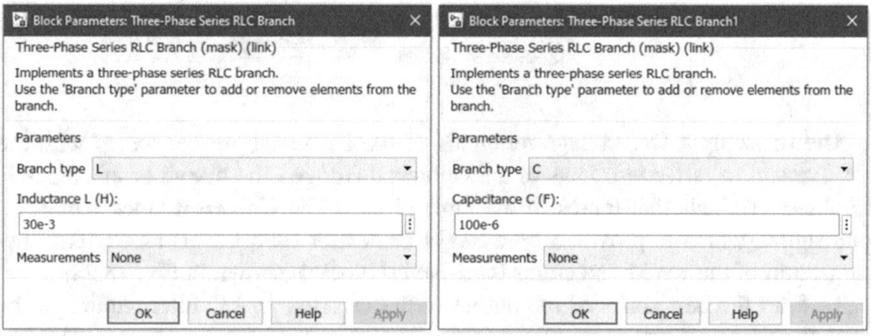

Fig. 18.23 Block parameters of the three-phase series RLC branch

The three-phase terminals of the AC voltage are connected with a subsystem named Grid. This subsystem incorporates the design of a Utility Grid. By double-clicking on the Subsystem, the original design of the Grid will appear in another subwindow. The overall design of this Grid is shown in the Fig. 18.25:

The Grid contains a Three-Phase Transformer (Two Windings) at the beginning, which is configured as a step-up transformer of 240 V/11 kV voltage. The nominal power of the transformer is set as 250 MVA. The first winding of the transformer is set to be connected in "Yg," while the other is assigned to be in the "Delta(D1)" connection. The parameter window of the Three-Phase Transformer (Two Windings) is shown in the Fig. 18.26:

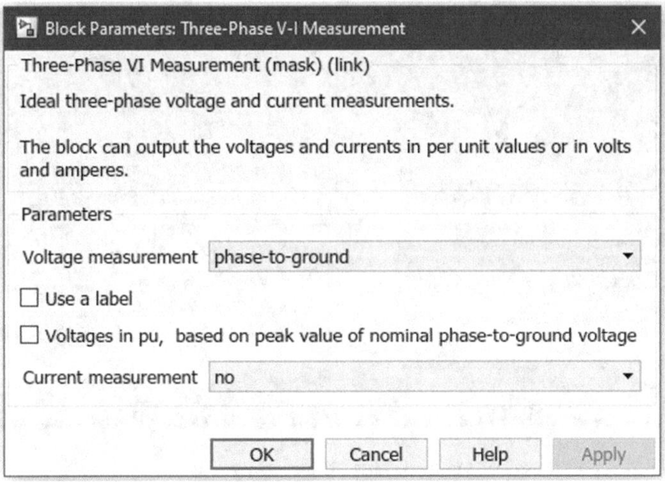

Fig. 18.24 Block parameters of the three-phase V-I measurement

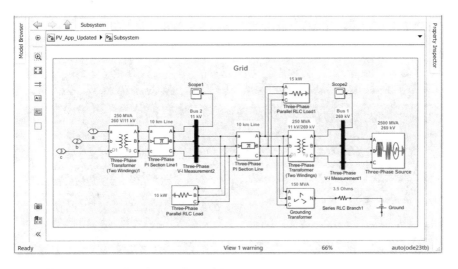

Fig. 18.25 Overall Simulink design of the grid

Two Three-Phase PI Section Line blocks customized to be of 10 km lines are connected with a Three-Phase V-I Measurement block in the middle. These two blocks are used to represent the transmission line of a grid system. The customized windows of these two blocks are illustrated in Fig. 18.27, and the parameter window of the Three-Phase V-I Measurement block is shown in Fig. 18.28.

A Scope is connected with the Three-Phase V-I Measurement block to observe the phase-to-ground three-phase voltages at this stage. The observed output voltages from the Scope1 window are shown in the Fig. 18.29:

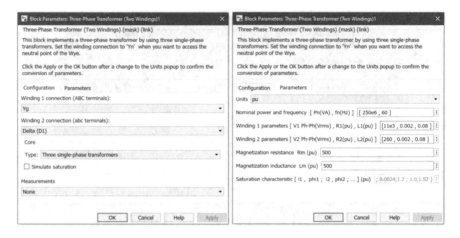

Fig. 18.26 Block parameters of the three-phase transformer

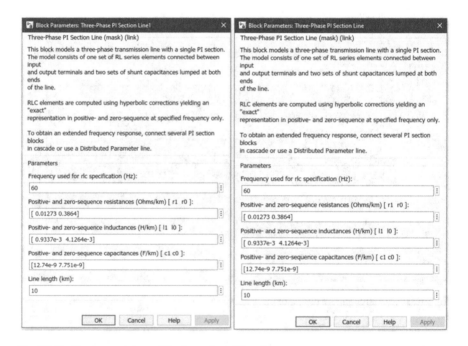

Fig. 18.27 Block parameters of the three-phase PI section

Before Bus2 of the design that is configured by the Three-Phase V-I Measurement2 Block, a load of 10 kW is connected. For representing the load, a Three-Phase Series RLC Load block is utilized. The customized parameter window of this block can be found in Fig. 18.30.

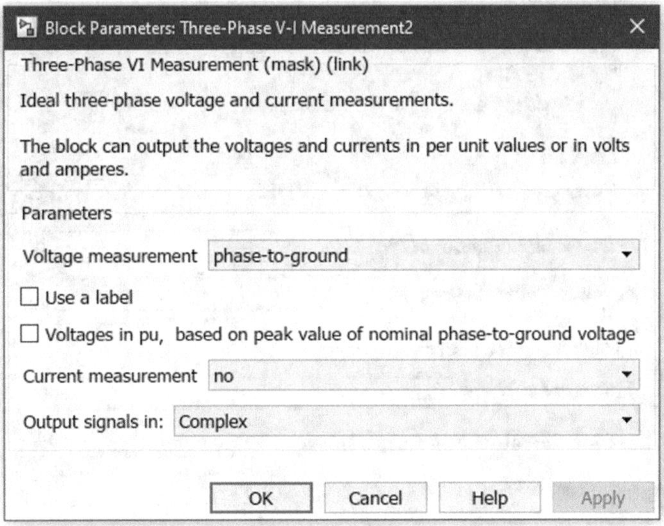

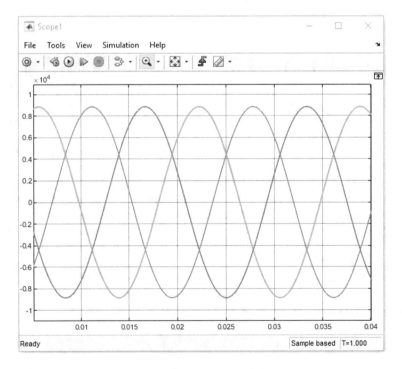

Fig. 18.28   Block parameters of the three-phase V-I measurement 2

Fig. 18.29   The output voltages observed from the scope window

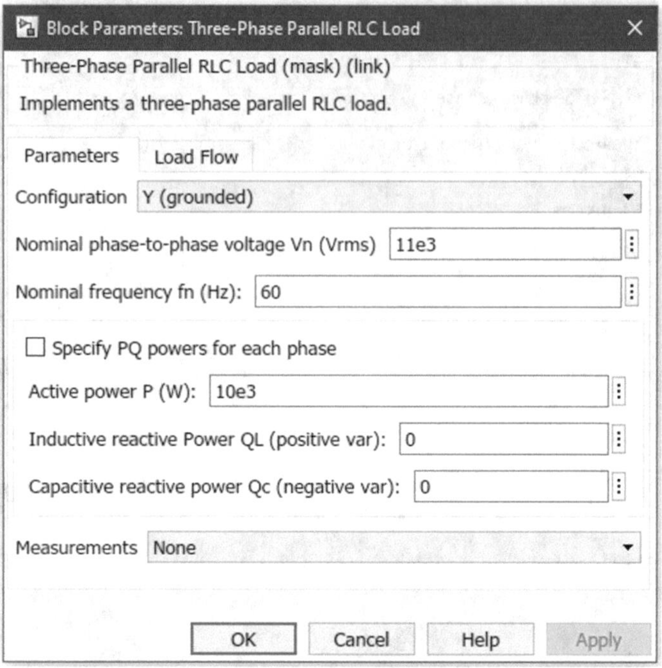

Fig. 18.30 Block parameters of the three-phase parallel RLC load

After the second bus of 11 kV, a Grounding Transformer block is connected with a resistance of 3.5 ohms. The parameter window of this Grounding Transformer block is shown in Fig. 18.31. A load of 25 MW is also connected here, which is represented by using the Three-Phase Series Load1 block. The parameter window of this block can be seen in Fig. 18.32.

A step-up transformer of 11 kV/269 kV is connected right after that to step up the voltages of the three phases. The Three-Phase Transformer (Two Windings)1 is customized on this account as follows (Fig. 18.33):

A Three-Phase V-I Measurement block is connected to observe the phase-to-ground voltage at this stage, which can be observed from the Scope2 window. It is also configured to represent Bus1. The customization of this block can be found in Fig. 18.34. The voltages of the three phases observed from the Scope2 window are shown in Fig. 18.35.

Finally, a Three-Phase AC Source of 2500 MVA and 269 kV RMS voltage is connected to complete the grid design. The customized parameter window of this block is presented in Fig. 18.36. With this, the entire design of a grid-connected PV panel model is completed to simulate.

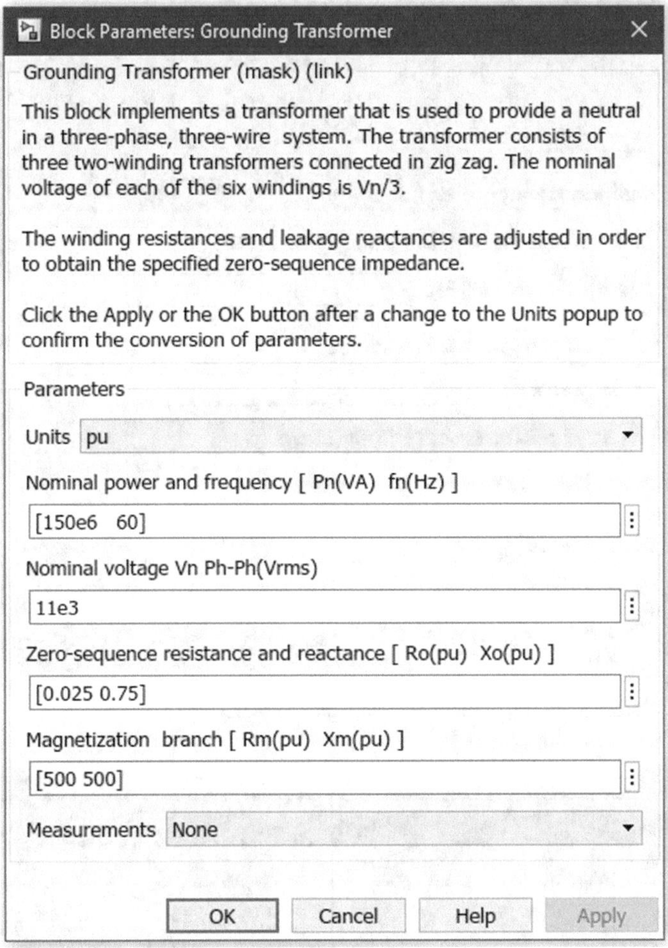

Fig. 18.31 Block parameters of the grounding transformer

18.2 Wind Turbine

Wind turbine is used to generate electricity by converting the kinetic energy developed from wind speed into electric energy. Similar to solar energy, the kinetic energy of the wind speed is significantly used as a renewable source for generating electricity. With the advent of modern technologies, the capacity of wind turbine-based generators has significantly increased. In Simulink, a wind turbine-based generator can be designed and modeled to explore its characteristics as well as to realize its implications in the power system domain. In this section, Wind Turbine Doubly-Fed Induction Generator (DFIG) will be utilized to explain the applications of the wind turbine in the power system domain.

Block Parameters: Three-Phase Parallel RLC Load1   ✕

Three-Phase Parallel RLC Load (mask) (link)

Implements a three-phase parallel RLC load.

Parameters	Load Flow

Configuration   Y (grounded)   ▼

Nominal phase-to-phase voltage Vn (Vrms)   11e3

Nominal frequency fn (Hz):   60

☐ Specify PQ powers for each phase

Active power P (W):   15e3

Inductive reactive Power QL (positive var):   0

Capacitive reactive power Qc (negative var):   0

Measurements   None   ▼

          OK   Cancel   Help   Apply

Fig. 18.32 Block parameters of the three-phase parallel RLC load 2

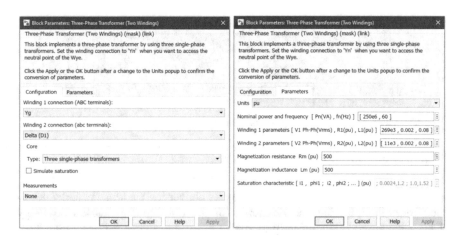

Fig. 18.33 Block parameters of the three-phase transformer (2 windings)

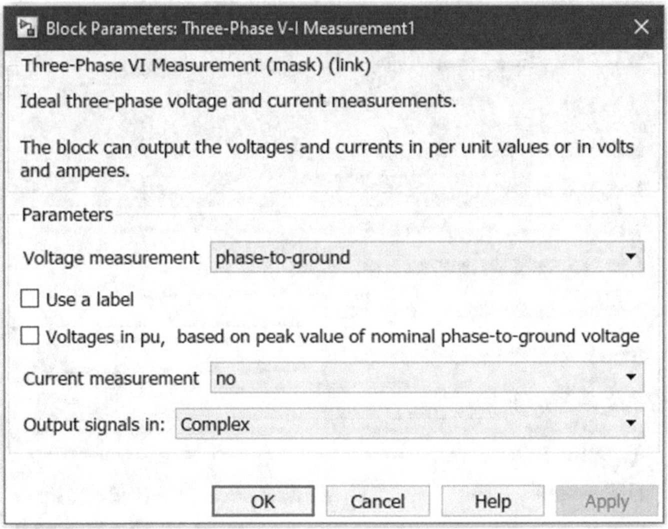

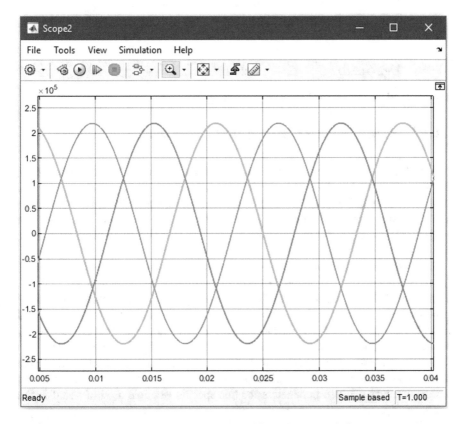

Fig. 18.34 Block parameters of the three-phase V-I measurement

Fig. 18.35 The voltages of the three phases observed from the scope window

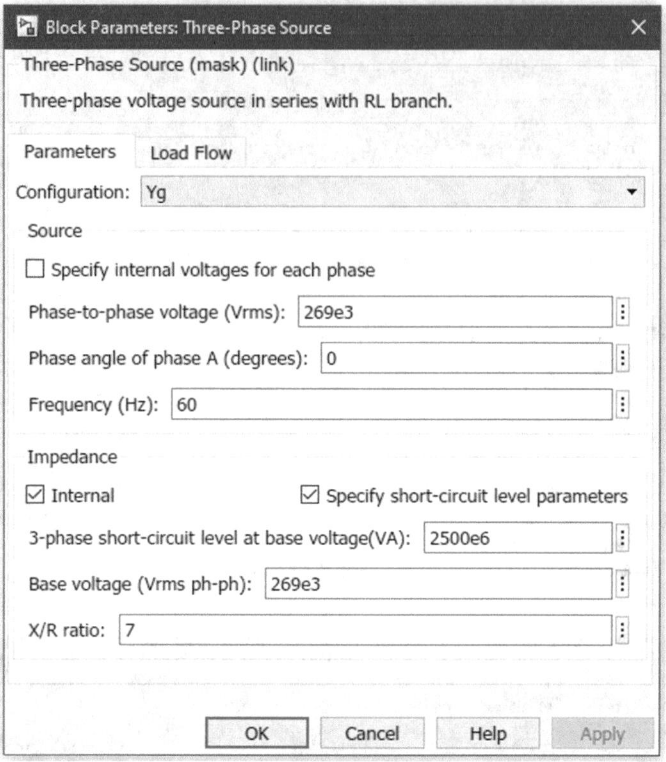

Fig. 18.36 Block parameters of the three-phase source

18.2.1   Model Wind Turbine-Based Generator in Simulink

In Simulink, Wind Turbine Doubly-Fed Induction Generator (Phasor Type) block is available that can be utilized to simulate a wind turbine-based model. A Simulink example is provided in Fig. 18.37, where a Three-Phase Source is connected with the Wind Turbine Doubly-Fed Generator with a given wind speed and trip logic. The utilized blocks on this Simulink design are given below with their respective navigation routes (Table 18.6):

The different customizable features of the parameter window of the Wind Turbine Doubly-Fed Induction Generator (Phasor Type) block are shown in Figs. 18.38, 18.39, 18.40, 18.41, and 18.42. In Fig. 18.38, the parameters that correspond to the generator data are shown, e.g., the nominal power, line-to-line voltage, stator and rotor impedances based on the ratings of the generator, etc. For this particular example, specifications of these parameters can be observed from this figure.

In Fig. 18.39, the parameters representing the Turbine data are shown, such as the mechanical output power of the wind turbine, tracking speeds, wind speed, etc. The assigned values of these parameters for this specific example are illustrated in the following figure:

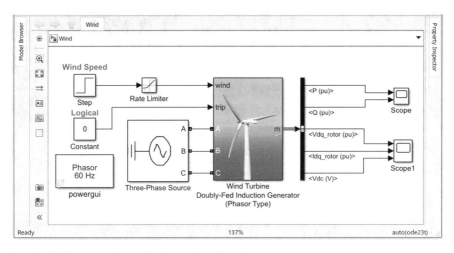

Fig. 18.37 Wind Turbine-based Generator in Simulink

Table 18.6 Blocks and navigation path for modeling wind turbine-based generator

Blocks	Navigation path on the Simulink Library Browser
Wind Turbine Doubly-Fed Induction Generator (Phasor Type)	Simscape → Electrical → Specialized Power Systems → Renewables → Wind Generation → Wind Turbine Doubly-Fed Induction Generator (Phasor Type)
Three-Phase Source	Simscape → Electrical → Specialized Power Systems → Fundamental Blocks → Electrical Sources → Three-Phase Source
Step	Simulink → Sources → Step
Constant	Simulink → Sources → Constant
Bus Selector	Simulink → Signal Routing → Bus Selector
Rate Limiter	Simulink → Discontinuities → Rate Limiter
Scope, Scope1	Simulink → Sinks → Scope
powergui	Simscape → Electrical → Specialized Power Systems → Fundamental Blocks → powergui

Based on the specification, the power characteristic curve of the wind turbine can be observed by clicking on the option named "Display wind turbine power characteristics" (Fig. 18.39), which will create the appearance of the following figure:

In Fig. 18.40, the output powers of the turbine are plotted against different turbine speeds. It can be observed from the graph that, up to point A, the turbine does not produce any output power. When the turbine speed exceeds 0.7 m/s, the output power of the turbine starts to increase gradually. After point D, the output power becomes almost stable.

The parameters that affect the converters can be customized from the Converters tab of the parameter window shown in Fig. 18.41. The specified values for such parameters for this example can be found from the following figure:

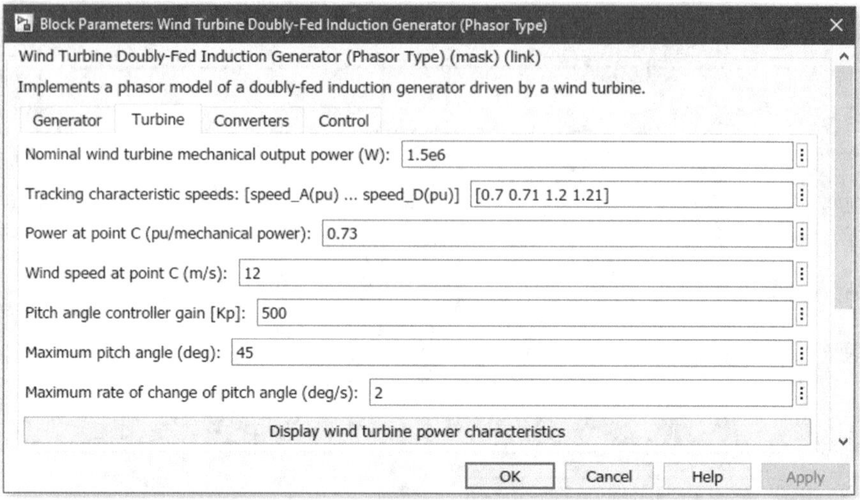

Block Parameters: Wind Turbine Doubly-Fed Induction Generator (Phasor Type)   ×

Wind Turbine Doubly-Fed Induction Generator (Phasor Type) (mask) (link)

Implements a phasor model of a doubly-fed induction generator driven by a wind turbine.

| Generator | Turbine | Converters | Control |

Nominal wind turbine mechanical output power (W): 1.5e6

Tracking characteristic speeds: [speed_A(pu) ... speed_D(pu)] [0.7 0.71 1.2 1.21]

Power at point C (pu/mechanical power): 0.73

Wind speed at point C (m/s): 12

Pitch angle controller gain [Kp]: 500

Maximum pitch angle (deg): 45

Maximum rate of change of pitch angle (deg/s): 2

Display wind turbine power characteristics

OK   Cancel   Help   Apply

Fig. 18.39 Block parameters of the Wind Turbine Doubly-Fed Induction Generator: Turbine

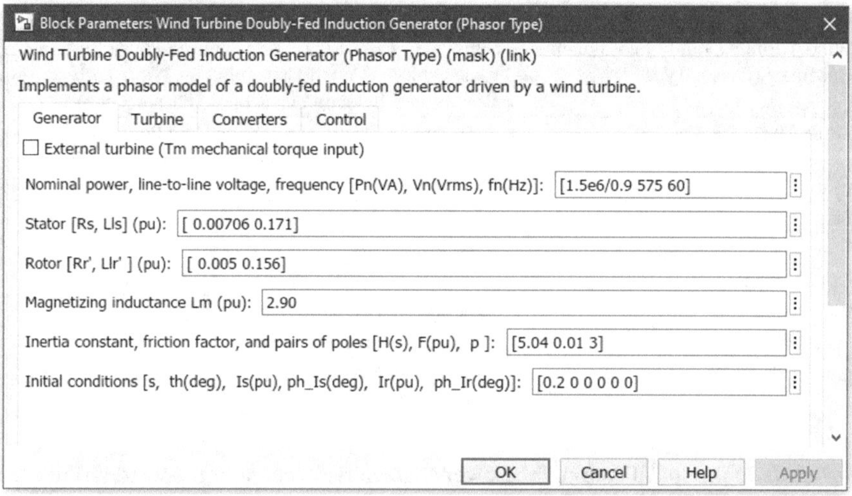

Block Parameters: Wind Turbine Doubly-Fed Induction Generator (Phasor Type)   ×

Wind Turbine Doubly-Fed Induction Generator (Phasor Type) (mask) (link)

Implements a phasor model of a doubly-fed induction generator driven by a wind turbine.

| Generator | Turbine | Converters | Control |

☐ External turbine (Tm mechanical torque input)

Nominal power, line-to-line voltage, frequency [Pn(VA), Vn(Vrms), fn(Hz)]: [1.5e6/0.9 575 60]

Stator [Rs, Lls] (pu): [0.00706 0.171]

Rotor [Rr', Llr'] (pu): [0.005 0.156]

Magnetizing inductance Lm (pu): 2.90

Inertia constant, friction factor, and pairs of poles [H(s), F(pu), p]: [5.04 0.01 3]

Initial conditions [s, th(deg), Is(pu), ph_Is(deg), Ir(pu), ph_Ir(deg)]: [0.2 0 0 0 0 0]

OK   Cancel   Help   Apply

Fig. 18.38 Block parameters of the Wind Turbine Doubly-Fed Induction Generator: Generator

The last tab of the parameter window of the Wind Turbine Doubly-Fed Induction Generator block is the Control tab (Fig. 18.42). From here, the mode of the control can be selected from two options—Voltage regulation or Var regulation. For this example, Voltage regulation mode is selected. Apart from this, several control parameters are available in this tab, which can be customized according to the user's preference. To know more about details regarding the different parameters, click on the Help button right below the parameter window.

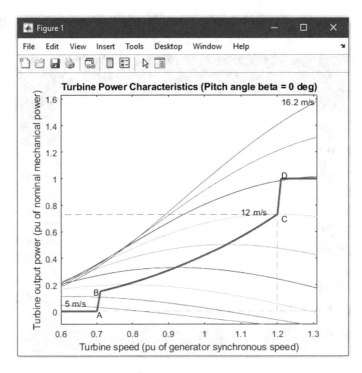

Fig. 18.40 Turbine power characteristics

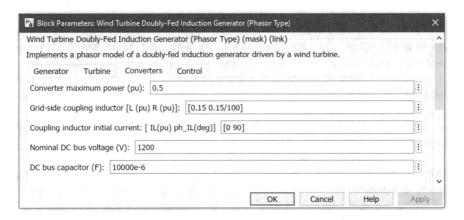

Fig. 18.41 Block parameters of the Wind Turbine Doubly-Fed Induction Generator: Converter

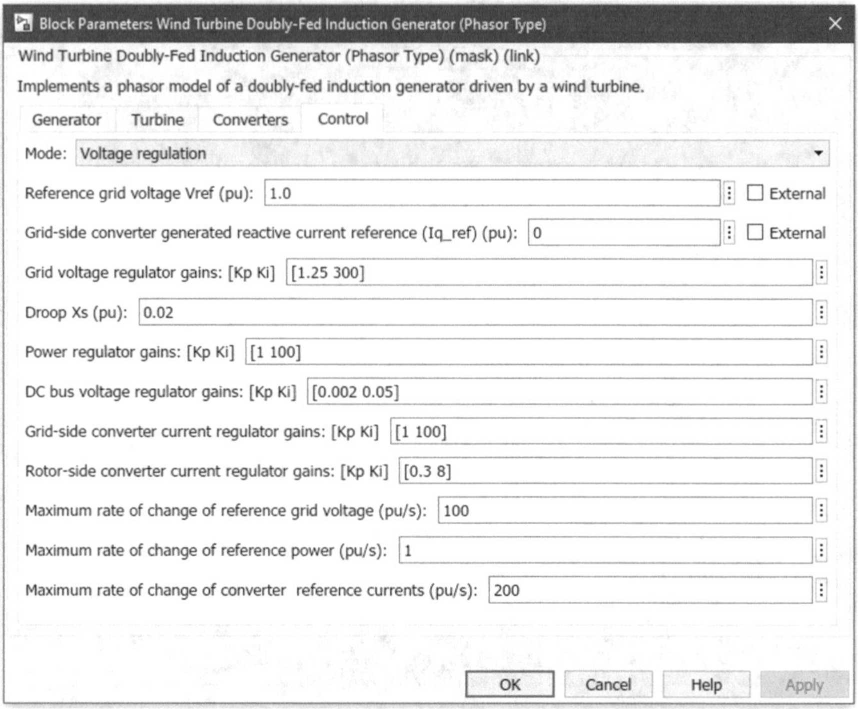

Fig. 18.42 Block parameters of the Wind Turbine Doubly-Fed Induction Generator: Control

The Wind Turbine Doubly-Fed Induction Generator (Phasor Type) has inputs named wind and trip. The wind takes wind speed as an input, while the trip takes any logical inputs. The wind speed is provided by utilizing two blocks—Step and Rate Limiter. The Step block creates a step response, whose initial and final values are defined as 6 and 16 with a stepping time of 4. As the wind speed is not constant or does not have a gradual increase or decrease properties, a Rate Limiter block is used to closely match with the nature of the wind speed. The customization of both of these blocks is shown in Fig. 18.43.

The Wind Turbine Doubly-Fed Induction Generator (Phasor Type) is connected with a Three-Phase Source, which is customized in the following manner from its parameter window (Fig. 18.44):

Finally, a Bus Selector is utilized to list all the output parameters of the Wind Turbine Doubly-Fed Generator and to select certain signals to be observed in the Scope window. After connecting the Bus Selector with the output port, double-click on the Bus Selector block, which will create the appearance of the parameter window as shown in Fig. 18.45. In the parameter window, the left box lists all the available output signals. To select any of these signals, click on a particular signal name and choose the "Select" option by clicking on it. The selected signal will be shifted to the right box. Repeat the same procedure to select multiple signals from the left box.

Fig. 18.43 Block parameters of the rate limiter and the step block

Fig. 18.44 Block
parameters of the three-
phase source

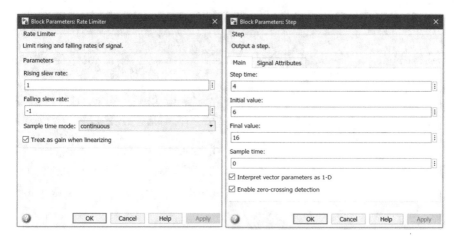

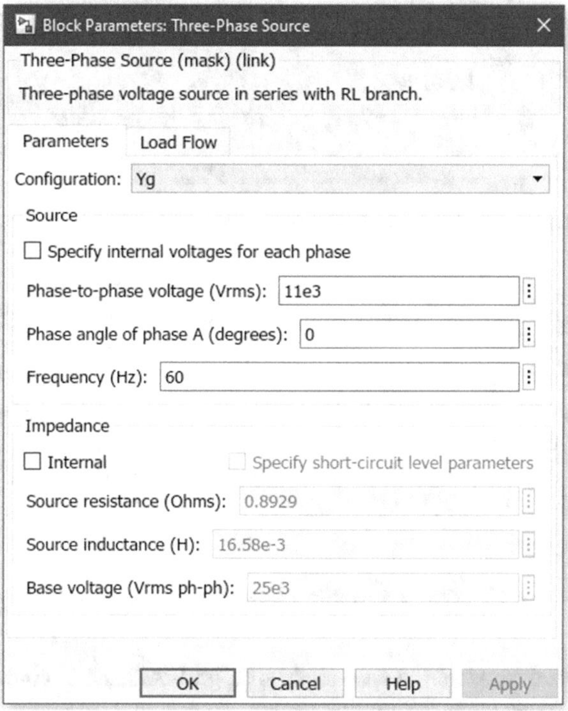

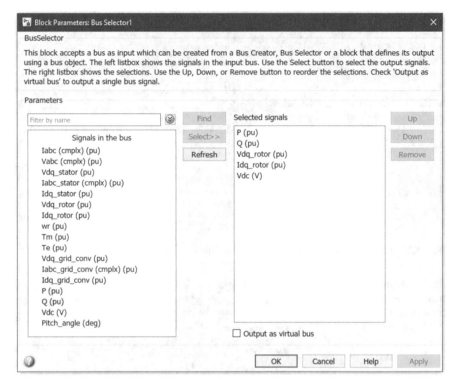

Fig. 18.45 Block parameters of the bus selector

Based on the number of signals selected to be on the right box, the number of output ports of the Bus Selector will be decided. For this example, five signals are made to be shifted to the right side, which automatically creates the appearance of five output ports of the Bus Selector block. Two separate Scope blocks are connected with the output signals to observe the graphs. The selected output signals can be observed from the Scope windows as shown in Figs. 18.46 and 18.47.

18.2.2   Case Study: Grid-Connected Wind Turbine Generator

In this section, a case study of a grid-connected wind turbine generator is demonstrated using a Simulink design. To design the complete model, the Simulink blocks that are utilized are summarized below with their navigation routes (Table 18.7):

The complete Simulink design of this case study is illustrated in Fig. 18.48, where a Wind Turbine Doubly-Fed Induction Generator is connected with a grid system. To design the grid, a Three-Phase Programmable Voltage Source is used at the beginning to represent a source of 230 kV voltage. The customization of this block

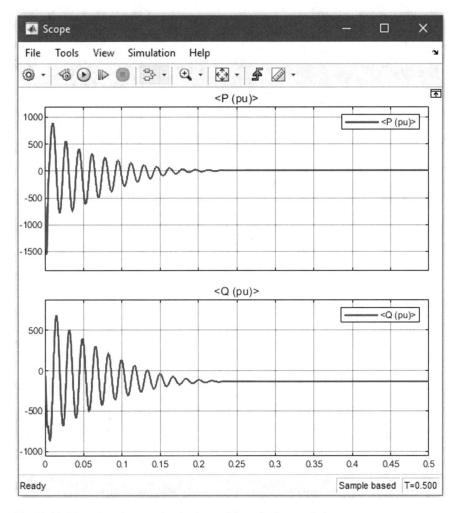

Fig. 18.46 The selected output signals observed from the Scope window

from its parameter window is shown in Fig. 18.49. In the parameter window, the amplitude of the phase-to-phase RMS voltage is assigned 230 kV with a frequency of 60 Hz.

With the Three-Phase Programmable Voltage Source, a Three-Phase Mutual Inductance Z1-Z0 block is connected in series, which is configured to be of 2500 MVA. The ratio of self and mutual reactance, $X0/X1$, is kept as a ratio of 2:1 for this instance. The parameter window of this block is demonstrated in the Fig. 18.50:

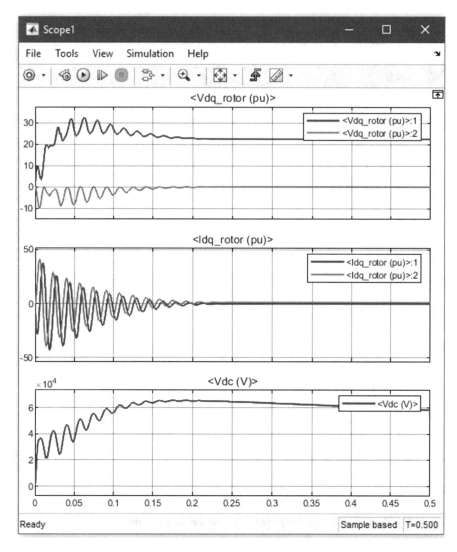

Fig. 18.47 The selected output signals observed from the Scope window

Afterward, a Three-Phase V-I Measurement block is utilized to act as Bus1. The format of the block is changed in this example, by right-clicking on the block and selecting Format → Background Color → Black. The customization of parameters of this block is shown below, where the measurements are checked as a label (Fig. 18.51).

The Bus1 that was defined previously by utilizing a Three-Phase Transformer is a 230 kV bus line. A step-up transformer is utilized to step up the voltage from 230 kV to 33 kV. The reason for doing so is to minimize the losses while transferring

Table 18.7 Blocks and navigation path for modeling grid-connected wind turbine generator

Blocks	Navigation path on the Simulink Library Browser
Wind Turbine Doubly-Fed Induction Generator (Phasor Type)	Simscape → Electrical → Specialized Power Systems → Renewables → Wind Generation → Wind Turbine Doubly-Fed Induction Generator (Phasor Type)
Three-Phase Programmable Voltage Source	Simscape → Electrical → Specialized Power Systems → Fundamental Blocks → Electrical Sources → Three-Phase Programmable Voltage Source
Three-Phase Mutual Inductance Z1-Z0	Simscape → Electrical → Specialized Power Systems → Fundamental Blocks → Elements → Three-Phase Mutual Inductance Z1-Z0
Three-Phase V-I Measurements, Three-Phase V-I Measurements1, Three-Phase V-I Measurements2	Simscape → Electrical → Specialized Power Systems → Fundamental Blocks → Measurements → Three-Phase V-I Measurements
Three-Phase PI Section Line, Three-Phase PI Section Line1	Simscape → Electrical → Specialized Power Systems → Fundamental Blocks → Elements → Three-Phase PI Section Line
Three-Phase Transformer (Two Windings), Three-Phase Transformer (Two Windings)1, Three-Phase Transformer (Two Windings)2	Simscape → Electrical → Specialized Power Systems → Fundamental Blocks → Elements → Three-Phase Transformer (Two Windings)
Three-Phase Series RLC Load	Simscape → Electrical → Specialized Power Systems → Fundamental Blocks → Elements → Three-Phase Series RLC Load
Series RLC Branch	Simscape → Electrical → Specialized Power Systems → Fundamental Blocks → Elements → Series RLC Branch
Step	Simulink → Sources → Step
Constant	Simulink → Sources → Constant
Bus Selector	Simulink → Signal Routing → Bus Selector
Rate Limiter	Simulink → Discontinuities → Rate Limiter
Scope, Scope1	Simulink → Sinks → Scope
powergui	Simscape → Electrical → Specialized Power Systems → Fundamental Blocks → powergui

electricity over a large distance by minimizing the current. As a step-up transformer, the Three-Phase Transformer (Two Windings) is utilized. For the first winding "Yg" connection and for the second winding "Delta(D1)" connection are configured for the transformer block. The 230 kV/33 kV configuration is set up from its Parameters option by updating the phase-to-phase RMS voltages of the two windings as shown in Fig. 18.52. The nominal power of the transformer is set as 250 MVA for this example.

Before introducing the transmission line, a Three-Phase Transformer (Two Windings) block is used to perform as a grounding transformer. Hence, the configuration of the first winding is set as "Yn," while for the second winding, the "Delta

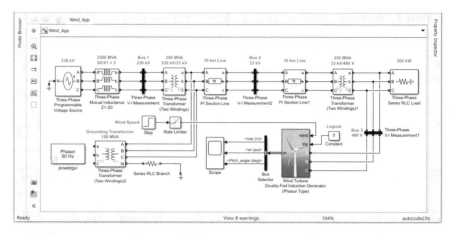

Fig. 18.48 Simulink diagram of a Grid Connected Wind Turbine Generator

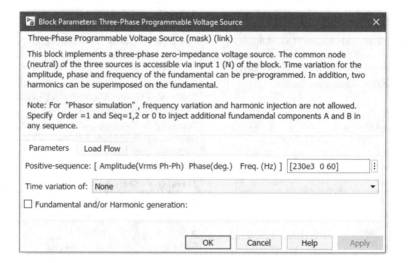

Fig. 18.49 Block parameters of the three-phase programmable voltage source

(D11)" connection is assigned. The nominal power of this transformer is set as 150 MVA. Both the magnetization resistance and inductance of this transformer are set as 500 per unit. The neutral port of the transformer is made grounded through a series-connected Series RLC Branch block, which is configured for 3.5 ohms resistance only. The customization of the 150 MVA grounding transformer is shown in the Fig. 18.53:

Two Three-Phase PI Section Line blocks are used to act as the transmission line linked with Bus2 in the middle. The Bus2 is modeled by using a formatted Three-Phase V-I Measurement block by following the same procedure as shown during the

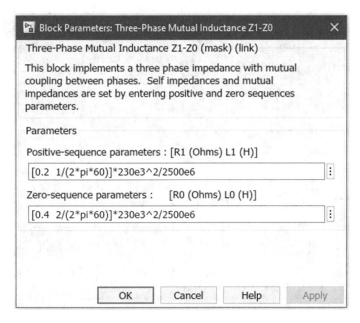

Fig. 18.50 Block parameters of the three-phase mutual inductance

formation of Bus1. The Three-Phase PI Section lines are configured for 10 km distance of lines. The parameter windows of both these line blocks are demonstrated in the Fig. 18.54:

The customization of the Three-Phase V-I Measurement2 block (Bus2) is shown in the left figure of Fig. 18.55. In this grid network, another 480 V bus line named Bus3 is modeled in the same manner. The configured parameter window of this Bus3 can be found on the right figure of Fig. 18.56, where the block is named Three-Phase V-I Measurement1.

Before the Bus3, the voltage of the 33 kV bus line is made to be stepped down by utilizing a step-down transformer to make the 33 kV/480 V conversion. A Three-Phase Transformer (Two Windings) is utilized on this account, and the customization of this block is shown in the following figure:

A 300 kW load is connected with the 480 V line, which is made by using the Three-Phase Series RLC Load. The active power of this block is set as 300 kW, and the nominal phase-to-phase RMS voltage is assigned as 480 V from its parameter window (Fig. 18.57).

With Bus3, finally, the Wind Turbine Doubly-Fed Induction Generator block is connected. As shown in Sect. 18.4.1, the Wind Turbine DFIG block requires wind speed and trip values as inputs. Therefore, a Step block is configured to act as the wind speed input of the Wind Turbine DFIG block with the association of the Rate Limiter block connected in series. For providing the trip input, a Constant block set as zero value is connected to perform as a logical zero input. The parameter windows of the Step and the Rate Limiter blocks are shown in Fig. 18.58.

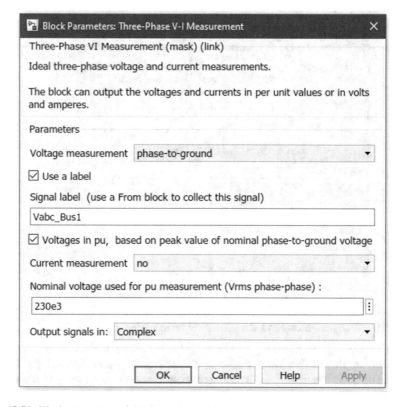

Fig. 18.51 Block parameters of the three-phase V-I measurement

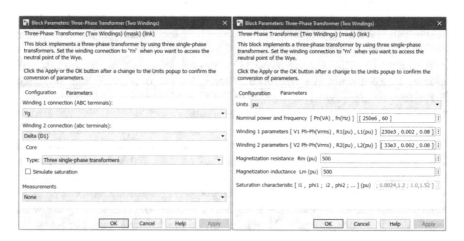

Fig. 18.52 Block parameters of the three-phase transformer

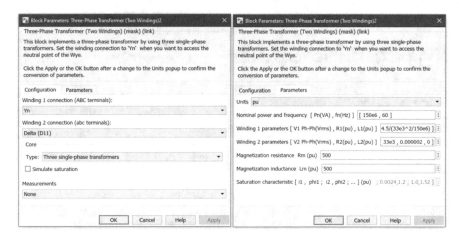

Fig. 18.53 Block parameters of the three-phase transformer 2

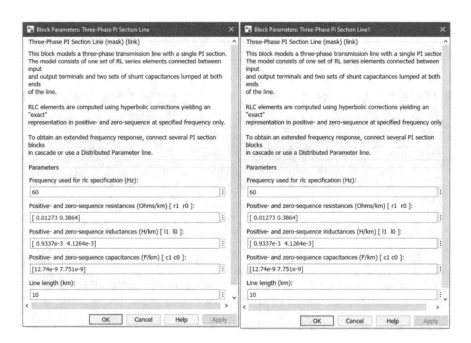

Fig. 18.54 Block parameters of the three-phase PI section line

With the output port of the Wind Turbine DFIG block, the Bus Selector block is connected to select the output signals that will be shown in the Scope window. The parameter window of the Bus Selector is shown in Fig. 18.59, where several output signals that can be observed and set as outputs can be seen from the left box. Out of these signals, three are chosen for this particular setup—Vdc (V), wr (pu),

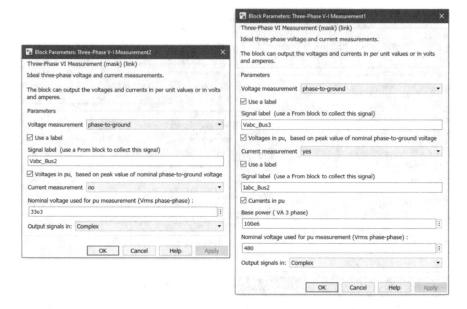

Fig. 18.55 Block parameters of the three-phase V-I measurement 2

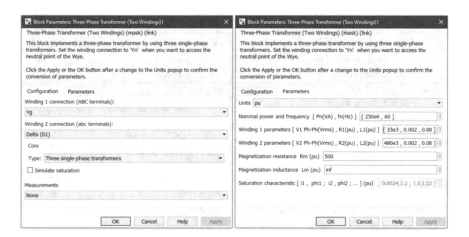

Fig. 18.56 Block parameters of the three-phase transformer

pitch_angle (deg). After selecting these three output signals from the Bus Selector, the number of output ports of this block is automatically configured as 3. A Scope block whose number of input ports is configured as 3 is connected with the three output ports of the Bus Selector.

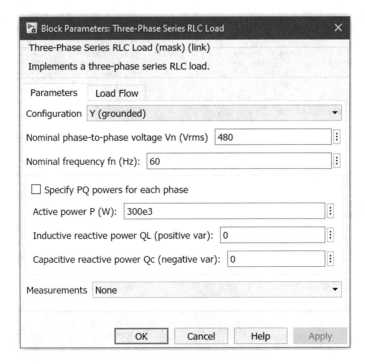

Fig. 18.57 Block parameters of the three-phase series RLC load

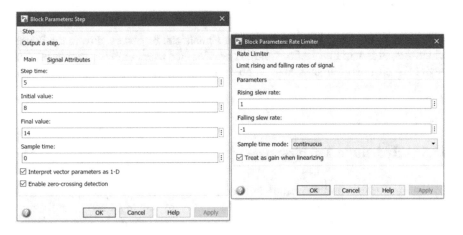

Fig. 18.58 Block parameters of the step and the rate limiter

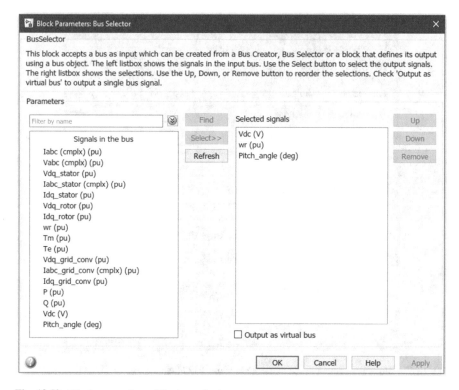

Fig. 18.59 Block parameters of the bus selector

After running the simulation, the output signals can be observed from the Scope window as shown in Fig. 18.60. Simulink provides the user flexibility to simulate the model for different output signals to observe and analyze in the Scope window as shown in this example.

18.3   Hydraulic Turbine

Hydraulic Turbine and Governor is a core component of a hydro-power-based electricity generator. In such a system, the energy of moving water is utilized to turn the blade of a hydro turbine. The spin of that rotating turbine, in other words, the mechanical energy, is converted into electrical energy. This is the basic principle of a hydraulic turbine-based power generator. In this section, a Hydraulic Turbine and Governor will be utilized to run a synchronous generator to produce electricity by utilizing the Simulink platform.

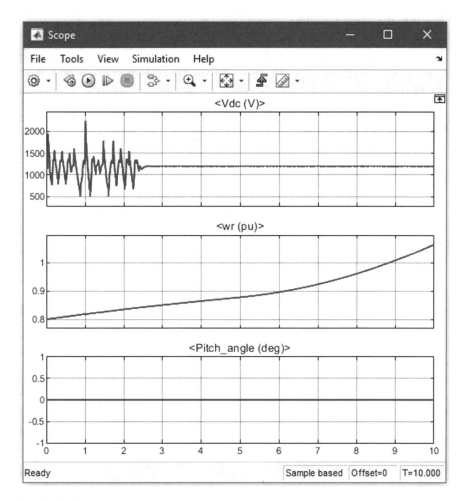

Fig. 18.60 The output signals observed from the scope window

18.3.1 Case Study: Hydro Turbine and Power Generator Model in Simulink

In this section, a Hydraulic Turbine and Governor block is utilized to power a synchronous generator with the assistance of an Excitation System. The overall design will be explained later with the necessary illustrations. The navigation routes of all the blocks that are used in this simulation are listed in the Table 18.8:

In Fig. 18.61, a Simulink design demonstrating the application of hydraulic turbines by powering up a synchronous generator is shown. In this design, the Hydraulic Turbine and Governor block is the core component and is customized from its parameter window. The parameter window of this block is shown in

Table 18.8 Blocks and navigation path for modeling hydro turbine and power generator

Blocks	Navigation path on the Simulink Library Browser
Hydraulic Turbine and Governor	Simscape → Electrical → Specialized Power Systems → Fundamental Blocks → Machines → Hydraulic Turbine and Governor
Excitation System	Simscape → Electrical → Specialized Power Systems → Fundamental Blocks → Machines → Excitation System
Synchronous Machine pu Standard	Simscape → Electrical → Specialized Power Systems → Fundamental Blocks → Machines → Synchronous Machine pu Standard
Three-Phase Transformer (Two Windings)	Simscape → Electrical → Specialized Power Systems → Fundamental Blocks → Elements → Three-Phase Transformer (Two Windings)
Three-Phase Source	Simscape → Electrical → Specialized Power Systems → Fundamental Blocks → Electrical Sources → Three-Phase Source
Three-Phase Series RLC Load, Three-Phase Series RLC Load1	Simscape → Electrical → Specialized Power Systems → Fundamental Blocks → Elements → Three-Phase Series RLC Load
Constant, Constant1, Constant2	Simulink → Sources → Constant
Bus Selector	Simulink → Signal Routing → Bus Selector
Ground	Simulink → Sources → Ground
Scope	Simulink → Sinks → Scope
powergui	Simscape → Electrical → Specialized Power Systems → Fundamental Blocks → powergui

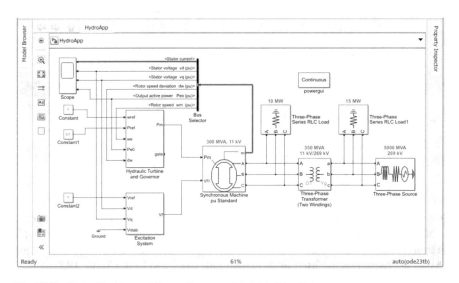

Fig. 18.61 Hydro Turbine and Power Generator Model in Simulink

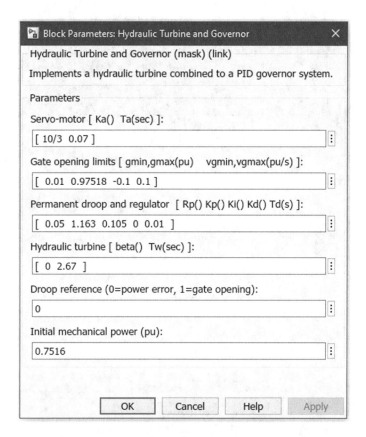

Fig. 18.62 Block parameters of the hydraulic turbine and governor

Fig. 18.62. There are several customizable parameters for this block, such as gain and time constant of the servo-motor, gate opening limits, etc. To learn about these parameters, click on the Help button of its parameter window. The Hydraulic Turbine and Governor block has five input ports—reference speed (wref), reference mechanical power (pref), actual speed of the machine (we), actual electrical power (Pe0), and speed deviation (dw). It is to be noted that all the inputs need to be provided in their respective per unit values.

An Excitation System block is required to provide the field voltage to the Synchronous Machine. The output port of the Excitation System block is the field voltage (V_f), in the per unit value. The output stator voltages V_d and V_q of the Synchronous Machine are the two inputs of the Excitation System. The reference voltage of the Excitation System is set as 1 by using a Constant block. The voltage stabilizer input of the Excitation System is made grounded in this example. The customized parameter window of this block is illustrated in Fig. 18.63.

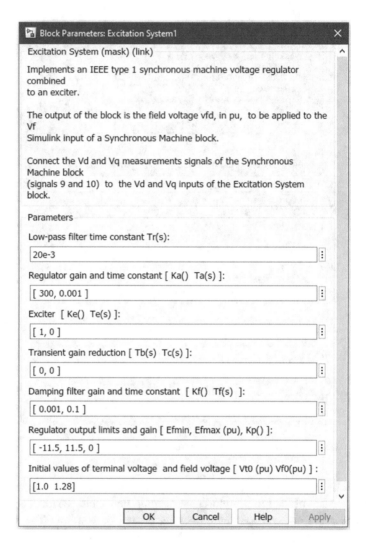

Fig. 18.63 Block parameters of the excitation system

The parameter window of the Synchronous Machine is provided in Figs. 18.64, 18.65, and 18.66. From this parameter window, the model of the Synchronous Machine can be selected from a list of preset models. In this example, no present model is selected. The nature of the mechanical input that is selected in this example is the mechanical power (P_m), which is provided via the Hydraulic Turbine and Governor block. The selected rotor type of the Synchronous Machine is the Salient-Pole type for this instance. After customizing the configuration of the Synchronous Machine (Fig. 18.64), go to the Parameters tab of its parameter window.

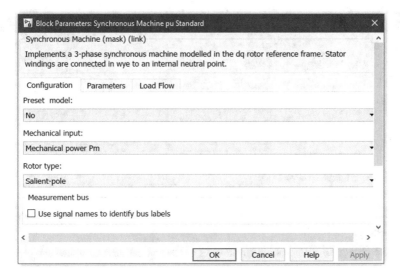

Fig. 18.64 Block parameters of the synchronous machine pu standard: Configuration

| Block Parameters: Synchronous Machine pu Standard | × |

Synchronous Machine (mask) (link)

Implements a 3-phase synchronous machine modelled in the dq rotor reference frame. Stator windings are connected in wye to an internal neutral point.

Configuration   **Parameters**   Load Flow

Nominal power, line-to-line voltage, frequency [Pn(VA) Vn(Vrms) fn(Hz)]:   [300E06 11E3 60]

Reactances [Xd Xd' Xd" Xq Xq" Xl] (pu):   [2.24 0.17 0.12 1.02 0.13 0.08]

Time constants

d axis: Short-circuit

q axis: Open-circuit

[Td' Td" Tqo"] (s):   [1.01, 0.053, 0.1]

Stator resistance Rs (pu):   0.037875

Inertia coefficient, friction factor, pole pairs [H(s) F(pu) p()]:   [0.1028 0.02056 2]

Initial conditions [dw(%) th(deg) ia,ib,ic(pu) pha,phb,phc(deg) Vf(pu)]:   [0 0 0 0 0 0 0 0 1]

☐ Simulate saturation   Plot

[ifd; vt] (pu): ,0.9214,0.9956,1.082,1.19,1.316,1.457;0.7,0.7698,0.8872,0.9466,0.9969,1.046,1.1,1.151,1.201]

| OK | Cancel | Help | Apply |

Fig. 18.65 Block parameters of the synchronous machine pu standard: Parameters

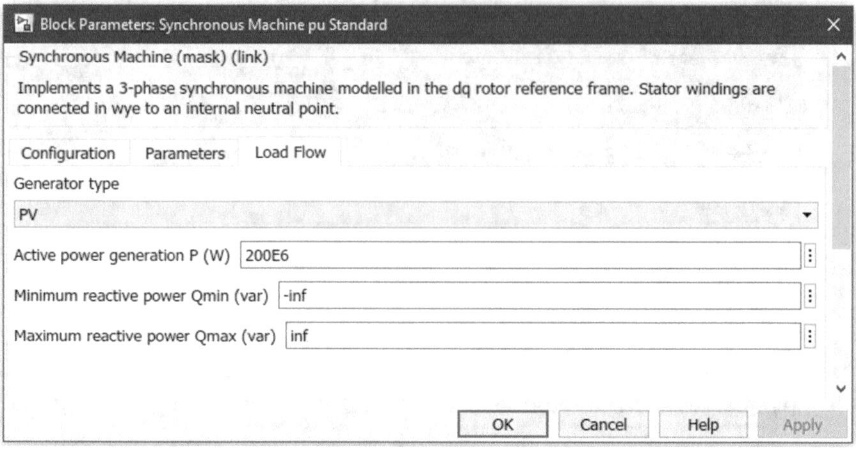

Fig. 18.66 Block parameters of the synchronous machine pu standard: Load flow

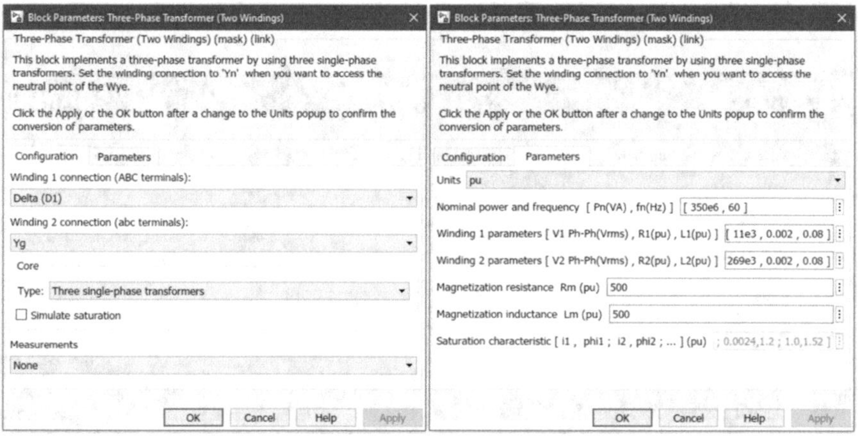

Fig. 18.67 Block parameters of the three-phase transformer

In Fig. 18.65, the customization of the parameters under the Parameters tab is shown. The nominal power of the Synchronous Machine is set as 300 MVA with the line-to-line RMS voltage of 11 kV. The d-axis is set in the short-circuit mode, while the q-axis is assigned in the open-circuit mode. All the other parameters that are defined for this example can be found in Fig. 18.65. The last remaining tab of the parameter window of the Synchronous Machine is called Load Flow. From this tab, the PV type generator is selected, and the active power of the generator is set as 200 MW, while the range of the reactive power is set as $-$inf to $+$inf (Fig. 18.66).

A Three-Phase Transformer is connected with the Synchronous Machine to step up the voltage from 11 kV to 269 kV. The configuration tab of the Three-Phase Transformer (Two Windings) is shown in the left figure of Fig. 18.67. The Delta

(D1) connection is selected for the first winding, whereas the "Yg" connection is chosen for the connection of the second winding. The Parameters tab of this block is shown on the right side of Fig. 18.67. The nominal power of the transformer is set as 350 MVA. As it will work as a step-up transformer, the RMS phase-to-phase voltage of the first winding is assigned as 11 kV, while for the second winding, the assigned voltage is 269 kV.

With the transformer a Three-Phase AC Source is connected, which is customized to have a phase-to-phase RMS voltage rating of 269 kV. The three-phase short-circuit power of the AC source at the 269 kV base voltage is set to be 5000 MVA as shown in Fig. 18.68. The X/R ratio of the source for this example is 7. A load of 10 MW is connected right after the Synchronous Machine, where the line voltage is 11 kV. Another load of 15 MW is connected with the 269 kV line after the step-up transformer. The customization of these two Three-Phase Series Load blocks is presented in Fig. 18.69.

The output port of the Synchronous Machine is connected with a Bus Selector block. The parameter window of the Bus Selector block is illustrated in Fig. 18.70, where the left box indicates all the available output signals, and the right box refers to the selected ones from the left box. The signal line of Mechanical Rotor speed

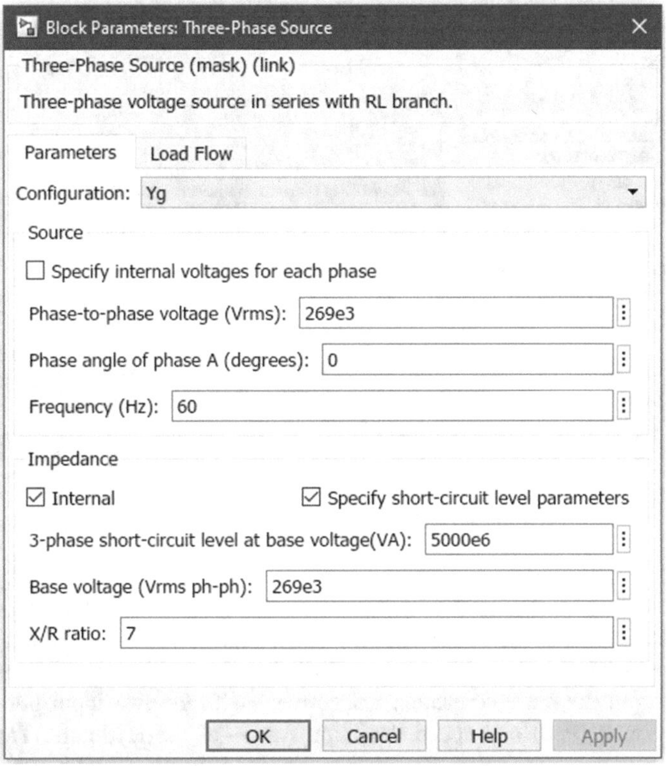

Fig. 18.68 Block parameters of the three-phase source

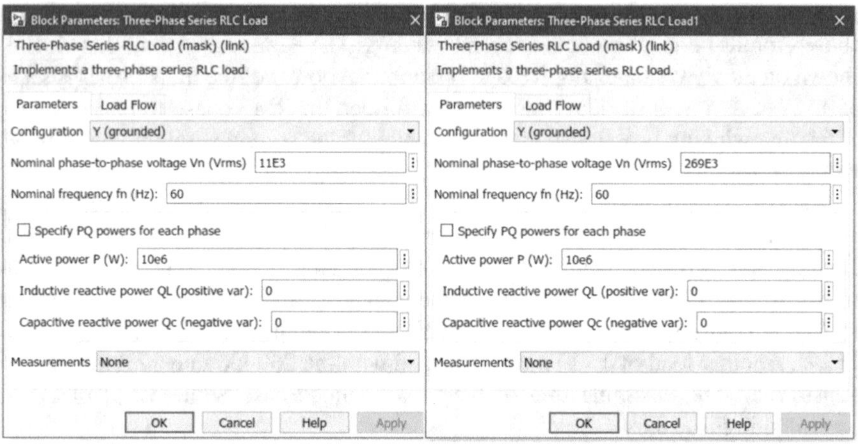

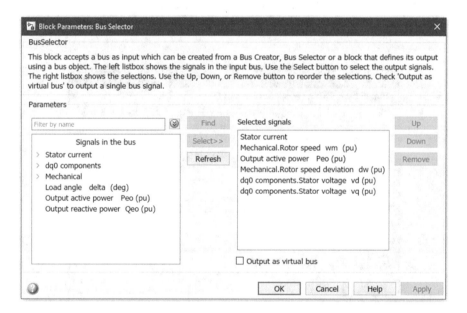

Fig. 18.69 Block parameters of the three-phase series RLC load

Fig. 18.70 Block parameters of the bus selector

(wm) is connected to one of the input ports of the Hydraulic Turbine and Governor block named actual speed of the machine (we). Similarly, the Output active power (wm) and the Mechanical Rotor speed deviation (dw) signal lines from the Bus Selector output ports are connected respectively with the two input ports—actual electrical power (Pe0) and speed deviation (dw)—of the Hydraulic Turbine and Governor. The two other selected signals in the Bus Selector v_d and v_q are connected with the two input ports of the Excitation System that match the name. A Scope is

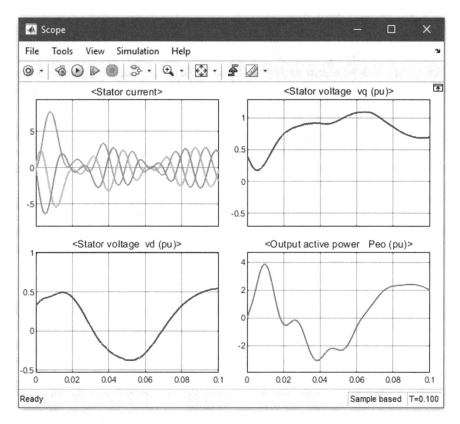

Fig. 18.71 The response observed from the scope window

connected with four output signals of the Bus Selector to realize the characteristics of the Synchronous Machine. The obtained simulated results observable in the Scope window are demonstrated in Fig. 18.71.

18.4 Battery

In the renewable energy-based power system, battery plays an important role. One of the disadvantages of renewable energy-based power systems is their intermittency problem. On this account, the battery system can be utilized to overcome the power intermittency by providing energy during the unavailability of the renewables. Due to the advancement of the current battery system, it can be employed as both temporary and permanent sources of power. In Simulink, a design can be modeled using a single battery cell, or by utilizing a battery pack created from multiple battery cells. Based on the structure of a battery cell, different types of batteries such as lead-acid batteries, Li-ion batteries, etc. are available. Simulink provides the opportunity to utilize any specific battery type to model a design.

18.4.1   Battery Cell Implementation in Simulink

In Simulink, the Battery block can be considered as a single battery cell, which can be used as a voltage source. In Fig. 18.72, a simple demonstration of the usage of a battery cell is made to supply current to a resistor. The utilized blocks in this example are listed below with their respective navigation routes (Table 18.9):

In Fig. 18.72, it can be observed that a single battery cell is connected with a resistor in series. To observe the voltage and current, a Voltage Sensor is connected in parallel across the resistor, and a Current Sensor is connected in series. As the utilized blocks are physical system blocks, a Solver Configuration block is required as explained in Sect. 12.5. Finally, the output of the voltage and current measurements are required to be displayed in the Display blocks. However, as the Display block can only display Simulink signals, the PS-Simulink Converters are required to convert the physical signals to the Simulink Signals. Hence, one PS-Simulink Converter is connected with the voltage signal line, whereas the other converter is connected with the current signal line as shown in the following figure:

In the parameter window of the Battery, three tabs are available. The Main tab contains the nominal voltage, internal resistance, and battery charge capacity parameters. The nominal voltage of the battery cell is assigned to be 10 V with an internal resistance (r) of 2.5 ohms (Fig. 18.73). The battery charge capacity is kept as infinite for this instance. The parameters belonging to the other tabs are kept in their default forms, such as no dynamics are chosen from the Dynamic tab of the parameter window.

The Resistor that is connected in series with the voltage source is customized as shown in Fig. 18.74. The value of the resistance is set as 10 ohms, while the tolerance is assigned to be 5%. All the other parameters in the remaining tabs are kept in their default forms or values.

For physical systems, the Solver Configuration block is used instead of the powergui block. The parameter window of the Solver Configuration block is shown in Fig. 18.75. To learn more about the Solver Configuration, click on the Help button, or get back to Sect. 12.5.

After simulating the model, it can be observed from the two Display blocks that the voltage across the load resistor is 8 V, and the current is 0.8 A. Due to the internal resistance of the battery cell and the load resistance, voltage drops occur, resulting in such a reduction of voltage across the load that is expected. Thus, a single battery cell can be utilized to provide the necessary power to a load.

18.4.2   Battery Modeling of Different Types in Simulink

In Simulink, different battery types can be customized by using Simulink block—Battery. The navigation route of this block can be found in the following table, which is different than the previous example. The previous Battery block represents

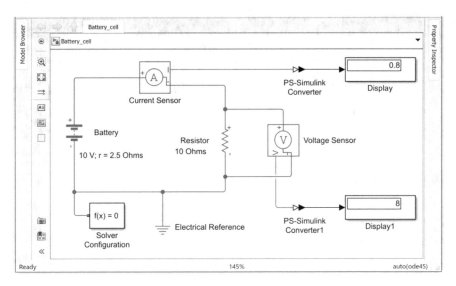

Fig. 18.72 Battery cell implementation in Simulink

Table 18.9 Blocks and navigation path for implementing battery cell model

Blocks	Navigation path on the Simulink Library Browser
Battery	Simscape → Electrical → Sources → Battery
Resistor	Simscape → Electrical → Passive → Resistor
Solver Configuration	Simscape → Utilities → Solver Configuration
PS-Simulink Converter, PS-Simulink Converter1	Simscape → Utilities → PS-Simulink Converter
Current Sensor	Simscape → Electrical → Sensors & Transducers → Current Sensor
Voltage Sensor	Simscape → Electrical → Sensors & Transducers → Voltage Sensor
Electrical Reference	Simscape → Foundation Library → Electrical → Electrical Elements → Electrical Reference
Display, Display1	Simulink → Sinks → Display

a single battery cell, whereas the Battery block within the Extra Sources is a generic battery model that can be customized for different types. In Fig. 18.76, a Lithium-Ion type battery is modeled by customizing its different parameters to simulate a design. The blocks that are used in this design are enlisted below along with their navigation routes (Table 18.10):

The Simulink design showed in Fig. 18.76 has a Battery connected with a Controlled Current Source. Here the Controlled Current Source acts as a constant load for the battery. A load resistor is also connected in parallel to the Controlled Current Source. To observe the different characteristic parameters of the battery, a Bus Selector with a Scope is connected with the output port of the Battery.

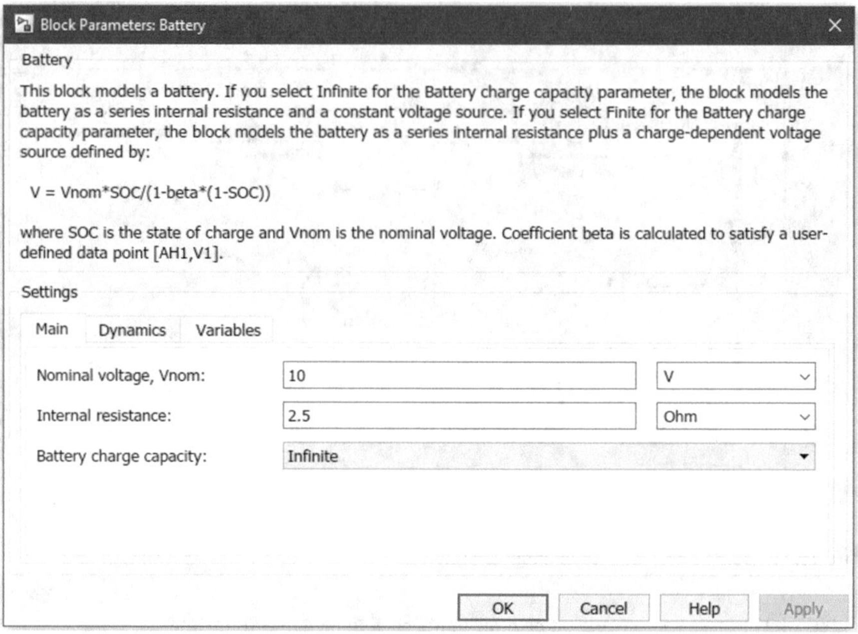

Fig. 18.73 Block parameters of the battery

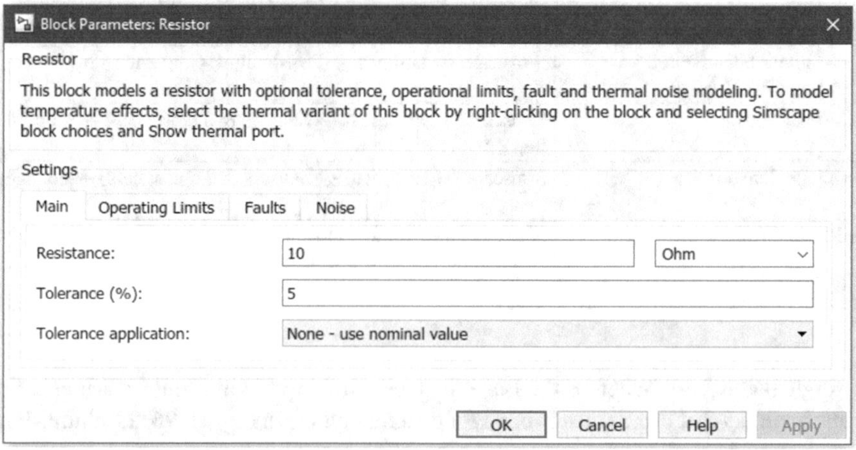

Fig. 18.74 Block parameters of the resistor

The Parameter window of the Battery is shown in Fig. 18.77, where two tabs named Parameters and Discharge can be found. Under the Parameters tab, the type of the battery can be selected out of four different options, which are Lithium-Ion Battery, Lead-Acid Battery, Nickel-Cadmium Battery, and Nickel-Metal-Hydride

Block Parameters: Solver Configuration   ×

Solver Configuration

Defines solver settings to use for simulation.

Parameters

☐ Start simulation from steady state

Consistency tolerance: | 1e-09 |

☐ Use local solver

    Solver type: | Backward Euler ▾ |

    Sample time: | 0.001 |

    Partition method: | Robust simulation ▾ |

    Partition storage method: | As needed ▾ |

    Partition memory budget [kB]: | 1024 |

☐ Use fixed-cost runtime consistency iterations

    Nonlinear iterations: | 3 |

    Mode iterations: | 2 |

Linear Algebra: | auto ▾ |

Equation formulation: | Time ▾ |

Delay memory budget [kB]: | 1024 |

☑ Apply filtering at 1-D/3-D connections when needed

Filtering time constant: | 0.001 |

| OK | | Cancel | | Help | | Apply |

Fig. 18.75 Block parameters of the solver configuration

Battery. For this example, the Lithium-Ion Battery type is chosen. The temperature and the aging effect of the battery can be defined while modeling such batteries by clicking the checkbox for each of these options. Here, for this particular example, both are kept unchecked. The nominal voltage of the battery is selected as 50 V with a rated capacity of 6.5 Ah. The initial State-of-Charge (SoC) of the battery is assigned to be 100%, and the response time is set as 30 s. The customization can be seen in the left figure of Fig. 18.77. Under the other tab named Discharge, several customizable parameters are available that can be seen in the right-side figure of

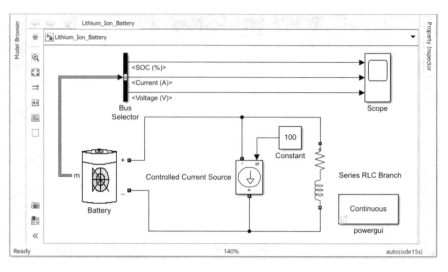

Fig. 18.76 Battery modeling of different types in Simulink

Table 18.10 Blocks and navigation path for modeling different types of battery

Blocks	Navigation path on the Simulink Library Browser
Battery	Simscape → Electrical → Specialized Power Systems → Electric Drives → Extra Sources → Battery
Resistor	Simscape → Electrical → Passive → Resistor
Controlled Current Source	Simscape → Foundation Library → Electrical → Electrical Sources → Controlled Current Source
Bus Selector	Simulink → Signal Routing → Bus Selector
Electrical Reference	Simscape → Foundation Library → Electrical → Electrical Elements → Electrical Reference
Constant	Simulink → Sources → Constant
Display	Simulink → Sinks → Display
Scope	Simulink → Sinks → Scope
powergui	Simscape → Electrical → Specialized Power Systems → Fundamental Blocks → powergui

Fig. 18.77. In the Discharge tab, all the parameters can be set automatically based on the nominal values assigned earlier for the battery by clicking the checkbox option that appears in the front. According to the users' preferences, the parameters can be set manually as well. At the bottom of that tab, a Plot option is available, which can be used to observe the discharge characteristics of the modeled battery. At the left side of the Plot button, the Unit option can be seen, which can be selected to observe the plot either in the time scale or in the ampere-hour scale along the *x*-axis. The discharge characteristic curve of the modeled battery in the time scale is shown in Fig. 18.78, which can be observed by double-clicking on the Plot button.

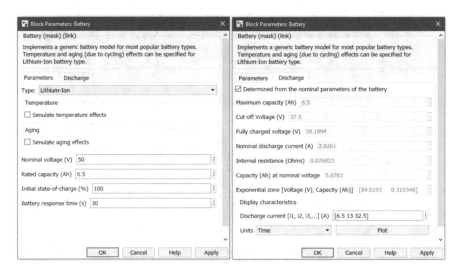

Fig. 18.77 Block parameters of the battery

The Controlled Current Source is customized by selecting the initial parameters as shown in Fig. 18.79. The initial amplitude, phase, and frequency are assigned to zero with a source type of AC. A Constant block is connected with its input port, which is customized to be a value of 100. A Series RLC Branch is connected in parallel, which is configured as a RL branch with a resistance of 10 ohms and inductance of 1 mH (Fig. 18.80).

Afterward, a Bus Selector is connected with the output port of the battery. From the parameter window of the Bus Selector (Fig. 18.81), three parameters can be observed in the left-side box—SOC (%), Current (A), and Voltage (V). All three of these parameters are selected to shift on the right-side box. A Scope customized with three input ports is connected with the three output signals of the Bus Selector. After simulating the model for 250 s, the observed graphs from the Scope window are shown in Fig. 18.82. From the graph, it can be observed that the State-of-Charge of the battery reduces over time from 100% to 0%. As the battery discharges, the current and voltage also reduce over the period.

18.4.3 Case Study: Battery Pack Design Using Battery Cells

In practical applications, it is always required to design a battery pack for certain rated values. A battery pack can be designed using multiple battery cells. In many applications, different rated voltages or ampere-hour ratings may be required. It is not efficient to purchase different rated batteries all the time based on different application requirements. Instead, same-rated multiple battery cells can be utilized

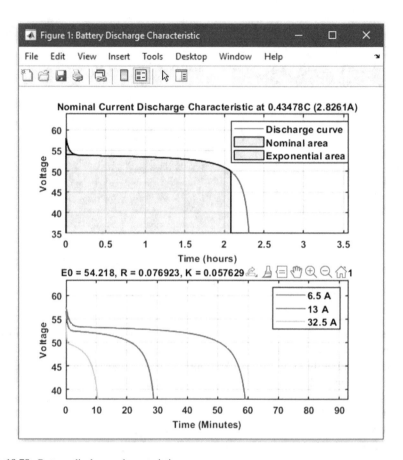

Fig. 18.78 Battery discharge characteristics

to design different battery packs of different sizes. In this case study, a battery pack will be designed to fulfill only the voltage requirement of a particular application.

In this case study, a battery pack of 15 V will be designed by utilizing multiple 4 V single battery cells with an internal resistance of 1 ohm. The designed model is shown in Fig. 18.83, where four single battery cells are connected in series with an external resistance of 60 ohm. The output voltage across the resistor is shown in the Display block that shows the exact 15 V voltage, as desired.

The four battery cells have the same rating and hence, are customized in the same manner. The parameter window of the Battery block is shown in Fig. 18.84, where it can be observed that the voltage and the internal resistance are assigned as 4 V and 1 ohm, respectively.

The external resistance is entered via the Resistor block, which is configured to have a resistance of 60 ohms, and tolerance of 5%, as shown in the Fig. 18.85:

All the utilized blocks in this design are summarized in the following table with their navigation routes (Table 18.11):

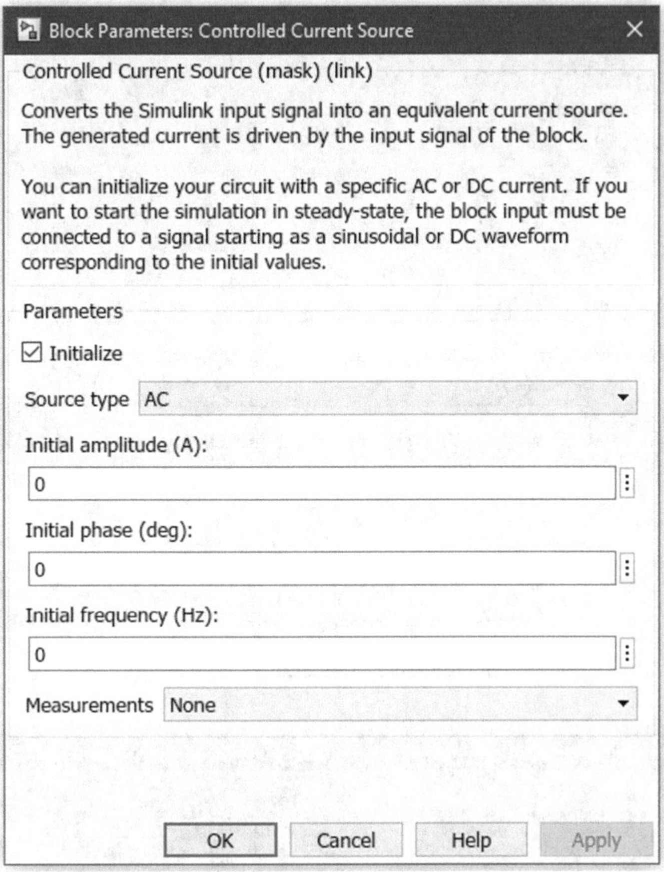

Fig. 18.79 Block parameters of the controlled current source

18.5   Conclusion

In this chapter, the modeling and applications of solar photovoltaics, wind turbines, hydraulic turbines, and batteries are demonstrated via Simulink. As solar photovoltaics are widely used renewable energy in the power system domain, this chapter comparatively emphasizes more on this content. The chapter initiates by creating a mathematical model of a PV cell and demonstrating the characteristic curves of the PV cell using different mathematical blocks of Simulink. Later, the same characteristic curves are generated by utilizing the Solar Cell graphical block of the Simulink. For demonstrating the applications of solar photovoltaics in the power system domain, it is imperative to introduce the PV Array to design a PV panel. Followed by the modeling of a PV panel by using PV Array, the chapter provides a case study that illustrates the grid connection of a PV panel to realize the practical application of

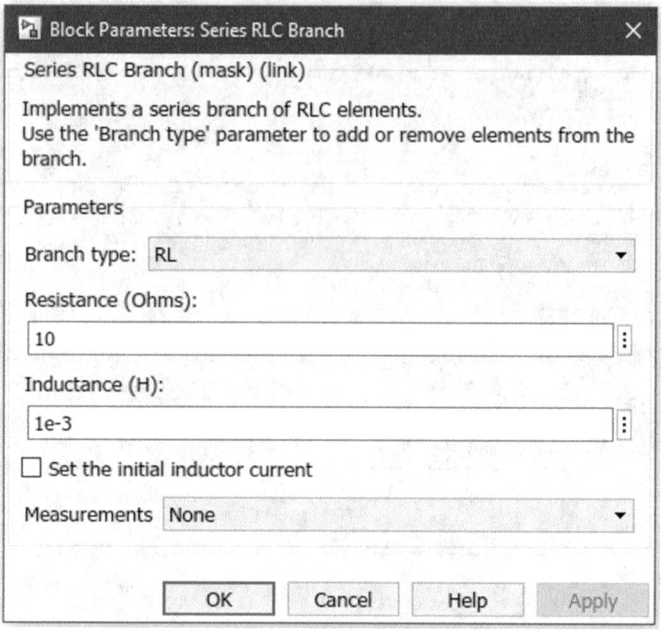

Fig. 18.80 Block parameters of the series RLC branch

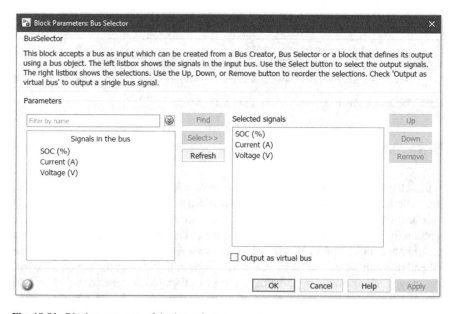

Fig. 18.81 Block parameters of the bus selector

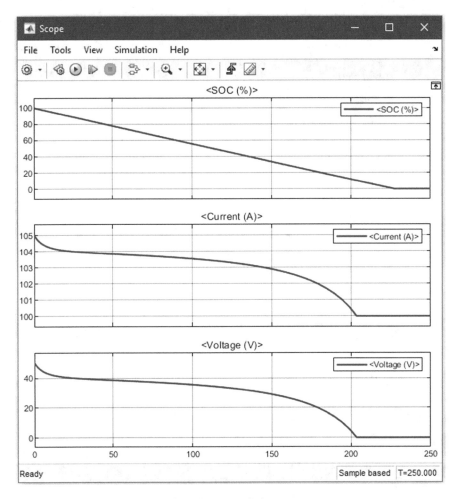

Fig. 18.82 The response observed from the scope window

solar photovoltaics. Afterward, the modeling and application of the wind turbine are showed with necessary illustrations in this chapter. Due to the recent popularity of hydraulic turbines, a case study of hydraulic turbines to run a synchronous generator is demonstrated. Finally, the modeling of both battery cells and different battery models is explained by using Simulink designs. A case study of creating battery packs from single battery cells is also covered in this chapter to make the reader familiarized with some of the practical aspects of batteries. This chapter will tremendously help a reader to grasp the concept of the applicability of renewable energies in the power system.

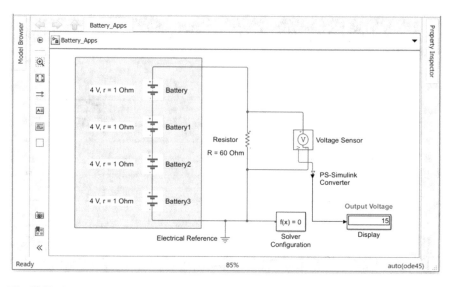

Fig. 18.83 Battery pack design using battery cells in Simulink

---

Block Parameters: Battery   ✕

Battery

This block models a battery. If you select Infinite for the Battery charge capacity parameter, the block models the battery as a series internal resistance and a constant voltage source. If you select Finite for the Battery charge capacity parameter, the block models the battery as a series internal resistance plus a charge-dependent voltage source defined by:

$V = Vnom*SOC/(1-beta*(1-SOC))$

where SOC is the state of charge and Vnom is the nominal voltage. Coefficient beta is calculated to satisfy a user-defined data point [AH1,V1].

Settings

| Main | Dynamics | Variables |

Nominal voltage, Vnom: | 4 | | V |

Internal resistance: | 1 | | Ohm |

Battery charge capacity: | Infinite |

                                        OK   Cancel   Help   Apply

Fig. 18.84 Block parameters of the battery

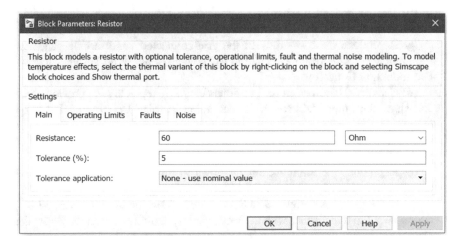

Fig. 18.85 Block parameters of the resistor

Table 18.11 Blocks and navigation path for designing battery pack using battery cells

Blocks	Navigation path on the Simulink Library Browser
Battery, Battery1, Battery2, Battery3	Simscape → Electrical → Sources → Battery
Resistor	Simscape → Electrical → Passive → Resistor
Solver Configuration	Simscape → Utilities → Solver Configuration
PS-Simulink Converter	Simscape → Utilities → PS-Simulink Converter
Voltage Sensor	Simscape → Electrical → Sensors & Transducers → Voltage Sensor
Electrical Reference	Simscape → Foundation Library → Electrical → Electrical Elements → Electrical Reference
Display	Simulink → Sinks → Display

Exercise 18

1. Why renewable energy-based technologies are important? Name some of the renewable energies that are used for electricity generation.
2. (a) Draw the simple circuit model of a single solar cell.

    (b) Write the equation of the output power of a single solar cell.

    (c) Design a Simulink model demonstrating the mathematical model of a single solar cell to generate both PV and VI characteristic curves. Use the same values of the parameters of Fig. 18.2.

    (d) Simulate the model for a temperature of 37°C and quality factor, $N = 2$. Show both PV and VI characteristic curves.

    (e) Design a three-phase inverter for an input DC voltage of 300 V.
3. (a) Write the navigation route of the Wind Turbine Doubly-Fed Induction Generator (Phasor Type).

(b) Design a wind turbine-based generator in the Simulink to display the output powers.

(c) Reproduce the case study of the grid-connected wind turbine generator shown in Fig. 18.48 to show the output DC voltage of the Wind Turbine Doubly-Fed Induction Generator.

(d) Use a 500 kW load in the previous design by replacing the previous load and show the output DC voltage of the Wind Turbine Doubly-Fed Induction Generator.

4. (a) What is the importance of the Excitation System block used in the design of Fig. 18.61.

(b) Reproduce the Simulink model shown in Fig. 18.61 to show the Output active power of the Synchronous Machine.

(c) In the previous design, use a Synchronous Machine of 300 MVA, 33 kV, and replace the transformer with a 33 kV/269 kV step-up transformer. Show the Output active power curve of the Synchronous Machine in the Scope window.

5. (a) Name some of the types of batteries.

(b) Design a Lead-Acid Battery model with a Controlled Current Source and an RL impedance in parallel. Show the discharge characteristic curve of the battery.

(c) Simulate the previous design to display the SOC (%) curve in the Scope window.

(d) Design a battery pack of 15 V by using single battery cells of 6 V and internal resistance of 2 ohms. Show the output voltage in a Display block.

Answer Keys to the End-of-Chapter Exercises

Chapter 1

1. See Sect. 1.4
2. See Sect. 1.8
3. Type "help <command/function name>" in MATLAB command window to learn more details about the commands/functions
4.
```
a=2*4^2;
b=(2*4)^2;
c=503+224-604;
d=(10^3)/(9*2);
e=6.25*0.42^3.56;
f='MATLAB is fun!';
whos
```

5.
```
num1=input('Enter num1:');
num2=input('Enter num2:');
result1=num1/num2;
result2=num1\num2;
fprintf('result1:%f\n',result1);
fprintf('result2:%f\n',result2);
diary('diaryFile.txt');
```

These operations do not produce the same result because the operator "/" calculates num1/num2, whereas "\" calculates the expression 1/(num1/num2)

© The Editor(s) (if applicable) and The Author(s), under exclusive license to
Springer Nature Switzerland AG 2022
E. Hossain, *MATLAB and Simulink Crash Course for Engineers*,
https://doi.org/10.1007/978-3-030-89762-8

Chapter 2

1   See Sect. 2.1
2   See Sect. 2.6
3   See Sect. 2.3
4
```
MatA=[4,7,1;7,2,3;5,5,9];
MatB=[6,0,4;9,8,1;7,5,2];
MatC=[2,5,3;0,17,9;8,0,1];
fprintf('(a):'); MatA+MatB
fprintf('(b):'); MatB-MatC
fprintf('(c):'); MatA/MatC
fprintf('(d):'); MatB'
fprintf('(e):'); det(MatC)
fprintf('(f):'); inv(MatA)
fprintf('(g):');[MatA,MatC]
fprintf('(h):');[MatC;MatA]
fprintf('(i):');
MatA*MatB
MatB*MatA
MatA.*MatB
```

5
```
a = linspace(2,20,100);
fprintf('Mean:'); mean(a)
fprintf('Variance:'); var(a)
fprintf('Standard deviation:'); std(a)
fprintf('Mode:'); mode(a)
```

6a
```
Serial_Number = {1;2;3;4;5};
Element_Name = {'Silicon'; 'Germanium'; 'Tin'; 'Carbon'; 'Tellurium'};
Element_Symbol = {'Si'; 'Ge'; 'Sn'; 'C'; 'Te'};
Bandgap = {1.12; 0.67; 0.08; 5.47; 0.33};
Table = table(Serial_Number, Element_Name, Element_Symbol, Bandgap);
disp(Table)
```

6b
```
Field1 = 'Serial_Number';
val_field1 = {1;2;3;4;5};
Field2='Element_Name';
val_field2 = {'Silicon'; 'Germanium'; 'Tin'; 'Carbon'; 'Tellurium'};
Field3= 'Element_Symbol';
val_field3= {'Si'; 'Ge'; 'Sn'; 'C'; 'Te'};
Field4 = 'Bandgap';
val_field4 = {1.12; 0.67; 0.08; 5.47; 0.33};
array = struct(Field1,val_field1,Field2,val_field2,Field3,val_field3,...
    Field4,val_field4);
disp(array(1))
disp(array(2))
disp(array(3))
disp(array(4))
array(3).Bandgap = 0.07;
disp(array(3))
```

Chapter 3

1   See Sect. 3.2

2   See Sect. 3.4

3
```matlab
a = input('Take user input:\n');
if(a<0 || a>100)
    fprintf('Outside range\n');
else
    fprintf('Inside range\n');
    if(a > 25 && a <= 50)
        fprintf('First half\n');
    elseif(a > 50 && a <=75)
        fprintf('Second half\n');
    end
end
```

4
```matlab
x = input('Enter the hexadecimal number:', 's');
switch x
    case '#FF0000'
        disp('Red');
    case '#00FF00'
        disp('Green');
    case '#0000FF'
        disp('Blue');
    case '#FFA500'
        disp('Orange');
    case '#FFFF00'
        disp('Yellow');
    case '#000000'
        disp('Black');
    case '#FFFFFF'
        disp('White');
    otherwise
        disp('The color code you entered is wrong/not available');
end
```

5
```matlab
function dist = distance(x1,y1,x2,y2)
dist = (((x1-x2)^2)+((y1-y2)^2))^0.5;
end
```

6
```matlab
function c = usercal(num1, num2, op)
if(op == 1)
    c = num1+num2;
elseif(op == 2)
    c = num1 - num2;
elseif(op == 3)
    c = num1 * num2;
else
    disp('wrong operation entered');
end
```

Chapter 4

1 See Sect. 4.1

2 Type "help <command/function name>" in MATLAB command window to learn more details about the commands/functions

3
```
a=input('Enter the real portion of the complex number:\n');
b=input('Enter the imagniary portion of the complex number:\n');
[m_angle,m_mag]=cart2pol(a,b);
m_angle=m_angle*(180/pi);
m=[m_mag,m_angle];
disp(m)
[n_angle,n_mag]=cart2pol(5,-1);
n_angle=n_angle*(180/pi);
n=[n_mag,n_angle];
e_mag=m_mag*n_mag;
e_angle=m_angle+n_angle;
disp(['Exponential form: ', num2str(e_mag),'exp(i*',num2str(e_angle),')'])
```

4 See Example 4.4, as the solution code closely matches this example

5 See Example 4.6, as the solution code closely matches this example

6 See Example 4.7, as the solution code closely matches this example

7
```
P=20; Q=35; t=86400; R=10; L=20*10^(-3); C=4*(10^-6); f=60;
X_L=2*pi*f*L;
X_C=-1/(2*pi*f*C);
Z_rec=R+i*(X_L-X_C);
disp('Impedance in rectangular form:');
disp(Z_rec);
Z_mag=abs(Z_rec);
Z_angle=angle(Z_rec)*(180/pi);
Z_polar=[Z_mag,Z_angle];
disp('Impedance in polar form: ');

disp(Z_polar);
S_rec=complex(P,Q);
disp('Apparent power in rectangular form');
disp(S_rec)
S_mag=abs(S_rec);
S_angle=angle(S_rec)*180/pi;
S_polar=[S_mag,S_angle];
disp('Apparent power in polar form');
disp(S_polar)
E=P*t; %Electrical energy
disp(['Electrical Energy: ',num2str(E), ' Joule']);
```

Chapter 5

1 See Sect. 5.1

2 Type "help <command/function name>" in MATLAB command window to learn more details about the commands/functions

3 See Examples 5.2 and 5.3

4 See Examples 5.4 and 5.5

5 See Examples 5.8 and 5.11

6 See Example 5.9

Chapter 6

1   See Sect. 6.2
2   See the examples that use the functions to solve equations
3
```
M=[-4,5;8,-11];
N=[0.33,1,3.3;0.5,0.45,-5.12;2,-2,0];
fprintf('(a)');
rank(M)
rank(N)
fprintf('(b)');
inv(M)
inv(N)
fprintf('(c)');

eig(M)
[vector1,lambda1]=eig(M);
disp('The eigenvalues of M:');
lamda1=sum(lambda1);
disp(lamda1);
disp('The eigenvector of M:');
disp(vector1)
[vector2,lambda2]=eig(N);
disp('The eigenvalues of N:');
lamda2=sum(lambda2);
disp(lamda2);
disp('The eigenvector of N:');
disp(vector2);
```

4   Only (a) and (d) are done since the others are similar to these solutions
    (a)

```
syms x;
x_val=solve(x^2+5*x+9==0,x);
disp('The solutions are:');
disp(x_val)
```

    (d)

```
syms x y;
[x_val,y_val]=solve(9*x^2+3*x*y-2==-3,4*x^2+7*x*y+(5/2)==0);
disp('The solutions are:');
disp('x=');
disp(x_val);
disp('y=');
disp(y_val);
```

5   See Example 6.6 for (a) and (b), Example 6.7 for (c) and (d), Example 6.8 for
    (e) and (f), and Example 6.9 for (g) and (h), and change the equations as suitable
6   Only (a)(i) has been done as the other one can be done similarly

```
syms x;
I1=int((log10(x))^2,x);
disp('The solution without limit:');
y_sol=solve(I1-2*x==0,x);
disp(y_sol);
%with limit of [0 2]
I2=int((log10(x))^2,x,0,1);
disp('The solution with limit: ')
y_sol=solve(I2-2*x==0,x);
disp(y_sol);
```

Chapter 7

1   See Sects. 7.2, 7.3, and 7.4 to summarize the basic steps
2   Gauss-Seidel method is used to solve a set of equations and for determining unknown variables, and Newton-Raphson method is used to approximate the root of a nonlinear function. On the other hand, Runge-Kutta method is used to solve ordinary differential equations
3a See Example 7.1 and change the equations as follows:

```
fx=@(x,y,z) (1/20).*(122+2*y+z);
fy=@(x,y,z) (1/-60).*(76-4*x-18*z);
fz=@(x,y,z) (1/35).*(50-2*x+15*y);
```

Result:

```
The solution after 7 th iteration:
x: 6.079227 y: -0.616263 z: 0.81707
```

3b For tolerance = 0.0001:   For tolerance = 0.001:

```
The solution after 6 th iteration:      The solution after 5 th iteration:
x: 6.079221 y: -0.616269 z: 0.81707     x: 6.079169 y: -0.616321 z: 0.81705
```

The results vary from the answer in question (a). It is seen that with less precision in tolerance level, the iteration decreases, hence the less converged solutions may be visible

3c For tolerance = 0.000001:   For tolerance = 0.0000001:

```
The solution after 8 th iteration:      The solution after 9 th iteration:
x: 6.079227 y: -0.616263 z: 0.81707     x: 6.079227 y: -0.616262 z: 0.81707
```

The result does not vary much from the answer in question (a) (apart from the value of y, which changes very little). It is seen that with more precision in tolerance level, the iteration increases, but since converged solutions have already been obtained, the solution can be considered as the final solution, and further iterations can be curtailed.

4a See Example 7.2 and change the equation as follows:

```
F=@(x) (x^5)-(x)-2;
```

Result:

```
Root of the equation after 4th iteration: 1.85110
```

4b
```
F=@(x) 3*x+2*cos(x)-5;
```

Result:

```
Root of the equation after 6th iteration: 1.26717
```

5a See Example 7.2 and change the equation as follows:

```
F=@(x,y) -4*(x^3)-6*(x^2)-(10*x)+2;
```

Result:

```
The final solution for x=2 is:-43.00000
```

5b
```
F=@(x,y) x*sin(y)+y*cos(x);
```

Result:

```
The final solution for x=2 is:13.11056
```

Chapter 8

1
```
R1=5;R2=7;R3=7;R4=7;R5=10;
eq_r= R1+(1/((1/R2)+(1/R3)+(1/R4)))+R5;
fprintf('Eqivalent resistance: %f\n', eq_r);
V=1:12;
I=V/eq_r;
plot(V,I,'o-b','Linewidth',1.2);
xlabel('Voltage, Volt');
ylabel('Current, Amp');
title('Verification of Ohm''s Law');
grid on;
```

2
```
R1=0.2;R2=0.5;R3=1;R4=0.8;R5=1.44;
Ry1=(R1*R2)/(R1+R2+R3);
Ry2=(R1*R3)/(R1+R2+R3);
Ry3=(R2*R3)/(R1+R2+R3);
Rs1=Ry2+R4;
Rs2=Ry3+R5;
Rp=(Rs1*Rs2)/(Rs1+Rs2);
Req=Ry1+Rp;
fprintf('Equivalent Resistance of the circuit: %f Ohms\n', Req);
V=6;
I=V/Req;
fprintf('Current: %f Ampere\n',I);
```

3a
```
function [v2,v3]=voltdiv(R1,R2,R3,E)
v2=(R2/(R1+R2+R3))*E;
v3=(R3/(R1+R2+R3))*E;
fprintf('Voltage across R2:%f\n',v2);
fprintf('Voltage across R3:%f\n',v3);
end
```

3b
```
function [i1,i2]=curdiv(R1,R2,I)
i1=(R2/(R1+R2))*I;
i2=(R1/(R1+R2))*I;
fprintf('Current through R1, I1:%f\n',i1);
fprintf('Current through R2, I2:%f\n',i2);
end
```

4
```
R1=4; R2=9; RL=5; V=12;
Vth=(R2/(R1+R2))*V;
Rth=1/((1/R1)+(1/R2));
fprintf('Thevenin Voltage: %.3f V\n', Vth);
fprintf('Thevenin equivalent resistance: %.3f ohms\n',Rth);
IRL=Vth/Rth+RL;
fprintf('Load current: %.3f A\n', IRL);
RL=1:1:20;
for i=1:1:20
    I(i)=Vth/(Rth+RL(i));
    Power(i)=I(i)^2*RL(i);
end

plot(RL,Power,'o-b','LineWidth',1.2);
xlabel('Load resistance, R_L (Ohms)');
ylabel('Output power, P (W)');
title('Maximum power transfer theorem');
grid on;
RL=5;
P_max=(Vth/(Rth+RL))^2*RL;
fprintf('Maximum output power=%.3f\n',P_max);
```

5 Replicate and run Example 8.13 by changing the value of *P* as 50 and *Q* as 13. Since (b) is similar in procedure with (a), the result of (a) is shown only

```
Positive reactive power
─────────────────────────
Apparent Power:
Apparent power in polar form:
|S|=51.662 VA     Power angle=143.842 degree
Power factor= 0.783; Lagging

Negative reactive power
─────────────────────────
Apparent Power:
Apparent power in polar form:
|S|=51.662 VA     Power angle=-143.842 degree
Power factor= 0.783; Leading

Zero reactive power
─────────────────────────
Apparent Power:
Apparent power in polar form:
|S|=50.000 VA     Power angle=0.000 degree
Power factor= 1.000; Unity
```

6a See and replicate Example 8.14 with the following equations:

```
V_AB=100*cos(0)+i*100*sin(0);
V_BC=110*cos(120*(pi/180))+i*110*sin(120*(pi/180));
V_CA=120*cos(240*(pi/180))+i*120*sin(240*(pi/180));
%Impedances
Z1=8*cos(25*(pi/180))+i*8*sin(25*(pi/180));
Z2=14*cos(55*(pi/180))+i*14*sin(55*(pi/180));
Z3=18*cos(-23*(pi/180))+i*18*sin(-23*(pi/180));
```

6b See and replicate Example 8.15 with the following equations:

```
%Line-to-line voltages
V_AB=100*cos(0)+i*100*sin(0);
V_BC=110*cos(120*(pi/180))+i*110*sin(120*(pi/180));
V_CA=120*cos(240*(pi/180))+i*120*sin(240*(pi/180));
%Impedances
Z=5*cos(30*(pi/180))+i*5*sin(30*(pi/180));
```

7a See Example 8.20 and change the values as per the question
7b See Example 8.21 and change the values as per the question
Result:

```
(a)Output voltage: -16.80 V
Gain: -0.70
(b)Output voltage: 28.80 V
Gain: 2.40
```

8a See Example 8.23 and change the values as per the question
8b See Example 8.24 and change the values as per the question

Chapter 9

1 See Sect. 9.2.2
2 See Sects. 9.2.3 and 9.2.4
3 Type "help <command/function name>" in MATLAB command window to learn more details about the commands/functions
4 Only (a) has been done as the others can be done similarly

```
syms s
H=@(s) (s-12)/(s^2-4*s+1);
G=tf([1,-12],[1,-4,1]);
N=[1,-12];
D=[1,-4,1];
disp(G);
fprintf('Pole:\n');
disp(pole(G));
fprintf('Zero:\n');
disp(zero(G));
pzmap(G);

fprintf('Inverse Laplace:\n');
A=ilaplace(H(s));
pretty(A);
fprintf('Laplace:\n');
B=laplace(A);
pretty(B);
DC_gain=limit(H(s),s,0);
fprintf('DC Gain:\n');
disp(DC_gain);
[r,p,k]=residue(N,D);
Expan=@(s) r(1)/(s-p(1)) + r(2)/(s-p(2));
fprintf('Partial Fraction Decomposition:\n');
disp(vpa(Expan(s),2));
```

5 Only (a) has been done as the others can be done similarly. The system is
   overdamped because zeta is greater than 1

```
k=2;
omega_n=3;
zeta=2;
s=tf('s');
fprintf('Transfer Function:\n');
G=(k*omega_n^2)/(s^2+2*zeta*omega_n*s+omega_n^2)
step(G);

grid on;
ylim([0 2.5]);
fprintf('Parameters:\n');
disp(stepinfo(G));
figure(2);
pzmap(G);
grid on;
```

6
```
R=2;L=1.5;C=0.6;
A=[0,1/C;-1/L,-R/L];
B=[0;1/L];
C=[-1,-R];
D=[1];
disp('State space representation:');
sys=ss(A,B,C,D)
[Num,Den]=ss2tf(A,B,C,D);
disp('Transfer function');
TF=tf([Num],[Den])
syms s;
I=eye(2);
G1=C*inv(s*I-A)*B+D;
disp('Tranfer function using formula');
disp(simplify(G1));
```

7 Only (a) has been done as the other one can be done similarly

```
Enter the coefficient:[1 3 27 45 -60]
      1     27    -60
      3     45      0
     12    -60      0
     60      0      0
    -60      0      0

The system is unstable

Verification
Poles:
  -0.6172 + 5.1632i
  -0.6172 - 5.1632i
  -2.6143 + 0.0000i
   0.8488 + 0.0000i
```

8 Only (a) has been done as the others can be done similarly
   (a) The system is stable

```
%root locus
sys1=tf([36],[2 14 61]);
figure(1);
rlocus(sys1);
hold on;
%bode plot
figure(2);

margin(sys1);
grid on;
%nyquist plot
figure(3);
nyquist(sys1);
hold off;
```

Chapter 10

1 See Sects. 10.2 and 10.3
2 Only (a) has been done as the others can be done similarly

```
syms x;
obj=@(x)6*x^4-11*x+10;
x_low=-12;
x_up=12;
[x, value]=fminbnd(obj,x_low,x_up);
fprintf('Optimized value of the decision variable: %.5f\n',x);
fprintf('Minimized value of the objective function: %.5f\n',value);
```

3a Use the functions that are coded in Example 10.2. Change the upper and lower limits
   in the script as follows:

```
x_low=[-4,-2,-1];
x_up=[7,9,10];
```

Result:

```
Local minimum found that satisfies the constraints.

Optimization completed because the objective function is non-decreasing in
feasible directions, to within the value of the optimality tolerance,
and constraints are satisfied to within the value of the constraint tolerance.

<stopping criteria details>
Optimized value of the decision variable:
x1: -1.49579
x2: 1.48717
x3: 2.74790

Minimized value of the objective function: 9.76389
```

3b
```
x_low=[1,-3,-7];
x_up=[4,3,-3];
```

Result:

```
Solver stopped prematurely.

fmincon stopped because it exceeded the function evaluation limit,
options.MaxFunctionEvaluations = 3.000000e+03.
```

4   See Example 10.3, as the code is similar. Only these values are needed to
be changed

4a
```
obj=[4 6 2];
x_low=[0,-3, 0];
x_up=[10,9,12];
A=[4 5 8;-7 12 3];
B=[30 65];
A_EQ=[2 3 5];
B_EQ=[11];
```

Result:

```
Optimal solution found.

Optimized value of the decision variable:
x1: 0.00000
x2: -3.00000
x3: 4.00000

Minimized value of the objective function: -10.00000
```

4b
```
obj=[5 7 -2];
x_low=[-3,-2,2];
x_up=[4,7,11];
A=[2 1 3;-4 2 0];
B=[20 10];
A_EQ=[3 1 -2];
B_EQ=[16];
```

Result:

```
No feasible solution found.

Linprog stopped because no point satisfies the constraints.
```

4c
```
obj=[4 9 1];
x_low=[2,-10, 0];
x_up=[6,10,22];
A=[1 1 1;0 8 -3];
B=[26 15];
A_EQ=[1 9 4];
B_EQ=[18];
```

Result:

```
Optimal solution found.

Optimized value of the decision variable:
x1: 2.00000
x2: -8.00000
x3: 22.00000

Minimized value of the objective function: -42.00000
```

5   Only (a) has been done as the other one can be done similarly

```
H=[10 3;3 7];
F=[-5;3];
x_low=[0,-5];
x_up=[8,5];
A=[4 5;3 -9];
B=[21 15];
A_EQ=[5 3];
B_EQ=[12];
[x,value]=quadprog(H,F,A,B,A_EQ,B_EQ,x_low,x_up);
fprintf('Optimized value of the decision variable:\n');
fprintf('x1:%.5f\n',x(1));
fprintf('x2:%.5f\n',x(2));
fprintf('Minimized value of the objective function:%.5f\n', value);
```

Result:

```
Minimum found that satisfies the constraints.

Optimization completed because the objective function is non-decreasing in
feasible directions, to within the value of the optimality tolerance,
and constraints are satisfied to within the value of the constraint tolerance

<stopping criteria details>
Optimized value of the decision variable:
x1:2.29714
x2:0.17143
Minimized value of the objective function:16.69714
```

Chapter 11

1 See Sect. 11.2.1
2 See the examples to infer some of the engineering applications
3 Step 1: Draw a "panel," an "Edit Field (Text)," and a "Button" from the component library, and arrange it according to the app shown
   Step 2: Rename the panel "Exercise 3," the edit field "Sentence," and the button "Click me"
   Step 3: Right-click on the button, and select "Callback" and "Go to ClickmeButtonPushed callback"
   Step 4: Write the following code in the function:

```
% Button pushed function: ClickmeButton
function ClickmeButtonPushed(app, event)
    app.SentenceEditField.Value = "AppDesigner is Fun";
end
```

Step 5: Save and Run

4 Step 1: While doing Example 11.1, add two more buttons named "POW" and "Z." Make the panel appear as the interface shown in the question. The component browser would look like this:

Step 2: While adding the callback functions for each of the buttons (first app. ADDButton, then for app.SUBButton, then for app.MULButton, then for app. DIVButton, then for app.POWButton, and finally for app.ZButton), add the following lines of code in their respective locations as shown as follows:

```
% Button pushed function: ADDButton
function ADDButtonPushed(app, event)
     app.out.Value=app.Num1.Value+app.Num2.Value;
end

% Button pushed function: SUBButton
function SUBButtonPushed(app, event)
     app.out.Value=app.Num1.Value-app.Num2.Value;
end

% Button pushed function: MULButton
function MULButtonPushed(app, event)
     app.out.Value=app.Num1.Value*app.Num2.Value;
end

% Button pushed function: DIVButton
function DIVButtonPushed(app, event)
     app.out.Value=app.Num1.Value/app.Num2.Value;
end

% Button pushed function: POWButton
function POWButtonPushed(app, event)
     app.out.Value=app.Num1.Value^app.Num2.Value;
end

% Button pushed function: ZButton
function ZButtonPushed(app, event)
     app.out.Value=(app.Num1.Value+app.Num2.Value)/2;
end
end
```

Step 3: Run the app.

5 Step 1: Drag and drop a panel, a knob (not the "Discrete Knob" used in Example 11.2), and five lamps, and rename the labels and titles as shown in the following figure:

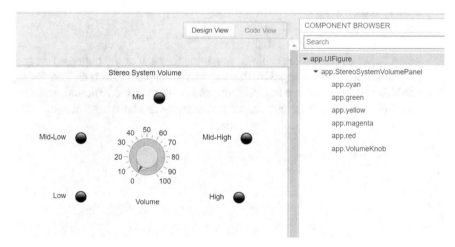

Step 2: Right-click on the app.VolumeKnob and select callbacks, and then VolumeKnobValueChanged() function. Enter the following code and save:

```
% Value changed function: VolumeKnob
function VolumeKnobValueChanged(app, event)
    value = app.VolumeKnob.Value;
    if value==0
        app.green.Color=[0 0 0];
        app.cyan.Color=[0 0 0];
        app.yellow.Color=[0 0 0];
        app.magenta.Color=[0 0 0];
        app.red.Color=[0 0 0];
    elseif value>0&&value<=20
        app.green.Color=[0 1 0];
        app.cyan.Color=[0 0 0];
        app.yellow.Color=[0 0 0];
        app.magenta.Color=[0 0 0];
        app.red.Color=[0 0 0];
    elseif value>20&&value<=40
        app.green.Color=[0 1 0];
        app.cyan.Color=[0 1 1];
        app.yellow.Color=[0 0 0];
        app.magenta.Color=[0 0 0];
        app.red.Color=[0 0 0];
    elseif value>40&&value<=60
        app.green.Color=[0 1 0];
        app.cyan.Color=[0 1 1];
        app.yellow.Color=[1 1 0];
        app.magenta.Color=[0 0 0];
        app.red.Color=[0 0 0];
    elseif value>60&&value<=80
        app.cyan.Color=[0 1 1];
        app.yellow.Color=[1 1 0];
        app.magenta.Color=[1 0 1];
        app.red.Color=[0 0 0];
    elseif value>80&&value<=100
        app.green.Color=[0 1 0];
        app.cyan.Color=[0 1 1];
        app.yellow.Color=[1 1 0];
        app.magenta.Color=[1 0 1];
        app.red.Color=[1 0 0];
    else
        app.green.Color=[0 0 0];
        app.cyan.Color=[0 0 0];
        app.yellow.Color=[0 0 0];
        app.magenta.Color=[0 0 0];
        app.red.Color=[0 0 0];
    end
end
end
```

Step 3: Run the program

Chapter 12

1   See Sect. 12.2
2   See Sects. 12.2–12.7. The final model would look like this:

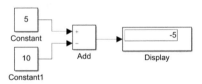

3   See Sect. 12.8. The function and model for (d)(i) is shown only since (d)(ii) can be done similarly
     Function:

```
function [Real, Imaginary] = Polar_to_Rectangular(Magnitude, Angle)
Real = Magnitude*cos(Angle);
Imaginary = Magnitude*sin(Angle);
```

Model:

Model:

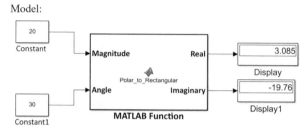

4a   Follow the example shown in Example 12.3 and replace the value of solar irradiance in the PS constant block into 1000

4b   Connect the power and voltage output only to the two inputs of the scope. The graph should look like this:

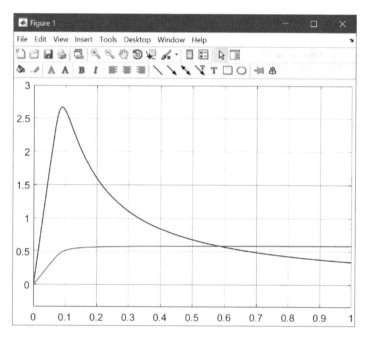

4c Connect the power and voltage output only to the two inputs of the XY graph.
    The graph should look like this:

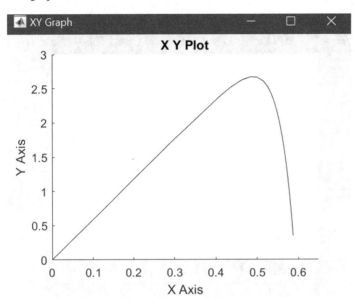

4d The graph should look as follows. The legend has been changed from North-
    west to Southwest for this solution

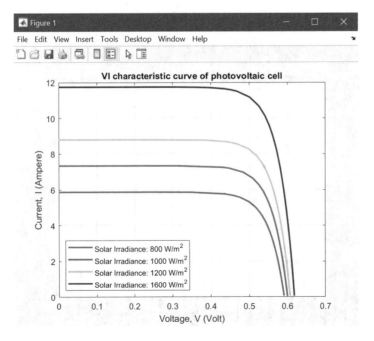

Chapter 13

1   Select any two blocks from the chapter and use corresponding examples
2a See Sect. 13.2
2b

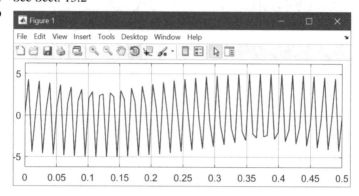

2c

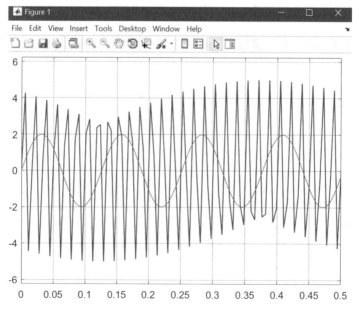

3a See Sects. 4.3 and 4.4 in Chap. 4 for the formula

3b

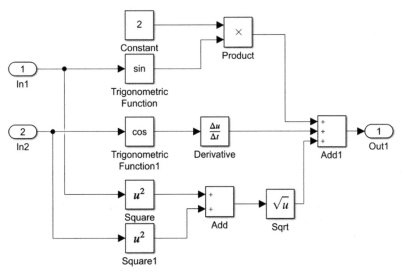

3c Use "Complex to Real-Imag," "Math Function," "Trigonometric Function," and other mathematical blocks to display the result

4a See Sect. 13.4

4b The subsystem is as follows:

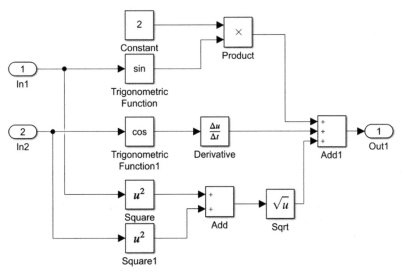

4c "Display" block can be used for demonstrating the result

5a See Sect. 13.5.3

5b See Sect. 13.5.4

5c The model should look as follows:

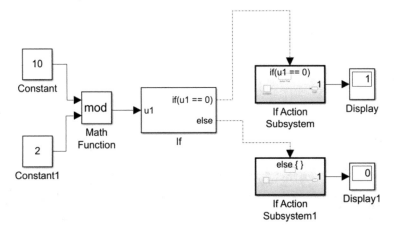

Chapter 14

1   See Sects. 14.1, 14.2, and 14.3
2a See Sect. 14.2
2b The response graph should look as follows:

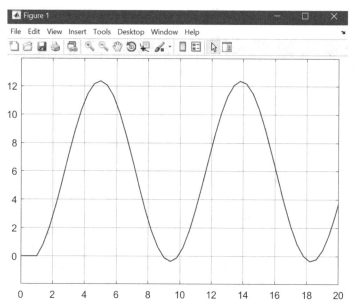

2c Change the values of the gain into 5, 8, and 12, respectively, in the same
    system. Describe the changes noticed in each waveform
3a See Sect. 14.3
3b

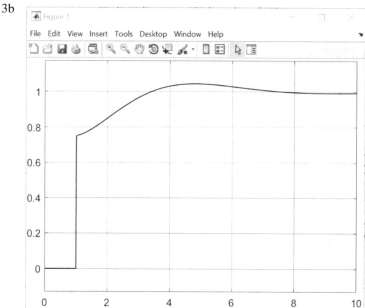

3c

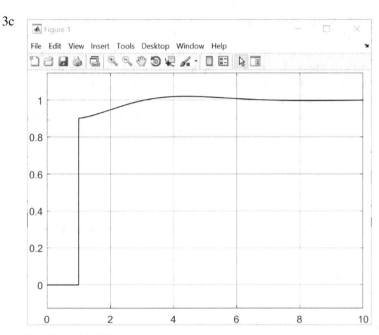

4a   See Sect. 14.5.2
4b   Follow the example shown in Sect. 14.5.2 to parameterize the PID controller
4c   The graph should provide the following response:

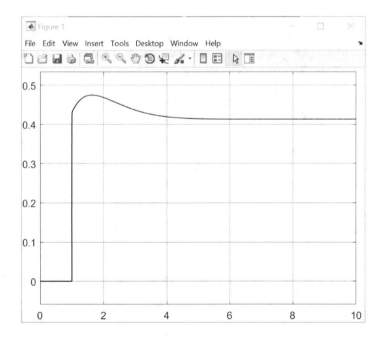

4d Describe the changes noticed from the scope of the PI and PD controller
5a See Sect. 14.6
5b Replicate example shown in Sect. 14.6.1 with the given transfer function. Double-click on the "Pole-Zero Plot" section and for pole-zero analysis and the "Gain and Phase Margin Plot" for observing the Nyquist plot
5c The system is unstable. See Sects. 14.6.1 and 14.6.2 for the characteristics of a stable and an unstable system. Compare the obtained graph with the characteristics and comment on the resembling features

Chapter 15

1   See Sect. 15.1
2a See Sect. 15.2 and corresponding examples in the section
2b See the example shown in Sect. 15.1.2. The resultant values are (i) 4.509, (ii) 12.75, and (iii) 4.06
3a See Sect. 15.2
3c Use Power Measurement block to measure real and reactive power and demonstrate the graph using a Scope block
4a See Sect. 15.1.2 for the definition of power factor
4b The value of the RMS current is 13.76 A
4c The value of the power factor is 0.1281

Chapter 16

1a See Sects. 16.1 and 16.2
1b, See the example shown in Sect. 16.2.1 and follow the model provided at the
c   right side of the example. In the Three-Phase Source block, the input voltage is given as RMS. Therefore, the input for (i) will be peak amplitude/$\sqrt{2}$. For (ii), the given parameter can be used as the RMS value is given
1d With the three-phase power source parameters in question 1b(ii), and default parameters of three-phase series RLC load, follow the example shown in Sect. 16.3 to measure voltage and current only
2a Follow the formula provided in Sect. 16.5
2b Replicate the example shown in Sect. 16.5, and replace the Three-Phase Source with the three-phase delta power source
2c Follow the measurement scheme used in the example in Sect. 16.5
3a See Sect. 16.6.1
3b, Follow the example shown in Sect. 16.6.1 and replace the load with balanced
c   three-phase delta load as shown in Sect. 16.6.3. Use the measurement scheme used in the example in Sect. 16.6.1
3d Comment on the basis of voltage, current, and impedances on each side and match with the individual characteristics of wye and delta connection
4a See the descriptions in Sects. 16.6.1 and 16.6.2

4b,   Follow the example shown in Sect. 16.6.4 and replace the load with unbal-
c   anced three-phase wye load as shown in Sect. 16.6.2. Use the measurement
    scheme used in the example in Sect. 16.6.4

4d   Comment on the basis of voltage, current, and impedances on each side and
    match with their individual characteristics of delta and wye connection

5a   See Sect. 16.7

5b,   See the example shown in Sect. 16.7.1 and replace the Preset model value as
c   per the question. Follow the measurement scheme shown in the example of
    Sect. 16.7.1

6a   See Sect. 16.7.2 for the types of rotors

6b,   See the example shown in Sect. 16.7.2 and replace the wye model with the
c   delta model from Sect. 16.2.2. Consider resistive loads only if RL loads do
    not work for the system

Chapter 17

1a   See Sect. 17.1

1b   See the explanation for each type of rectifier in Sects. 17.1.2 and 17.1.3

1c   Replicate the example in Sect. 17.1.3 and change the value of the transformer
    winding parameters for 220 V input voltage and 24 V output voltage

1d   Replicate the design of three-phase rectifier from Sect. 17.1.4. Change the load
    from R to RL and RC. Change the value of L and C as shown in Sects. 17.1.2.2
    and 17.1.2.3 and see the explanation to justify the observed changes

2a   See Sects. 17.1 and 17.2

2b   Replicate two examples shown in Sect. 17.2.1 for NPN and PNP transistors,
    and change the value of the DC current source into 0.004 V (for NPN
    transistor) and -0.004 V (for PNP transistor), and change the parameters of the
    XY Graph for better visualization

3a   See Sects. 17.3.3 and 17.3.4 for the formula

3b   See Sect. 17.3.1. With the given value of input and output voltages, determine
    the ratio of R_2/R_1. Pick appropriate values of R_1 and R_2 to match the ratio. Then
    replicate the example in the section with those resistor values to obtain the
    desired input and output voltage

3c   See Sect. 17.3.2. With the given value of input and output voltages, determine
    the ratio of R_2/R_1. Pick appropriate values of R_1 and R_2 to match the ratio. Then
    replicate the example in the section with those resistor values to obtain the
    desired input and output voltage

4a   See the initial explanation in Sect. 17.4

4b   (i) Replicate the example of MOSFET in Sect. 17.4.2 and change the phase
    delay first into (45*(1/60)*(1/360)) and then into (90*(1/60)*(1/360)) in Pulse
    Generator block. Compare the output and explain the changes with the help of
    Sect. 17.4.2
    (ii) Replicate the example of GTO in Sect. 17.4.3 and change the pulse width to
    10 and phase delay for (45*(1/60)*(1/360)) in the Pulse Generator block. Then

change the pulse width to 25 and phase delay for (90*(1/60)*(1/360)). Compare the output and explain the changes with the help of Sect. 17.4.3

5a Replicate the example shown in Sect. 17.5.1. Add an alpha-beta-zero to dq0 block from Simulink, and provide the alpha-beta-zero signal as the input. Compare the dq0 output from this block with the one in the example

5b See Sect. 17.6.1.1 and calculate the value of the duty cycle from the source voltage and output voltage. Change the duty cycle in the Pulse Generator block and change the value of resistor, inductor, and capacitor if needed to obtain a graph with less ripple and oscillation as much as possible

5c See Sect. 17.6.1.2 and calculate the value of the duty cycle from the source voltage and output voltage. Change the duty cycle in the Pulse Generator block and change the value of resistor, inductor, and capacitor if needed to obtain a graph with less ripple and oscillation as much as possible

6a See Sect. 17.6.3

6b Replicate the three-phase full-wave converter shown in Sect. 17.6.2.3 with DC voltage of 24 V. Measure the phase-to-phase voltages using Voltage Measurement blocks and demonstrate the output in a Scope block

6c Replicate the example cycloconverter designed in Sect. 17.6.4 with the same parameters. Provide a period of 0.0668 s at the "Pulse Generator3" block

Chapter 18

1   See the abstract of the chapter

2a See Fig. 18.1

2b See Sect. 18.1.1

2c See Fig. 18.2

2d See Fig. 18.2. Update the value, $T = 310$ and $N = 2$. Simulate the model for showing the characteristic curves

2e See Fig. 18.18. Use the Second Area box for the three-phase inverter design. Replace the first PV panel Area box with a DC voltage source of 300 V

3a See Sect. 18.2.1

3b See Fig. 18.37. From the Bus Selector, only select P (pu) and Q (pu) to show them in the Scope

3c See Fig. 18.48. From the Bus Selector, only select Vdc (V) to show it in the Scope

3d See Fig. 18.48. Update the active power (W) of the Three-Phase Series RLC Load from its parameter window (see Fig. 18.57). From the Bus Selector, only select Vdc (V) to show it in the Scope

4a See Sect. 18.3.1

4b See Fig. 18.61. From the Bus Selector, only select Output active power, Pe0 (pu), to show it in the Scope

4c See Fig. 18.65 to update the nominal power and line-to-line voltage of the Synchronous Machine

See Fig. 18.67 to update the two winding voltages of the Three-Phase

Transformer (Two Windings)
From the Bus Selector, only select Output active power, Pe0 (pu), to show it in the Scope
5a See Sect. 18.4.2 to find out the types of batteries
5b See Fig. 18.76 to design the model. See Fig. 16.77 to select the battery type and to show the discharge characteristic curve
5c See Fig. 18.76. From the Bus Selector, only select SOC (%) to show it in the Scope
5d See Fig. 18.83. The resultant output voltage = 15 V. The design may vary. One of the acceptable designs is as follows:

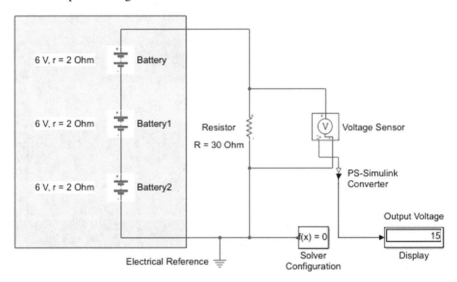

Printed in the United States
by Baker & Taylor Publisher Services